This book will provide an up-to-date and comprehensive basis for undergraduate courses in structural concrete and concrete materials taken by students in civil and structural engineering. Its numerous design tables and graphs, together with extracts from BS 8110 and the Department of the Environment publication *Design of Normal Concrete Mixes*, will also make it an important book for all practising professionals and those involved in graduate-level studies.

A quick glance at the author's contents list will reveal the comprehensive nature of the book. In particular the book covers the philosophy and practice of structural design, concrete material properties and the design of concrete mixes, reinforced concrete beams and slabs (including the Johansen and Hillerborg methods of slab analysis and design), concrete columns and walls, and reinforced concrete frames. The final three chapters cover very important material on prestressed concrete, shell roofs and folded plate roofs. An appendix also provides many useful tables from BS 8110, in addition to those included in the text.

Professor Wilby's qualifications, experience and reputation make him an ideal author, and in fact the present volume builds upon, updates and expands his previous book *Structural Concrete*. He has personally taught much of the material covered, and provides a great number of worked examples. The student is shown how to create and design practical structures, and how to set out calculations as required in practice. All the author's work is in accordance with BS 8110. Undergraduates, graduate students and professionals alike will benefit greatly from this book.

Concrete Materials and Structures

A UNIVERSITY CIVIL ENGINEERING TEXT

Concrete Materials and Structures

C. B. WILBY, PhD, BSc, CEng, FICE, FIStructE
Emeritus Professor, University of Bradford and
Consultant to Robinson Consulting Engineers
Bradford, West Yorkshire, England

CAMBRIDGE UNIVERSITY PRESS
Cambridge
New York   Port Chester   Melbourne   Sydney

Published by the Press Syndicate of the University of Cambridge
The Pitt Building, Trumpington Street, Cambridge CB2 1RP
40 West 20th Street, New York, NY 10011-4211, USA
10 Stamford Road, Oakleigh, Melbourne 3166, Australia

First published in 1983 by Butterworths as *Structural concrete* and © Butterworth & Co. (Publishers) Ltd 1983
This revised edition published in 1991 by Cambridge University Press as *Concrete materials and structures* and © Cambridge University Press 1991

Printed in Great Britain at the University Press, Cambridge

British Library cataloguing in publication data

Wilby, C. B. (Charles Bryan)
Concrete materials and structures.
1. Construction materials: Concrete
I. Title
624.1834

Library of Congress cataloguing in publication data

Wilby, C. B. (Charles Bryan)
Concrete materials and structures: a university civil engineering text/C. B. Wilby.
p. cm.
Includes bibliographical references and index.
ISBN 0-521-37334-4 (HC). – ISBN 0-521-37706-4 (PB)
1. Concrete construction. 2. Concrete. I. Title.
TA681.W644 1991
624.1′834–dc20   90-45681 CIP

ISBN 0 521 37334 4 hardback
ISBN 0 521 37706 4 paperback

UP

To
**Jean,
Charles (Anthony),
Chris and
Mark**

CONTENTS

Preface xv

1 Serviceability and safety 1
1.1 Serviceability and safety 1
1.2 Elastic theory of design 2
1.3 Load factor method of design 3
1.4 BS 8110 philosophy of design 5
1.4.1 *Summary of BS 8110 philosophy of design* 11
1.4.2 *Simplified statement of BS 8110 philosophy of design* 11
  References 12

2 Properties of materials and mix design 13
2.1 Cement 13
2.2 Aggregates 21
2.3 Concrete 24
2.3.1 *Workability* 25
2.3.2 *Water-to-cement ratio and strength of concrete* 26
2.3.3 *Strength tests of concrete* 28
2.3.4 *Vacuum concrete* 28
2.3.5 *Vibrated concrete and pressure compaction* 29
2.3.6 *Gap graded concrete* 29
2.3.7 *No fines concrete* 31
2.3.8 *Curing of concrete* 32
2.3.9 *Design of concrete mixes using Road Note 4* 33
2.3.10 *Design of concrete mix of given mean (or average) strength using Road Note 4* 36
2.3.11 *Combining aggregates to obtain a grading for the Road Note 4 mix design method* 39
2.3.12 *Design of concrete mixes and the D.O.E. method* 41
2.3.13 *D.O.E. mix design method* 41
2.3.14 *Quantities of materials required to make 1 m^3 of concrete* 49

2.3.15	*Prescribed mixes* 50	
2.3.16	*Shrinkage* 50	
2.3.17	*Relationship between stress and strain for concrete* 54	
2.3.18	*Tensile strength of concrete and granolithic toppings* 57	
2.4	Types of reinforcement 61	
2.5	Practical use, creation and economics of structural concrete 63	
2.6	'Bond' between concrete and steel 65	
2.6.1	*Anchorage or bond length* 68	
2.6.2	*End anchorages* 70	
2.6.3	*Laps in reinforcement* 71	
2.6.4	*Curtailment of reinforcement in beams* 74	
2.6.5	*Anchorage of bent-up shear bars* 77	
2.6.6	*Bearing stresses inside bends* 78	
2.6.7	*Anchorage of stirrups (or links)* 79	
2.6.8	*Splitting effects of bar anchorages* 80	
2.6.9	*Anchorage lengths based on elastic analysis* 81	
2.7	Corrosion of reinforcement, carbonisation and 'cathodic protection' 82	
	References 83	
3	**Reinforced concrete beams** 85	
3.1	Design 85	
3.2	Elastic analysis for bending moments 86	
3.2.1	*Assumptions made in the elastic design of reinforced concrete* 86	
3.2.2	*Moment of inertia of a reinforced concrete section* 88	
3.2.3	*Method for tabulating calculations for x and I* 91	
3.2.4	*Popular formulae for slabs and rectangular beams (elastic theory)* 95	
3.3	Elastic theory for shear stresses 100	
3.4	Shear reinforcement 105	
3.4.1	*Design of shear reinforcement by BS 8110 truss analogy* 106	
3.5	'Bond' stresses due to shear (or flexural bond) 112	
3.6	Torsion 113	
3.7	Plastic analysis 118	
3.7.1	*Assumptions of plastic design methods* 118	
3.7.2	*Plastic design in bending* 119	
3.7.3	*Plastic design of 'under-reinforced' rectangular sections* 120	
3.7.4	*'Balanced' plastic design of rectangular sections* 123	
3.7.5	*Plastic design of any shape of 'under-reinforced' section* 126	
3.7.6	*'Balanced' plastic design of any shape of section* 126	
3.7.7	*Plastic design of any shape of 'under-reinforced' section containing compression steel* 127	
3.7.8	*'Balanced' plastic design for any shape of section containing compression steel* 128	

Contents   xi

- 3.7.9 Design of compression steel for a rectangular section   129
- 3.7.10 Compression steel near to neutral axis   130
- 3.7.11 Further points about compression steel   131
- 3.8 Limit state of deflection   131
- 3.9 Limit state of cracking   132
  - References   132

4 Reinforced concrete slabs   134
- 4.1 Slabs spanning 'one-way'   134
- 4.2 Slabs spanning 'two-ways'   134
- 4.2.1 General discussion of design of two-way spanning slabs   135
- 4.2.2 Design tables for two-way slabs   139
- 4.3 Flat slabs   141
- 4.4 Yield-line theory of slab analysis   141
- 4.4.1 Reinforcement   144
- 4.4.2 Further points on yield-line analyses   146
- 4.4.3 'Upper-bound' and 'lower-bound' solutions   147
- 4.4.4 Further consideration of the 'equilibrium method'   148
- 4.4.5 Further consideration of the virtual-work method   153
- 4.4.6 Combination of equilibrium and virtual-work methods   154
- 4.4.7 Affine slab transformations   162
- 4.5 Hillerborg's strip method of slab design   166
- 4.5.1 Further points on Hillerborg's strip method   167
- 4.6 BS 8110 and CP 110 and yield-line and strip methods   168
- 4.7 Hillerborg's advanced method   169
- 4.8 Slab with hole using Hillerborg's strip method   190
- 4.9 Hillerborg elements with shear forces along their edges   196
- 4.10 Traditional U.K. design office methods   201
- 4.11 General discussion of design methods for two-way and flat slabs   201
  - References   203

5 Columns and walls   205
- 5.1 General   205
- 5.2 Slender and short columns   205
- 5.3 Axially loaded short columns   205
- 5.4 Plastic analysis for eccentrically loaded short columns   207
- 5.4.1 Design of eccentrically loaded columns   212
- 5.5 Reinforced concrete walls   212
- 5.6 Design of columns to frameworks   213
- 5.7 Very slender columns   214
  - References   215

6 Reinforced concrete frames and continuous beams and slabs   216
- 6.1 Introduction   216

xii *Contents*

  6.2 Frames 216
  6.3 Continuous beams and slabs 218
    References 218

7 Design of structures 219
  7.1 Design of an *in-situ* R.C. framed building 219
 7.1.1 *Floor slab* 221
 7.1.2 *Beams of 7 m span* 223
 7.1.3 *External columns between ground and first floor* 227
 7.1.4 *Bases* 230
 7.1.5 *Anchorage of column bars into bases* 233
 7.1.6 *Design calculations* 234
 7.1.7 *Student design office exercise* 234
 7.1.8 *Floor of a building (two-way and flat slabs)* 234
  7.2 Design tables 238
  7.3 Creation (design or selection) of structural system 242
  7.4 Creation of structures generally 249
    References 250

8 Prestressed concrete 252
  8.1 Prestressing 252
 8.1.1 *Advantages and disadvantages of prestressing* 253
  8.2 Materials 255
 8.2.1 *Stress corrosion* 255
  8.3 Losses of prestress 255
  8.4 Limit state design of members 259
 8.4.1 *Simple assessment of size of prestressed members* 261
 8.4.2 *Assumptions for elastic design* 262
 8.4.3 *Limit states of stresses and deflections* 263
 8.4.4 *Simplified elastic design of prestressed concrete beams* 264
 8.4.5 *Ultimate limit state due to flexure (bonded tendons)* 274
 8.4.6 *Additional untensioned steel (bonded tendons)* 281
 8.4.7 *Top non-prestressed steel* 282
 8.4.8 *Ultimate limit moment due to flexure (unbonded tendons)* 283
 8.4.9 *Prestressed columns* 283
 8.4.10 *Prestressed ties* 283
 8.4.11 *Shear resistance of prestressed concrete beams* 284
 8.4.12 *Inclined tendons* 287
 8.4.13 *Composite construction* 287
 8.4.14 *Continuity* 294
 8.4.15 *End splitting forces* 295
 8.4.16 *Prestressed concrete tanks, pipes, domes, shells and piles* 295
 8.4.17 *Torsional resistance* 296
  8.5 Load balancing 296

8.6	Post-compressing 299	
	References 299	
9	**Shell roofs** 301	
9.1	Notation 301	
9.2	Introduction 303	
9.2.1	*Types of shells* 303	
9.2.2	*Designation of shells* 310	
9.2.3	*Applications* 311	
9.3	Economics 313	
9.4	Design 319	
9.5	Analysis 321	
9.6	Construction 324	
9.7	Generalized membrane analysis 332	
9.7.1	*Cylindrical shells* 332	
9.7.2	*Shells of revolution* 341	
9.7.3	*Shells of arbitrary shape* 347	
9.8	Cylindrical shells 347	
9.8.1	*Design tables and graphs* 348	
9.8.2	*Analogue computer* 350	
9.8.3	*Instability* 350	
9.9	Domes 351	
9.10	Hypars 352	
9.11	Conoids 354	
9.12	Method of finite elements 355	
9.13	Computer programs 356	
9.14	Design practicalities 356	
	References 357	
10	**Folded plate roofs** 360	
10.1	Folded plates 360	
10.2	Design and analysis of folded plates 361	
10.3	Analysis of folded plates 361	
10.4	Analysis due to Parme 363	
	References 378	

Appendix 1: Tables and graphs for design 379
Appendix 2: Units and Greek symbols 383
Appendix 3: Nomenclature 385
Appendix 4: Extracts from British Standard BS 8110: 1985 392
Index 406

PREFACE

This book has been written as a good course for undergraduate students of Civil and/or Structural Engineering and for concrete work included in Master Degree and Post-graduate Diploma Courses. It has everything and more than required for lower than Batchelor Degree courses in Civil and/or Structural Engineering and for degree and lower courses in Architecture and Building. It should also be useful to practising professionals to update their knowledge.

It should be useful in design offices because of its numerous design tables and graphs, and extracts from BS 8110 and the Department of the Environment publication *Design of Normal Concrete Mixes*.

Chapter 1 deals with the general philosophy of design for serviceability and safety, describing plastic and elastic methods of design and the BS 8110 recommendations.

Chapter 2 describes the manufacture of various cements, the various aggregates available and the ways of classifying aggregates by grading, etc. Various types of concrete and methods of designing concrete mixes are described. Shrinkage, strain and creep of concrete are discussed. Problems concerned with granolithic flooring and the recent development from this technology of the flexural tensile strength concrete design for slabs on solid, and with A.S.R., H.A.C., corrosion due to chloride ion, cathodic protection of reinforcement, etc., are described. Not many books of this type consider all these problems. The reinforcements which can be used are described and methods are described for determining anchorage or bond length of reinforcement, lap lengths, bearing stresses on the concrete inside reinforcement bends, etc.

Chapter 3 describes all that is necessary for the design of cross sections of reinforced concrete beams or slabs.

Chapter 4 deals with the design and analysis of one-way and two-way spanning and flat R.C. slabs. Johansen's Yield Line analyses, both 'equilibrium' and 'virtual work' methods, are gone into more exhaustively than is done in many books of this kind. Hillerborg methods (and the author is grateful for discussions with Professor Hillerborg of Sweden and his help) both 'strip' and 'advanced' methods, are also gone into more exhaustively than is done in most books of this kind. Professor Hillerborg produced the difficult design of the complex slab of Fig. 4.24 posed to him by the author. Hillerborg methods have most useful design, analytical and detailing advantages over Johansen's methods in the author's opinion.

Chapter 5 deals with the design and analysis of columns and walls.

Chapter 6 deals with the design and analysis of frames and continuous beams and slabs.

Chapter 8 deals with pretensioned and post-tensioned prestressed concrete, and composite construction. In addition it describes Reiffenstuhl's methods of what the author calls 'post-compressing' as opposed to post-tensioning (and the author is grateful for discussions with Professor Reiffenstuhl of Austria and his help). Professor Reiffenstuhl has designed important bridges with very shallow depths relative to their spans by this method. Perhaps the presentation of his methods in the English language in this book will result in their being taken advantage of where beneficial in the English-speaking world. An introduction is also given to Professor Lin's (University of Berkeley, U.S.A.) load-balancing method, which he has used for designing important continuously spanning prestressed concrete bridges. Professor Reiffenstuhl's work seems to have been neglected by books of this kind and Professor Lin's work seems to have been neglected by British books of this kind. Several examples take the reader through the design, using rapid approximate methods, of a prestressed concrete beam and then the refinement of the design using more accurate quantities for the losses of prestressing forces in the tendons. Losses are dealt with fairly thoroughly. This is not done by many similar books.

Chapter 9 considers the design, analysis, construction, economics, applications, etc., of shell roofs. It includes roofs of cylindrical shells, hypars, conoids and domes. Domes in R.C. have been used internationally since about 1920 for covering service reservoirs and for libraries, mosques and synagogues. Shells are always popular for buildings for Olympic Games, etc. and for impressive buildings.

Preface xvii

Chapter 10 deals with Folded Plate roofs and takes the reader through two different practical examples.

Appendix 4 gives useful extracts for much commonplace design from BS 8110.

In the book and Appendix 4 there are many useful Tables for designers. So that these can be located easily they are listed in Appendix 1.

Chapter 7 designs an *in situ* R.C. framed building. The beginner and the experienced designer have to create the disposition of the beams, slabs and columns, preferably so that subsequent analysis and design will not require outline drawings to be altered. Also these latter are often used by contractors and quantity surveyors for estimating and by architects for incorporating in their drawings, long before the elaborate analysis, design and detailing is produced. So the author describes how to create the geometry of the structure by rapid reliable calculations. Subsequently the design has been set out as a designer would for submission for approval. Many books show how to check the adequacy of the design of concrete members already disposed and sized and to design the reinforcement. But the student/beginner/designer needs to know, indeed he wonders how the outline drawings were obtained.

Again regarding creating structures Section 2.5 generally discusses the practical use, creation and economics of R.C. and prestressed concrete structures. Section 8.1.1 discusses the advantages and disadvantages of using prestressed concrete. Sections 9.2, 9.3, 10.1 and 10.2 discuss the applications and economics of shell roof and folded plate roof construction.

Section 7.4 takes the opportunity of discussing the creation of structures generally, that is how different types of steelwork, R.C., prestressed concrete, and shell and folded plate roofs are related in usefulness for various purposes.

The book deals with basics and principles and develops analyses and design methods and in this respect it is of value internationally. When these lead to code recommendations then instead of reporting code regulations internationally, the current British Codes and Standards and Ministry recommendations are quoted. For example when a particular loss of prestressing force is described, the book (as opposed to a research treatise) does not reproduce recommendations applying to numerous countries but quotes the current British ones. When design examples are given to illustrate the practical applications of the principles then current

British recommendations are used (e.g. BS 8110). British Codes and Standards are used in many countries internationally, but in those countries where other recommendations are in use (e.g. the U.S.A.) the reader will need to refer to his own country's recommendations but the main principles, basics etc. of this book should still be relevant.

A student at University has the opportunity of designing and analysing complex and intellectually stimulating structures. Subsequently in practice he may only spend a modest time on design and as a newcomer to the design office may mainly work on simple, perhaps highly profitable, structures required urgently. The most interesting and complex structures will probably be tackled by a certain few designers of considerable experience and ability. Time at University is at a premium and a final year design class on concrete (as designs have to be made in other materials) may probably be no more than 6 hours per week for 10 weeks, i.e. a total of 60 hr, which is only like working for about $1\frac{1}{2}$ weeks in a professional design office most probably under tighter supervision than can be effected in a University as the work has to be 100% correct and a beginners work would be supervised and checked meticulously. A student can learn R.C. detailing on a shell or folded plate roof or on a prestressed concrete beam just the same as he could on mundane simple members and structures. For example on the MSc course (at Bradford) some have detailing experience and some have none, so we have had them designing with tables and detailing conoidal shell roofs, which none of them have done before. Also as an example, at Bradford University, in the final year of the Batchelor degree and on the MSc course we have had students designing and detailing folded plates as per Chapter 10.

The author has personally taught much of the work in this book to final year Batchelor and Master Degree and post-graduate Diploma students. This has led to some improvements in certain explanations.

References have been kept to a minimum because of the huge number available. The author cites ones best known to himself and apologises to numerous academics, researchers and practicing engineers internationally for not including or selecting their meritorious works. Sometimes books are quoted rather than old original papers, when they are considered easier to obtain by the reader from libraries. References in the English language are always used in preference to those in other languages, again apologies to the great many concerned.

The work in this book has benefited by the work in the author's books *Structural Concrete* and *Concrete for Structural Engineers* and the author

Preface xix

thanks the publishers Butterworth & Co. (Publishers) Ltd for their kind cooperation. This material had some degree of indebtedness generously given by each in the following list (in alphabetical order), which also includes those who have kindly helped with this present book:

1. Professor A. L. L. Baker, formerly of London University.
2. Dr A. Beeby, a member of the BS 8110 Committee and a considerable contributor to BS 8110, formerly of C.&C.A.
3. Dr E. Bennett, Leeds University
4. Bradford University (at some time or other): Mr W. Appleyard, Dr J. C. Boot, Professor John Christian (Canada), Dr A. Dracos (Robinson's Consultants, Bradford), Mr D. Franklin, Mr N. K. Gogoi, Mr P. Gregory, Dr I. Hussain (Canada), Dr I. Khwaja, Dr M. Naqvi (Rolls-Royce, Derby), Dr Noor (Hatfield Polytechnic), Dr V. R. Pancholi, Dr D. H. Schofield, Mr D. Walker (P.S.A.) and Mr R. Westbrook.
5. Mr George Brown, Managing Director, (and Len Reed), Kontrad Ltd, Manchester Airport.
6. B.S.I.: Extracts from the D.O.E.'s *Design of Normal Concrete Mixes* included in Chapter 2 are contributed by courtesy of the Director, Building Research Establishment. Crown Copyright reproduced by permission of the Controller H.M.S.O., and extracts from BS 8110 in Appendix 4 and elsewhere in this book, are included by kind permission of the British Standards Institution, 2 Park Street, London W1A 2BS from whom complete copies of the documents can be obtained.
7. Mr A. T. Corish, Blue Circle Cement, London.
8. Professor A. Hillerborg, Lund Institute of Technology, University of Lund, Sweden.
9. Professor L. L. Jones, Loughborough University.
10. Mrs H. Mahony, formerly of C.&C.A.
11. Dr A. L. L. Parme, U.S.A.
12. Professor H. Reiffenstuhl, University of Technology, Vienna.
13. Dr R. E. Rowe, formerly of C.&C.A.
14. Professor R. H. Evans, University of Leeds.
15. Mr C. A. Wilby, P.A.F.E.C., Nottingham.
16. Mr Chris B. Wilby, Kirklees M.D.C.
17. Mr M. Stainburn Wilby, Leeds.

The work in Chapter 9 has benefited by the work in the author's

Chapter 32 of *Handbook of Structural Concrete* published by Pitman who are now owned by Longman Group UK Ltd. and the author thanks the latter and Professors F. K. Kong (Newcastle University) and R. H. Evans (Leeds University) for their kind cooperation.

1

Serviceability and safety

1.1 Serviceability and safety

A structure or any part of it, such as a beam, column, slab, etc., must be serviceable in use and safe against collapse. Serviceability requires that, at the kind of loads likely to occur during use, everything will be satisfactory, for example, deflections will be adequately small, vibrations will be tolerable, the maximum width of cracks will be no greater than specified, etc. For example, for prestressed concrete no cracks may be specified whatsoever, whilst for reinforced concrete design the maximum size of crack might be specified as small enough not to admit rainwater (about 0.25 mm) or, if inside a building, not to be visually unacceptable.

Safety requires that the strength of a structure or any part of it be adequate to withstand the kind of loads reasonably considered to be most critical as regards collapse.

In assessing the requirements for serviceability and safety just described, it is necessary to assess, for example, deflections and ultimate strengths. Deflections require assessments of stress/strain, creep/time and shrinkage/time relationships. Safety requires assessments of ultimate strengths. These properties vary to some extent for any material used. For example, if one cast a large number of concrete cubes and endeavoured to make them identical so that they all had the same strength, on crushing these cubes one would obtain a result like the graph of Figure 2.4. One can hardly assume that this particular concrete can be assumed to have a strength equal to say its mean strength of 35 N/mm^2 as shown on this graph because one or two cubes out of this very large number have failed at near to 15 N/mm^2. Also it is not economic to try and assume this particular concrete to have the strength of the weakest cube tested. So a compromise based on experience, and involving a decision on chance with regard to safety, has to be made by any code committee. The tensile strength of specimens of steel reinforcement all thought to be the same, would give a graph similar to Figure 2.4 except that the range and standard deviation of the histogram would be very much less.

Again, in assessing the previously described requirements for serviceability and safety, it is necessary to decide upon loads which may have to be carried during use and occasionally sustained to prevent collapse. It may well be impractical to consider the worst possible event which could ever occur, for example, a nuclear holocaust coinciding with an earthquake and a hurricane – the client has to be able to afford the building for his planned use. So a compromise based on experience, and probability with regard to serviceability and safety, has to be made by any code committee.

1.2   Elastic theory of design

This method (also called permissible stress method) of design is based on the assumptions described in Section 3.2.1.

The loading which has to be carried in use, or when working, is assessed and known as the 'working load'. Then using the elastic theory, sections of members are designed so that the maximum 'working stresses' in the concrete and reinforcement are not greater than certain 'permissible stresses' or 'allowable working stresses'. A permissible stress is restricted by a 'factor of safety' to be sufficiently below the ultimate stress of the material, to be well within the limit of proportionality of the steel reinforcement and sufficiently low to be within the initial fairly linear portion of the stress/strain curve for the concrete (see Figure 2.10). The 'factor of safety' times the permissible stress is equal to either the yield or 0.2% proof stress for steel reinforcement or the cube strength for concrete. Codes used to make the factor of safety greater for concrete than steel because of the approximate linearity of the stress/strain curve for concrete not extending to much of a proportion of its ultimate stress. Subsequently with the arrival of recent codes of practice in the U.K. and U.S.A. the term 'factor of safety' almost requires definition each time it is used, so for any particular code the definition needs to be carefully studied. For example, the term 'factor of safety' as used in this section is not the same as the term 'partial safety factor' used in BS 8110 (see later).

In the case of frames and continuous beams and slabs an elastic theory has been used (sometimes modified slightly in later years) for evaluating bending moments and shear forces.

In the early days of (reasonable) structural concrete design, the elastic theory was well established and had proved reliable for designing steel structures. It therefore seemed to be the most reliable, sensible and indeed only theory to use for designing structural concrete since concrete appeared to have a fairly linear stress/strain relationship up to the stresses likely to be permissible. The permissible stress method was used in the U.K. and U.S.A., prior to 1957 and 1963, respectively. After these dates an alternative 'load factor' method (see later) was recommended by the respective British and A.C.I. codes. With regard to prestressed concrete the first national (previously private ones existed) code of practice CP 115[1] was published in 1959 and required both permissible stress and load factor designs to be made. The present British Code BS 8110[2] does not principally use the permissible stress method for reinforced concrete design but uses it for the limit states of stress and deflection (see Section 8.4) for prestressed concrete. It also uses it for the calculation of more precise evaluations of crack widths and deflections for reinforced concrete. In 1989 the Institution of Structural Engineers was obliged, because of a majority vote of its members, to endeavour to produce a Code based on permissible stresses like CP 114[3]. The British Code BS 8007[4] for designing water-retaining structures recommends the elastic theory for assessing crack widths at the limit state of serviceability.

Permissible stress design has certainly been very satisfactory for a long time.

1.3 Load factor method of design

When it was eventually considered that the ultimate moments of resistance of sections could be reasonably reliably assessed, the elastic theory for designing sections was thought to be basically uneconomic because of its inability to predict collapse or 'ultimate loads'. The theories for assessing ultimate bending moments made use of the plastic action of concrete, that is the behaviour at higher stresses when stress is not directly proportional to strain (see Figure 2.10) and peak stresses calculated by elastic theory are relieved by plastic action. Thus the load factor method is based on 'plastic theory' and is sometimes called 'plastic design' (see Section 3.7.2). The ratio of the ultimate load to the working load is called the 'load factor'.

In a structure, sections designed by elastic theory would have different load factors. It can be seen from Figure 3.6 how the distribution of concrete stress in the upper part of a beam alters from that shown in Figure 3.6(a) for working stresses to that shown in Figure 3.6(c) just before failure. The reinforcement, if of mild steel, would have a stress/strain curve like curve 11 on Figure 8.4. The stress in it would therefore increase linearly with increase in bending moment from Figure 3.6(a) to Figure 3.6(c), if the 'moment or lever arm' (see dimension z in Figures 3.2(d) and 3.7), remained constant. From Figures 3.2(d), 3.6 and 3.7 it can be seen that the moment arm reduces slightly towards failure. Thus if one designed a section of a beam by elastic theory, even if the same factors of safety for concrete and steel reinforcement were used, the load factor would not be the same as the factor of safety. This is made more so if the code used for elastic design uses different factors of safety for concrete and steel. As the elastic design requirements of CP 114[3] consider that the strength of concrete is less reliable, because of its method of manufacture, than the strength of steel, a greater factor of safety for concrete than steel is used. In other words, designing sections of different members such as beams, slabs and columns and various types of all these in a structure, by say using the elastic theory requirements of CP 114, results in these sections possessing differing load factors.

The advocates of load factor design considered a constant load factor desirable for economy and that this should take priority over permissible stress design. Now the latter did limit stresses and therefore strains and thus crack widths and deflections at working loads, whereas a load factor design did not. To endeavour to overcome this, and to not make radically different sized members from previously, the load factor design recommendations of CP 114 were more conservative. As the permissible stresses in CP 114: 1957 were increased from previously, greater deflections would occur so Table 7.1 was introduced to endeavour to limit deflections (unfortunately it does not include loading which of course affects deflection).

In the early days of prestressed concrete design in the U.K., structural concrete members were being made considerably smaller than ordinary reinforced concrete members and contained thin wires instead of robust bars. Prior to code CP 115 they were designed by the permissible stress method, sometimes without checking the load factor. When CP 115 was introduced it required a load factor of 2 but this could be less if the member would fail at a load not less than the sum of 1.5 times the dead load plus 2.5 times the imposed, or live, load. This introduced the concept of what has subsequently been called 'partial safety factors' for loads in BS 8110. The imposed load may increase by accident. For example, a flat roof may be designed for occasional access but while a procession was passing by it might become packed tight with spectators. The dead load cannot increase unless, for example, the finishes to a roof or floor are renewed or changed, in which case the client would usually seek or encounter some building advice. Thus the load factor used for the imposed load part of the loading must be greater than that used for the dead load part of the loading.

The illogicality that existed after the publication of CP 114 was that, for example, individual ordinary reinforced concrete sections of a frame, or continuous beam or slab, could be designed to have a constant load factor but the distribution of bending moments was obtained by elastic analysis. The ideas of plastic collapse mechanisms (see Chapter 6), first developed for steelwork structures, had not been established well enough for inclusion in CP 114 in any greater way than allowing bending moments obtained by elastic analysis at supports to be increased or decreased by up to 15% provided that these modified moments were used for the calculation of the corresponding moments in the spans.

Still most analyses used would give bending moments at sections which would not increase in direct proportion to the loading towards failure, so to design sections of indeterminate structures with a constant load factor seemed pointless. Also the load factor method, with a general conservatism incorporated, only indirectly controlled crack widths and deflections compared to the permissible stress design method. Historically, however, a start presumably had to be made somewhere and somehow with the introduction of methods endeavouring to gain extra economy by the use of load factor methods.

To summarise, when the load factor method of CP 114 was used for sections, crack size was limited by incorporating conservatism into the formulae (in effect limiting the tensile stress in the reinforcement) and deflection was limited by the use of Table 7.1. Of course in important cases the designer could use the elastic methods of CP 114 and calculate deflections.

The book by Evans and Wilby[5] gives considerable description and many examples on the elastic and plastic methods of CP 114 and the plastic method of the A.C.I.[6] code of practice.

1.4 BS 8110 philosophy of design

The European Concrete Committee (abbreviated to C.E.B., the initials of the Committee in French) introduced the concept of probability and used statistics in connection with the strengths of materials, loadings and safety and produced recommendations[7] for a code of practice for reinforced concrete. The underlying philosophy involved has been used as a basis for the present British BS 8110.

With regard to concrete strength, the previous British practice was essentially to specify a minimum concrete strength below which no cubes should fail. This meant that the contractor needed to decide upon the quality of his control (see Table 2.2) to be able to calculate the average strength of the concrete he should endeavour to make. Then he designed his mix for this mean strength as in Section 2.3.10. When on the site, if any of the concrete cubes tested failed below the minimum strength then the concrete was either removed or cores of the concrete taken and tested or a load test was performed to see if the extra age had increased the strength and if the general monolithic construction (sometimes permitted to receive help from, for example, surrounding brickwork if any) was such that the construction could be considered to be safe. The BS 8110 philosophy was to specify, not a minimum concrete strength as previously, but a strength which 5% of the cubes would not achieve, called the 'characteristic strength'. This involved the use of statistics and is explained in Section 2.3.9. The idea of accepting a strength below that at which some cubes would fail was hard for many British engineers to accept, because of their being brought up to think and desire that their designs should be very safe – failure was out of the question.

With regard to loading, the previous British practice was to assess the load which would be unlikely to be exceeded in use, and this would be called the 'working load'. Then if the CP 114 load factor method of design was used, sections would be designed to have a factor of safety of 1.8 against an ultimate load which would be taken as 1.8 times the working load. Now the BS 8110 philosophy was not to assess the maximum load for the working load as previously but was to assess a load which, in effect, only 5% of occurrences of loading would exceed, called the 'characteristic load'. This involved the use of statistics as is explained in Section 2.3.9. The idea of seemingly now accepting a working load which was planned to be sometimes exceeded was again hard for many British engineers to accept. Then, as if to make it more difficult for engineers to accept, BS 8110 introduced the idea of probability of characteristic strengths and loads being variable.

British engineers had always prided themselves on designing structures which in their opinion could never fail. Well, of course, scientific reality cannot be ignored, materials do vary and probability does exist. Apart

BS 8110 philosophy of design

from negligence and natural catastrophes, the most likely cause of failure of a structure, or inadequacy at working loads (that is cracks or deflections being unacceptable), is the coincidental occurrence of both overload and excessive weakness at a critical section.

The probability of failure, for example, could involve the concept of an accident rate intuitively accepted for a given type of structure. For example, how often are crane gantries liable to fail by overload? The probability of failure could also involve economy, for example a reduced probability of failure will require a stronger structure at an increased cost.

Discussions of probability of failure become very emotive because of probable loss of life. A possible analogy is a motor coach full of passengers because if it crashes loss of life is also involved. There is a certain statistical level of probability of hitting a lamp standard or telegraph post, of running into a ditch or river, of rolling over, of hitting another vehicle head on, etc. The designers would not dream of designing the motor coach so that no lives would be lost, or even that no parts of the coach would fail, under all these eventualities. It would not be economically desirable even if possible with brilliant engineering design. On the other hand one would expect the coach floor not to fail due to a suitcase dropping from a luggage rack. One would expect the walls and floor not to fail due to unequal loading of passengers or even a fight amongst some passengers. So with structures a compromise has to be reached between practicality, economy and probability of failure. A jetty designed for a certain use, namely a ship being piloted up to it by a skilled skipper cannot economically be designed to withstand the fairly remote probability of say a drunken skipper sailing a large ship at full speed at right angles to and into the side of the jetty. In such a case it would be argued that the damage and loss of life to anyone on the jetty was the responsibility of the skipper and it was not the responsibility of the owners of the jetty to build it strong enough for this eventuality.

The BS 8110 use of probability manifests itself in the use of 'partial safety factors'. The word 'partial' is used as each part of the problem may have a different safety factor. The characteristic strength of a material permits 5% of the control specimens to be inadequately strong. Dividing the characteristic strength by a partial safety factor (a number greater than unity) means that less specimens will be below the resulting 'design strength' used. The characteristic loading is such that it will only be exceeded on 5% of occasions. Multiplying the characteristic load by a partial safety factor (a number mainly greater than unity) means that the resulting 'design load' should be exceeded on less than 5% of occasions. Thus these partial safety factors are intended to reduce the probability of failure towards zero.

BS 8110 also introduced the concept of 'limit state design'. In design everything that matters as regards the strength and serviceability of a structure is limited or restricted to a satisfactory amount. The condition of a structure or part of it, when it becomes unfit for use, is called a 'limit state'. We can categorise these limit states into two broad divisions, namely 'limit states of serviceability' and 'ultimate limit states'. 'Limit states of serviceability' include:

1. Deflection: This must not impair the appearance or efficiency of the structure.

2. Cracking: Cracks must not adversely affect the durability or appearance of a structure although the latter does not seem to matter in some parts of the world. In Britain there is a practice of generally limiting cracks and this was often done no less for a hidden and protected member than for one that is seen in a building or exposed. This uneconomic and inefficient practice was established in previous codes. The limit state design of BS 8110 now gives opportunities of using different limit states for different members whereas CP 114 did not.

3. Vibration: This must not cause unpleasantness or alarm to the occupants, damage to fixtures, fittings and services (such as water pipes), etc.

4. Other limit states: BS 8110 requires consideration of any other limit states considered necessary by the engineer.

'Ultimate limit state' requires that the strength of the structure should be adequate to withstand the design loads with due consideration being given where appropriate to buckling and the general overall stability. Ultimate limit states may need to be assessed for the following:

(A) Flexural or compression failure at any critical sections
(B) Shear failure
(C) Torsion failure
(D) Bond or anchorage failure of reinforcement
(E) Instability of a member
(F) General instability (for example overturning)
(G) Bearing failure at a support or under a concentrated load or at bends or hooks in tension reinforcement
(H) Bursting of prestressed concrete end blocks

(I) Failure of connections (for example between precast concrete elements or in composite construction).

For the ultimate limit state Tables 1.1 and 1.2 summarise the partial safety factors for loads and strengths, respectively, as recommended by BS 8110. For example from Table 1.1 if one is designing for the ultimate limit state and considers the combination of loading (1) for a simply supported beam then, using BS 8110 symbols (Appendix 3)

Design load = sum of γ_f times each characteristic load
= $1.4G_k + 1.6Q_k$

Table 1.1. *Load combinations*

Load combination	Load type					
	Dead		Imposed		Earth and water pressure	Wind
	Adverse	Beneficial	Adverse	Beneficial		
1. Dead and imposed (and earth and water pressure)	1.4	1.0	1.6	0	1.4	—
2. Dead and wind (and earth and water pressure)	1.4	1.0	—	—	1.4	1.4
3. Dead and wind and imposed (and earth and water pressure)	1.2	1.2	1.2	1.2	1.2	1.2

Table 1.2. *Values of γ_m for the ultimate limit state*

Reinforcement	1.15
Concrete in flexure or axial load	1.50
Shear strength without shear reinforcement	1.25
Bond strength	1.4
Others (e.g. bearing stress)	$\geqslant 1.5$

The γ_f is smaller for the dead load because there is less likelihood of the dead load being increased (for example, a small increase can be due to members being cast slightly oversize) whereas the γ_f for the imposed load is greater because the imposed load can experience an overload.

Again from Table 1.1 if one is designing, say a simply supported beam, for serviceability limit state for the combination of loading (3) then

$$\text{Design load} = 1.2G_k + 1.2Q_k + 1.2W_k$$

The values of γ_f are smaller than the 1.4 and 1.6 values in the previous example because it is a fairly remote possibility that full imposed and wind loading will occur together.

In Table 1.2 it will be noticed for example that γ_m is less for the ultimate limit state for steel than it is for concrete. This is because the control in the manufacture of steel is considered to be better than it is for concrete.

An example will illustrate the meaning of the word 'beneficial' in Table 1.1. If one is calculating the maximum mid-span bending moment in the centre span of a three span continuous beam, one should take a maximum loading of $1.4G_k + 1.6Q_k$ on the centre span and minimum loadings of $1.0G_k$ on the other two spans.

With regard to the serviceability limit states of deflection and cracking, BS 8110 allows these to be dealt with as follows:

1. Deflections are controlled by limiting span to depth ratios. But if one wishes to make more accurate calculations it recommends a method based upon the elastic theory.
2. Crack widths are controlled by attention to detailing of the reinforcement. For example small bars closely spaced result in a greater number of cracks of smaller widths than larger bars spaced wider apart, which produce fewer wider cracks, and it is the maximum crack width which matters as regards appearance and allowing the entry of corrosive elements such as acidic rainwater. If the detailing recommendations of BS 8110 are altered in any way, and there is not normally any reason to do so, or if there is special cause for concern, then BS 8110 requires calculations to be made in a way it recommends of maximum crack widths.

BS 8110 philosophy of design

1.4.1 Summary of BS 8110 philosophy of design

(a) A 'limit state' is a condition of a structure at which it ceases to function in the manner for which it was designed. Limit states can be classified as follows:

1. 'Ultimate limit state' refers to failure.
2. 'Serviceability limit states' refer to conditions in normal use. The main ones are deflection, cracking, vibration, fatigue, durability and fire resistance.

(b) Materials:

1. 'Characteristic strength' is the strength below which only 5% of test specimens will fail (see Figure 2.4).
2. 'Partial factor of safety', γ_m, is given by

$$\text{`Design strength'} = \frac{\text{characteristic strength}}{\gamma_m}$$

and this is applied to each of concrete and steel, that is the parts involved. For example, γ_m is normally 1.5 for concrete and 1.15 for steel for assessing ultimate limit state. Refer to Table 1.2.

(c) Loads:

1. 'Characteristic load' is the load which is expected to be exceeded on, in effect, only 5% of occasions.
2. 'Partial factor of safety' for ultimate loads, γ_f, is a factor by which each part (dead, imposed, wind) of the loading is multiplied so as to obtain the 'design load', that is the load to be designed against.

The serviceability limit state of deflection is dealt with by controlling span to depth ratios and of cracking by careful detailing of the reinforcement.

1.4.2 Simplified statement of BS 8110 philosophy of design

An attempt to summarise the whole process of BS 8110 design is now made. Essentially 'characteristic loads' are determined. There are usually three: namely for dead, imposed and wind loadings. These are then, for design, considered in what are thought to be the most critical combinations for causing failure ('ultimate limit state') by using multipliers ('partial safety factors'), to give various 'design loadings'.

The resistance to these various load combinations is calculated using

12 *Serviceability and safety*

'design strengths' for concrete and steel obtained by dividing 'characteristic strengths' for concrete and steel by their respective 'partial safety factors'.

References
1. *Code of Practice for the Structural Use of Prestressed Concrete in Buildings, CP 115 (1959)*, last reprint 1969. British Standards Institution, London.
2. *Structural use of Concrete, BS 8110 (1985)*. British Standards Institution, London.
3. *Code of Practice for the Structural Use of Reinforced Concrete in Buildings, CP 114: Part 2 (1957)*, last reprint 1969. British Standards Institution, London.
4. *Design of Concrete Structures for Retaining Aqueous Liquids, BS 8007 (1987)*. British Standards Institution, London.
5. Evans, R. H. and Wilby, C. B., *Concrete – Plain, Reinforced, Prestressed and Shell*. Edward Arnold (1963).
6. *Building Code Requirements for Reinforced Concrete, A.C.I. 318–56*. American Concrete Institute, Detroit, Michigan, U.S.A. (1956).
7. C.E.B., *Recommendations for an International Code of Practice for Reinforced Concrete*. English edition. American Concrete Institute and Cement and Concrete Association (1964).

2

Properties of materials and mix design

2.1 Cement

Cement is the most important and expensive ingredient of concrete, on a price per tonne of material basis (dependent upon the mix, the aggregates can sometimes cost more than the cement in a cubic metre of concrete). It was patented by J. Aspdin in the U.K. in 1824 and he called his product *Portland Cement* because the 'artificial stone' (concrete) made with it resembled Portland stone. He had cement works in Yorkshire (Leeds and Sheffield) and then London. In the U.K., others made and had made previously, similar cements, for example the Romans in various baths and villas in England. He was, however, well ahead of his competitors at the time.

Portland cement is made by grinding together its principal raw materials, which are (a) argillaceous, for example silicates of alumina in the form of clays and shales, and (b) calcareous, for example calcium carbonate in the form of limestone, chalk, and marl which is a mixture of clay and calcium carbonate. The mixture is then burned in a rotary kiln (shaft kilns are still used for works with small outputs and there is an interest in their installation in developing countries) at a temperature between 1400 and 1500 °C; pulverised coal, gas or oil is the fuel. The material partially fuses into a clinker which is taken from the kilns, cooled and then passed on to ball mills where gypsum is added and it is ground to the requisite fineness. The resulting cement is allowed to contain small strictly limited percentages of materials not required, some disadvantageous for some uses, such as iron oxide and sulphur trioxide. A general idea of the composition of cement is indicated by the following oxide composition ranges for Portland cements: lime (CaO) 60–67%, silica (SiO_2) 17–25%, alumina (Al_2O_3) 3–8%, iron oxide (Fe_2O_3) 0.5–6%, magnesia (MgO) 0.1–4%, sulphur trioxide (SO_3) 1–3%, soda (Na_2O) and/or potash (K_2O) 0.5–1.3%.

The constituents forming the raw materials used in the manufacture of Portland cement combine to form compounds, sometimes called Bogue[1] compounds, in the finished product. The following four compounds are regarded as the major constituents of cement: tricalcium silicate ($3CaO.SiO_2$ or C_3S), dicalcium silicate ($2CaO.SiO_2$ or C_2S), tricalcium aluminate ($3CaO.Al_2O_3$ or C_3A) and tetracalcium aluminoferrite ($4CaO.Al_2O_3.Fe_2O_3$ or C_4AF).

A cement works is usually sited near to its raw materials. These sites vary and consequently cements from different works vary within permissible limitations. In the U.K. this variation seems to have an insignificant effect upon concrete. However, research by the author and others indicates that the asbestos cement manufacturing process is sensitive to the percentage of C_3S, which varies significantly with cements from different works in the U.K. Examples of other sensitivities: pipe spinners sometimes request coarse ground cement, aerated block manufacturers sometimes request cements with high total silicates, roof tile manufacturers sometimes prefer cements with higher alkalis (for the associated high strengths at early ages), floor layers dislike cements with short or long setting times, etc.

High alumina cement was first made by J. Bied for the French Lafarge Company in 1908, and named *Ciment Fondu*. This discovery was made whilst searching for a cement which liberated no free hydrated lime upon setting. Portland cement liberates free hydrated lime upon hydration and this in the resulting concrete is very vulnerable to attack from mineral sulphates, dilute acids and other agents.

When cement is hydrated, lime and alumina are liberated. The lime combines with the alumina and in the case of Portland cement an excess of lime results, whereas in the case of high alumina cement an excess of alumina results. Bearing this in mind, the properties of these two fundamentally different cements can often be predicted. For example, when these cements are mixed together and hydrated, the respective excesses of lime and alumina react chemically with one another and a *flash set* (almost instantaneous setting) can result. This can be useful for caulking small leakages in cofferdams and water-retaining structures. The flash set phenomenon is, however, a reason for new *Ciment Fondu* concrete not being suitable for jointing to new Portland cement concrete, and vice versa. Time limits have to elapse so that there is no danger of unhydrated Portland cement coming into contact with unhydrated high alumina cement and causing a flash set with accompanying local excessive thermal and shrinkage movement disrupting the concrete locally. The concrete which is to be extended should be 24 hours old if it is *Ciment Fondu* concrete, 2 days old if rapid hardening Portland cement, and 7 days old if ordinary Portland cement.

When cement is hydrated the terms *initial setting time*, *final setting time* and *rate of hardening* are used, often loosely. However, the first two are defined for cement by BS 12, 915 and 1370. Other tests of cement for soundness, tensile and compressive strength, chemical composition, fineness of grinding, etc., are described in BS 12. The definitions of *initial set* and *final set* unfortunately bear no precise relationship to practice. They do, however, enable the properties of different cements to be compared for their setting qualities. It can loosely be said that it is good practice not to disturb concrete after its initial set, and the initial setting time is normally not less than half an hour. There are exceptions to this rule in practice, however, since such operations as the trowelling of concrete floors and granolithic finishes, for example, usually need to be performed after the initial set, but before the final set has taken place. The final setting time is not usually more than ten hours.

If one imagines say a sewn-up sheep's bladder (a colloidal membrane) containing a solution, immersed in a similar solution of greater dilution, then water travels through the very fine pores in the bladder so that a pressure (an osmotic pressure) is developed in the bladder. This pressure continues to increase until the solutions on either side of the colloidal membrane have the same dilution. This is a very simple description of colloidal chemistry relative to the hydration of cement. Upon hydration the surface of a small portion of cement forms crystalline substances, which can be observed with an electron microscope.[2] These form a colloidal membrane, surrounding the portion of cement, called *tobermorite gel*[3] (a calcium silicate hydrate). As indicated previously, water travels through the membrane to dilute the solution of hydrating cement compounds within the membrane. This causes a pressure inside the membrane and hence expansion of the concrete or mortar. Conversely, drying of the cement after hydration causes *shrinkage* of the concrete. However, the amount of shrinkage caused by complete drying out of hydrated cement paste is not completely recovered by subsequent wetting.

If water is in contact with concrete, for example the wall of a basement, water can travel through the concrete not only via any cracks, construction joints, or voids, but also via the colloidal membranes. The water endeavours to pass through adjacent colloidal membranes, until all solutions surrounded by colloidal membranes have reached the same dilution. Thus water, or dampness, can be transmitted through a basement wall of sound concrete. Hence the desirability of 'tanking' (providing an impervious membrane) even if the concrete is very good.

The strength of a cement paste depends greatly upon the bonds formed between the very small particles of its cement gel. Generally the greater the number of these particles and the denser the gel structure, the stronger the gel mass.[2] The water-to-cement ratio used for a cement paste is related to its strength.

There are several types of cement available to the engineer, for example, as follows:

1. *Ordinary Portland cement.* This is the most inexpensive cement and is consequently widely used.

2. *Rapid hardening Portland cement.* As the name implies, concrete made with this cement hardens more rapidly than concrete made with ordinary Portland cement. Such a property enables early stripping of concrete formwork, especially advantageous for precast work where repeated uses are made of the same shutter. *Extra rapid hardening* cements can be obtained for special purposes. These two cements are of the same material as ordinary Portland cement except more finely ground. Over the years in the U.K. ordinary Portland cement has been more finely ground resulting in it being more rapid hardening than the latter was previously. Whilst this has allowed earlier stripping of formwork it has also been disadvantageous for example to the laying of granolithic toppings.

3. *High alumina cement (H.A.C.).* This cement is not classed as a Portland cement. It hardens much more rapidly than any other commercial cement, and it has the further advantage of being sufficiently immune, for practical purposes, to attack from several important chemicals. Some examples are: many of the sulphates present in subsoil waters and in sewage; sulphur compounds formed from the combustion of coal and oil; carbonic acid as experienced in subsoil waters from moorland areas; many of the chemicals contained in sea water; chemicals which attack Portland cement and which are present in important industries such as lactic acid (associated with milk), tar oil, cottonseed oil, beer, and sugar juices. H.A.C. was excluded from CP 110 by the August 1974 amendment, and is not mentioned in BS 8110, but was previously

allowed to be used when high strength was required urgently, for example on maritime structures when it was necessary to have a reasonably hard concrete before high tide; for the sealing of water leaks in emergencies when excavating in water-bearing ground; for structural work which was required to be in use within, say, 24 hours; for structural work where formwork was required to be stripped early or where it was required to prop further shutters from the members cast as soon as possible; for prestressed concrete, especially pretensioned concrete, where economy required release of the wires and removal of the members from the prestressing beds as early as the strength of the concrete permitted. The high early strength is obtained to some extent because the chemical reaction of the cement with water is very exothermic. To avoid the ills of overheating (see item 7 of this Section) it is desirable to have a low water-to-cement ratio (to reduce the rate of chemical activity), to cast at an ambient temperature of not more than about 20 °C, not to allow the internal temperature of the concrete to be more than 30 °C for more than 24 hours after casting, to cure with water or similar, and certainly not to steam cure.

The greatest disadvantage of high alumina cement was its cost, which made it prohibitive for many purposes. Another economic disadvantage was the necessity of curing with water or dampness. Concrete using this cement was nevertheless quoted as being more economical than steam cured Portland cement concrete for prestressed concrete work.

H.A.C. with suitable aggregate can be used as a refractory concrete or mortar for fireclay bricks and is suitable for temperatures up to about 1300 °C. High climatic temperatures in combination with high humidities as experienced in the tropics were found to reduce the strength of concrete made with H.A.C. rather alarmingly.[4] The chemical conversion of certain crystalline compounds having certain numbers of elements of water of crystallisation to other crystalline compounds with different numbers of elements of water of crystallisation could cause an internal volume change in the concrete with a consequent disruption and weakening of the concrete. The shape of crystals changes from hexagonal to cubic. Neville[4] claimed that this chemical conversion could also eventually occur with aging in the cool damp U.K. climate, although CP 110 prior to the August 1974 amendment, regarded this effect as negligible for properly cured concrete. It might be thought that high alumina cement concrete could be used in structures protected from moisture, which is the case with many buildings, without worrying about chemical conversion. Yet even indoors, with central heating and solar gain through large glass windows, temperatures can be high and it is argued that there is always water in some form inside the concrete, and the humidity of the atmosphere can be high in the U.K. and this air is not normally dried before entering buildings. After full conversion, concrete strength increases with age.

Although the dangers of conversion became rather catastrophically experienced about 1961, seemingly inadequate notice was taken of this subsequently, until about 1974 when there was considerable alarm concerning lack of reliable knowledge of when high alumina cement could be used. Inadequate notice was taken of work by Bolomey[5] of France in 1927 and Davey[5] in the U.K. in 1937; both demonstrated that high alumina cement concretes, hardened under good conditions, subsequently lost up to 40% of their strength permanently, due to curing in warm water, and experienced the colour change to yellow-brown, which we now know to be due to conversion.

In the case of the most publicised failure in the U.K., the prestressed concrete beams were over a swimming pool and experienced warmth, moisture from condensation and roof leaks, sulphate attack from the plaster, and possibly had poor concrete and support seatings. The other few failures in the U.K. seem to have had more than just conversion as a weakness. Subsequently most high alumina cement work has been tested in the U.K. and most of it found to be safe at the time of testing. Some structures have been strengthened against possible future weakness due to conversion. The author has tested roofs to a building with up to 95% conversion and found them very safe over several years. There is no doubt that steam constantly directed on to high alumina cement beams can cause them to disintegrate.

4. *Cement for use in cold weather.* Such cements (manufacture was discontinued in the U.K. some years ago) were usually achieved by adding about 1.5% of calcium chloride to rapid hardening Portland cement. The calcium chloride generates heat by reacting with the water used in mixing the concrete. This also enhances the rapid hardening qualities. Because of the heat evolved, these cements were very often profitably used in cold weather to allow concreting operations to continue. The high early strength properties were advantageous for allowing early stripping, and, in the case of precast concrete, handling. The chloride ion aggravates (is a catalyst to) the corrosion of steel (this is particularly well known in the case of NaCl, which is used for salting in winter to de-ice roads and has caused considerable corrosion to reinforcement near the joints of many

bridges, particularly in the Birmingham area of England). Hence if water and oxygen ion can penetrate to the reinforcement through pores and/or cracks in the concrete, the calcium chloride will increase the rate of corrosion of this reinforcement. It is interesting that in the case of water-retaining structures and underground pipelines, if the water is in contact with concrete containing very fine cracks which penetrate to the reinforcement, it is possible for corrosion to occur even though many would not imagine that air could penetrate through the crack. This is because the oxygen ion of air dissolved in the water is easily carried in the water penetrating the crack to the steel – refer to the theory of notch corrosion. The author came across this solution when consulted about various troubles with a prestressed concrete pipeline conducting water under a pressure of 24 metres of water, a distance of about 100 km. CP 110 prohibited calcium chloride in prestressed pretensioned concrete, and restricted it to not more than 1.5% by weight of the cement in reinforced concrete. Subsequent amendments effectively banned calcium chloride in reinforced concrete; theoretically a small amount could be used but this was too small to be effective practically as an accelerator. The author was consulted about a building where $CaCl_2$ used in the concrete had caused unbelievable corrosion to the reinforcement and as the volume of rust is many many times the volume of steel from which it is formed this had caused destructive damage to the concrete. In addition cables, each of eight wires, were completely severed with corrosion. Admittedly this was over a period of more than two decades but the concrete had been protected from the weather. It is considered by a certain chemist that the humidity in the U.K. atmosphere is sufficient to aid the corrosion from inside a building even if the concrete is protected adequately from rainfall. $CaCl_2$ is not allowed by BS 8110 in any concrete which is in contact with steel. It would seem that it is still useful for winter working for plain concrete construction such as certain paths, drives, roads and parking areas, providing these use no steel reinforcement. Non-chloride accelerators are now being used to some extent to aid winter working.

    5. *Sulphate-resisting cement*. This cement is made specifically to resist the attack of sulphates. Underground structures can experience sulphate attack from the soil, back-fill or ground water. There is a cement known as *super sulphated cement* which is sometimes claimed to be better when the sulphates are acid in nature.

    6. *Cements with a low coefficient of shrinkage* can be specifically devised

for highways, dams, water-retaining structures, etc., to reduce the magnitude of cracks caused by shrinkage. Such a cement, which also had low heat of setting, was devised and used for the mass concreting to the Boulder Dam, U.S.A. There are cements which are claimed to expand, but they do not always do so if the concrete subsequently dries out.

7. *Low heat Portland cements* generate less heat upon reacting with water than normally experienced with other cements and are thus suitable for mass concrete work. The heat generated with Portland cement in mass concrete work can literally boil off the water required for the necessary chemical reaction, the steam causing flash setting of some of the cement and also disruption and voids in the resulting concrete.

8. *Portland-pozzolana cements.* Fly ash (pulverised fuel ash, P.F.A., or pozzolana) is sometimes substituted for 15–35% (one cement manufactured in the U.K. uses 28%) by mass of the ordinary Portland cement to achieve low heat of setting and reduced shrinkage without reducing the 28-day strength of the concrete, but the early rate of hardening is reduced. About 1970 this idea was used for a gravity dam in Yorkshire, England and to help further, the concrete mix had a low cement content and used a large size of aggregate. Unfortunately fly ash contains a small amount of sulphate. BS 5328:1981 restricts the total sulphate content of a mix expressed as SO_3 to not more than 4% by mass of the cement. So far, in the U.K., in practice fly ash has had no difficulty in complying with this restriction. To use fly ash refer to BS 6588.

9. *Coloured cements* are used for reconstructed stones, renderings, and the like. Because of the high cost of these cements, coloured artificial stones usually have a facing about 38 mm thick made with the coloured cement, and a backing made with ordinary Portland cement. Coloured cements can be obtained by adding the following pigments to Portland cement: yellow ochre (yellow), brown oxide of iron (brown), green oxide of chromium (green), red oxide of iron (red), manganese black (black). The weight of the pigment should not exceed 10% of the weight of the cement, otherwise the strength will be impaired. White cements are popular and require to be specially manufactured. The colour of a concrete can be improved and will wear better if the aggregates also are of a colour similar to the coloured cement. For several years the manufacture of coloured cements has been discontinued in the U.K., except for white cement.

10. *Portland blast furnace cement* is obtained by grinding granulated blast furnace slag with the clinker which is normally ground down to make

Aggregates

ordinary Portland cement. It has a slightly lower heat of hydration than ordinary Portland cement, is slightly more resistant to sulphate attack, and is slower to develop its early strength.

11. *Water-repellent cements.* Certain ones are most effective in sealing leakages in water-retaining structures.

12. *Low-alkali cements, alkali–silica reaction and petrographic analysis of concrete.* Since about 1983 it has been realised that certain cracking has been due to alkali–silica reaction (ASR). This is caused by silica in certain aggregates reacting chemically with alkali in the cement and water is involved, provided by concrete retaining a high moisture content. This chemical reaction gives compounds which are greater in volume than the alkali and silica and therefore causes expansion. This disruption and internal expansion can cause severe and extensive cracking. To reduce this problem low-alkali Portland cements are available under BS 4027. The other solution is to use other aggregates. It is not uncommon for ASR to occur with many aggregates to such a minor extent that no significantly undesirable cracking occurs. This can be observed by a *petrographic analysis of concrete* – an impregnated thin section (e.g. measuring 75 mm by 50 mm by 30 microns) is cut from the concrete and examined under a polarising microscope at magnifications up to × 1000 in accordance with ASTM C856-83.

2.2 Aggregates

Aggregates are classed as *fine aggregates* and *coarse aggregates*. Generally, various sands are used as fine aggregates, and coarse aggregates are either water-worn gravels, sometimes larger pieces are crushed to smaller sizes, or crushed rocks. The aggregates chosen are usually the most inexpensive to give the requisite quality of concrete. The engineer must, however, be satisfied that the source selected will consistently supply the quality of aggregate which he has approved because the supply can vary in different parts of the same quarry. This can be difficult for certain special requirements. Sometimes the engineer requires stockpiles at the suppliers' works to meet with his approval. These are then drawn upon exclusively for the concreting operations.

Aggregates for normal concreting work are a fairly inexpensive commodity at the quarry and thus transport charges substantially influence their overall cost. Local aggregates are therefore generally employed, but an expensive type of aggregate may warrant greater transport costs if the necessary stone does not occur locally. Examples of more expensive stones are: granites for granolithic finishes; various types of coloured aggregates for artificial (reconstructed) stones (usually used for the surface layer of the stone only); and vermiculite (imported into the U.K.) for lightweight finishes.

Reference should be made to the British Standards 882, 1199, and 1200, which recommend various gradings of the particle sizes for both fine and coarse aggregates. These enable standardisation and control but are not necessarily ideal gradings for concrete. The standards quoted specify tests of other relevant qualities of the aggregates, namely specific gravity, water absorption, bulk density, organic impurities, and crushing strength. Figure 2.1 shows four gradings, upon which the mix designs of the D.S.I.R. *Road Note No. 4*[8] are based, for 19.05 mm ($\frac{3}{4}$ in) and down aggregates, and one average grading curve for 9.52 mm ($\frac{3}{8}$ in) aggregate. The grading of a 19.05 mm ($\frac{3}{4}$ in) aggregate should lie within the curves 1 and 4 and preferably within the curves 2 and 3 if this method of mix design is to be used.

Coarse aggregates can be classified according to shape (BS 812) as follows:

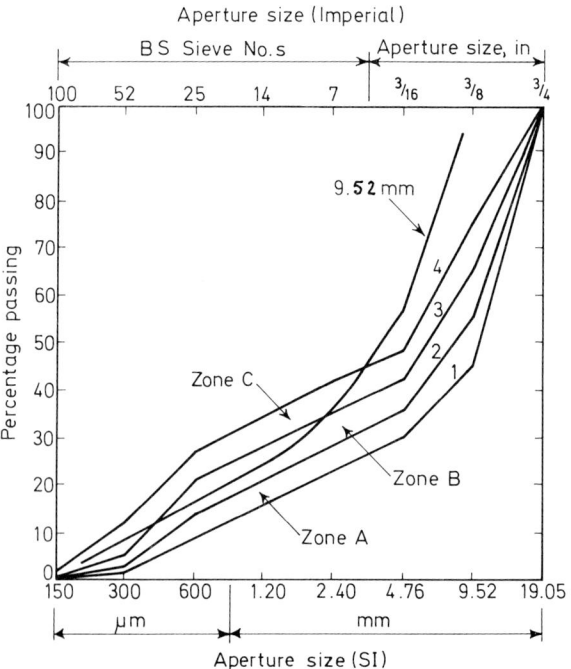

Figure 2.1

Aggregates

1. *Rounded aggregates*, for example beach and other well worn gravels.
2. *Irregular aggregates*, for example water worn river gravels.
3. *Angular aggregates*, for example crushed rock or manufactured materials. These are commonly granites, limestones, basalts, quartzites, flints, pumice, broken bricks, foamed slag, blast furnace slag, sometimes a strong sandstone, vermiculite and duromit, etc.

The grading, shape, porosity and surface texture of the aggregates can affect the workability and consequently the strength of concrete.

When a concrete is required to be lightweight, and to have a good resistance to heat transmission, and a high strength is not required, special lightweight aggregates are often used, such as vermiculite, foamed slag, clinker, breeze, pumice, wood wool and expanded shales.

If water is added to 1 m³ of sand, the gross volume of this sand increases until it occupies about 1.25 m³. After this volume is attained the addition of further water decreases the gross or bulk volume until when the sand is finally saturated the volume has returned to 1 m³. When concrete is 'batched' by volume (that is the ingredients measured by volume) the water content of the sand greatly influences the quality of the resulting concrete. Consider a 1(cement):2(sand):4(gravel) mix, the ratios referring to dry volumes of the respective materials (as is standard practice). If we were using a sand experiencing its maximum amount of 'bulking' of, say, 25%, then the mix actually produced in terms of dry volumes would be $1:2/1\frac{1}{4}:4$ or $1:1.6:4$.

If water is added to 1 kg of sand, the gross weight is increased by the weight of the water added to about 1.1 kg upon saturation. Hence, if the batching of concrete were by weight, the water content of the sand would still be troublesome but not to as great an extent as by volume. Consider again a 1:2:4 mix and let the sand be increased in weight by its maximum amount of say 10% due to its water content. Then the mix actually produced in terms of dry volumes would be $1:2/1.1:4$ or $1:1.818:4$. For illustrative purposes it has been assumed that the bulk densities of the dry materials are the same. Thus the inaccuracy of batching by weight is basically not as great as batching by volume. This reasoning ignores the fact that the same phenomenon also affects coarse aggregates, but to a far lesser extent. Several devices are available for measuring the water content of the aggregates, so that the mix can be adjusted accordingly. The water content often varies from place to place in a stockpile. When a large concreting programme is being conducted, sometimes the stockpiles will be insufficient (especially on congested sites) and sand which arrives during the course of the concreting operations will have a different water content to the sand in stock. Aggregates are commonly exposed to the weather so that the water content will vary with the rainfall. One needs to be vigilant therefore to allow for the errors in batching caused by the water content of the aggregates.

2.3 Concrete

Coarse aggregate, fine aggregate, cement and water are mixed together in suitable proportions, and this mixture, placed and compacted wherever required, solidifies after a lapse of time into what is known as concrete.

The mixes of concrete commonly used (CP 114) for structural purposes were 1 part (by dry volume) of cement:2 parts (by dry volume) of fine aggregate:4 parts (by dry volume) of coarse aggregate, and similarly, $1:1\frac{1}{2}:3$ and $1:1:2$. CP 110 calls such mixes 'prescribed mixes' and specifies them for various grades of concrete in terms of weights of cement and total dry aggregates with percentages by weight of fine aggregate in total dry aggregates. BS 8110 refers to BS 5328 to recommend *prescribed mixes*; see Section 2.3.15. It is now more common in the U.K. to design mixes to specified grades or strengths.

Many investigators have proved that most of the qualities desired of concrete (the materials being kept constant) benefit by increased compressive crushing strength, for example, strength in tension, shear, and resistance to weathering, abrasion and wear, and impermeability. Exceptions to this rule are lightness (in density), and thermal insulation.

The factors which have the greatest effect upon the strength of concrete are the cement-to-aggregate ratio, the compaction, the water-to-cement ratio of the mix, and the method of curing.

It is easy to imagine that the strength of concrete depends upon the absence of voids, or in other words, upon the final density after setting and maturing. For example, 5% of air voids can give a loss in strength of 30%, 10% of voids can give a loss in strength of 60% and 25% of voids can give a loss in strength of 90%. Compaction of the concrete is therefore extremely important, and this is dependent upon the 'workability' of the concrete.

2.3.1 Workability

Workability is the ease with which concrete can be placed in moulds, compacted around reinforcement and screeded to a level. Many tests have been devised for measuring this property, and all have been subjected to adverse criticism. The test which has possibly been condemned the most, namely the *slump test*, is and has been the most commonly used in the U.K., and is referred to by BS 5328:1981. The nature and the grading of the aggregates considerably affect the slump. Thus specifying the slump can ensure uniformity in the consistency of concrete during the progress of work only if the materials are of constant quality.

Other well known tests of workability are the *compacting factor test* and the *VB consistometer test*. The former was developed as an improvement upon the slump test in attempting to measure workability. The latter became useful in the U.K. when drier concretes than previously became necessary for prestressed concrete work, as it can distinguish between

Table 2.1. *Uses of concrete of different degrees of workability (Road Note No. 4)*

Degree of workability	Slump, mm	Compacting factor		Use for which concrete is suitable
		Small apparatus	Large apparatus	
Very low	0–25	0.78	0.80	Vibrated concrete in roads or other large sections
Low	25–50	0.85	0.87	Mass concrete foundations without vibration. Simple reinforced sections with vibration
Medium	50–100	0.92	0.935	Normal reinforced work without vibration and heavily reinforced sections with vibration
High	100–180	0.95	0.96	Sections with congested reinforcement. Not normally suitable for vibration

various concretes having virtually zero slump. It is also better for very dry mixes than the compacting factor test.

Table 2.1 recommends suitable approximate workabilities of concrete for various uses.

Good compaction of the concrete, and hence a high strength concrete with a good finish, can be obtained by manipulation of the grading and type of the aggregates, the use of additives to reduce the surface tension of the water, employment of vibration and/or pressure, and use of a high water content (but this latter acts contrary to strength; requiring more cement to compensate).

The additives are *plasticisers* and *super-plasticisers* comprising soaps, detergents, or resins. Essentially they reduce the surface tension of the water, that is the water wets the particles more easily, increasing workability. They allow the water-to-cement ratio to be reduced for no decrease in workability, thus giving a stronger concrete. Some entrain finely dispersed air bubbles sufficiently for the concrete to have increased frost resistance for little decrease of strength – used for roads in cold countries and called *air entrained concrete*.

The use of a high water content must be avoided as much as possible as it also decreases the strength of the concrete, as explained later. It can however be used with advantage when combined with a vacuum process (see Section 2.3.4).

2.3.2 Water-to-cement ratio and strength of concrete

A high strength concrete requires to be as free from voids as possible. If water in excess of the amount required for the chemical reaction with the cement is present in the mix, this water remains in a free state and the concrete sets around the drops of water. Such particles of water form pores and voids in the concrete, resulting in weakness and permeability. Dependent upon curing conditions they may freeze and expand, cause corrosion and/or eventually evaporate into the atmosphere.

The important effect of the water-to-cement ratio, by weights, on the strength of concrete was published in 1918 by D. Abrams of Chicago, who stated that the strength of any *workable* concrete, of constant materials other than water, was dependent upon the water-to-cement ratio alone, assuming the same cement and degree of compaction are used and the conditions of curing and age at comparison of strengths are constants. The statement applies whatever the types of aggregates used provided the concrete does not fail by the fracture of such aggregates. The workabilities of different mixes having the same water-to-cement ratios can be considerably different; for example a lean (low proportion of cement) mix might need vibration to obtain the same compaction as a richer (in cement) mix placed by hand.

The strength of concrete increases as the water-to-cement ratio decreases, provided the water present is sufficient to allow the full chemical reaction to occur with the cement. If the water is less than this amount, a decrease in strength is experienced. Figure 2.2 shows the relationship between the average ultimate compressive stress (or crushing strength) and the water-to-cement ratio for 150 mm cubes of fully compacted concrete for mixes of various proportions. Over the years U.K. manufacturers have altered the former ordinary Portland cement to what was rapid hardening cement by finer grinding.

Only the compressive strength of concrete has been considered so far. It is generally accepted that this is a fairly reliable guide to the tensile and

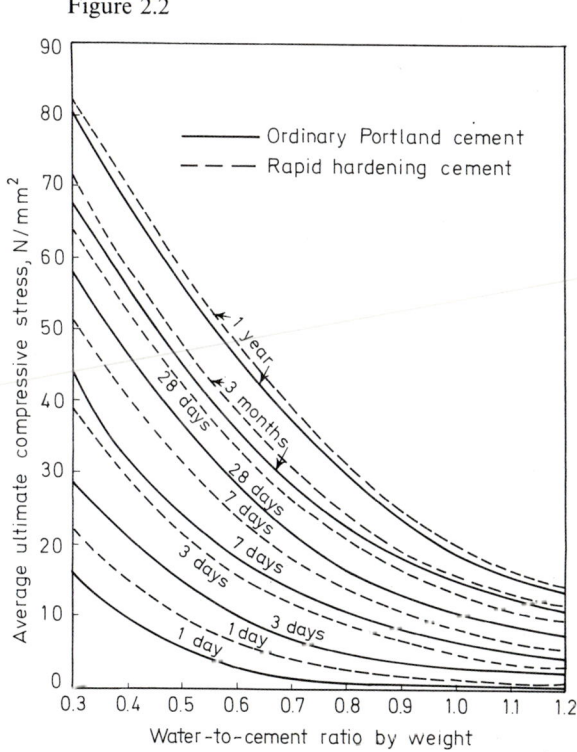

Figure 2.2

shear strengths, the modulus of rupture, the resistance to abrasion and wear, durability to the weather, density, porosity and watertightness for constant materials. For durability, cement content is also important and minima are specified for various conditions in BS 8110.

2.3.3   Strength tests of concrete

BS 1881 recommends a standard compressive test, and also a standard test for the modulus of rupture. The latter flexural tensile test gives greater values than those obtained from tension tests made on standard briquettes (BS 12). The cross section of the briquette which is tested in tension is 25 mm square, the specimen being primarily designed for testing cements by determining the strengths of their cement/sand mortars. Larger specimens should be used for tension tests when the maximum size of the aggregate is greater than 9 mm. The cylinder splitting test has become popular as a tensile test of concrete. Unfortunately it is an indirect test of tension and assumes an elastic theory to calculate the distribution of ultimate stress.

Shear in concrete beams is thought of in terms of diagonal tension and consequently the tensile strength of concrete is more relevant than the shearing strength. The shearing strength can be obtained from torsion tests of cylinders of concrete. The distribution of shear stress in such tests, however, is not the same as experienced in, say, a punching shear test.

With all the tests mentioned, size and shape of specimen matter, and thus empirical factors are usually required to relate these indicative control tests to the behaviour in the structural member.

2.3.4   Vacuum concrete

The concrete is made sufficiently wet to be placed and compacted easily and then the vacuum process removes water from the concrete, so that it finally has a low water-to-cement ratio. The water is extracted through mats placed in contact with the concrete. These mats are such that only water, and no cement, or fines (out of the aggregates) can be sucked from the concrete by the vacuum pump. Side shutters can usually be removed immediately afterwards if desired, as the concrete has almost zero slump. In the U.K. the vacuum concrete process is used by certain, but not all, firms making pavement flag stones.

2.3.5 *Vibrated concrete and pressure compaction*

Concretes with low water-to-cement ratios can be placed and compacted by internal or external vibrators. External vibrators usually consist of motors with heavy cams on their shafts, and are fastened to a mould. Internal vibrators are of a poker type and can be held in the hand and immersed in the concrete where required. Again the vibration is by cams on a shaft. These are inside the poker and are rotated by compressed air or by a flexible drive shaft driven with an electric motor; the latter type of poker vibrator can usually be of a smaller diameter (say 25 mm as opposed to 50 mm). They are the more efficient for compaction and do not require the strong moulds often necessary for the external vibrators. If sufficiently dry mixes are used, the sides of the moulds can be removed immediately after vibration. There are in fact beam-making machines where the concrete is compacted by vibration, the sides removed immediately, and the beam on its pallet dragged away along skids. Most block-making machines employ pressure as well as vibration. Here again, solid and hollow blocks can be removed immediately from block-making machines on their pallets.

Workmen, when not strictly supervised, tend to make concrete extremely wet, often to make for easy placing and sometimes to avoid the mix drying out if placing is delayed by problems on the site or in transit in the case of ready mixed concrete. Vibration does not increase the workability of such concrete and can be detrimental by causing segregation of the constituents of the concrete, the gravel tending to sink to the bottom, and the sand and cement to float to the top of the concrete. Such segregation can also occur with dry mixes if the vibration is sustained for a long enough period. The vibration employed with an apparently dry mix should be *only just sufficient* to make the concrete flow into the sharp arrises of the mould and around the reinforcement. Poker vibrators should not be removed rapidly or they can leave voids behind them.

Essentially compaction by pressure and/or vibration enables drier concretes to be satisfactorily compacted to make stronger concretes.

2.3.6 *Gap graded concrete*

The principle of this method is to omit from the gradings, such as those of Figure 2.1, certain undesirable sizes of aggregates. Undesirable sizes are those which prevent the efficient packing of the other sizes. If desired the smaller sizes of the coarse aggregate can be omitted, or one or two sizes only of coarse aggregate can be used.

The general aim of gap grading is to achieve strength from the efficient packing of the aggregate. This saves cement and allows aggregate suppliers to supply larger aggregate, less expensive to crush, which suits them also because there is normally a large demand for small aggregate for throwing with salt onto winter roads in the U.K. By careful packing of stones, a strong wall can be built without using any cement. If a cement paste were to fill all the minor voids in such a wall, then a very strong construction would result, and this would be the ideal aimed at by the advocates of the gap grading of concrete.

A multitude of spheres of diameter D have a rhombohedral form of packing. These can be termed *major spheres*, and spheres of diameter $0.414D$, known as *major occupational spheres*, can fit into the voids between the major spheres. These spheres could, mathematically, constitute our coarse aggregate. The fine aggregate would then consist mathematically of *minor occupational spheres* of diameter $0.225D$, which would fit into the remaining voids. The voids now remaining can be fitted by *admittance spheres* of diameter $0.155D$, and these could also be provided by the fine aggregate. Cement would then occupy the remaining voids and a mathematically perfect compact mix would result. Such a mix, however, could not normally be mixed and cast in this ideal fashion and consequently some authorities[6] consider that only the major and admittance spheres are of practical value in designing a mix.

Mixes therefore are often designed with one size of coarse aggregate (for example 20 mm) and a sand, all the particles of which can pass through the voids in the compacted coarse aggregate. The sand is designed to fill the voids in the coarse aggregate and the cement is designed to fill all the remaining voids. The particles of sand must not be smaller than necessary, as this will increase the total surface area to be wetted with water and cement, and consequently a wetter mix (giving a weaker concrete) would be required for any requisite workability. Irrespective of the calculation just suggested, the sand should be sufficient to distribute itself uniformly throughout the mix under practical conditions of mixing and placing. When the sand is less than 18% of the mix it is difficult to obtain uniformity even under laboratory conditions. Mixes are often designed and then modified to suit the particular site conditions of mixing and compacting. Of course allowance has to be made in the calculations when the particles of aggregate are not true spheres, indeed they can be very angular.

To increase workability it is advantageous to reduce the surface area of all the aggregates in a unit volume. This can be done by using larger particles. The largest aggregate possible should therefore be used, consistent with the minimum clearances allowed between and around reinforcement.

Gap grading enables leaner and drier mixes to be used, the absence of many intermediate sizes of aggregates having reduced the specific surface area of the aggregates and therefore having increased workability. The lean mixes usually utilised, however, make good mixing and vibration almost essential. Such concrete, being made of leaner and drier mixes than a conventional concrete of equivalent strength, will therefore experience less shrinkage and hence possess better weathering qualities. Compressive forces on the gap graded concrete described are ideally transmitted from particle to particle of the coarse aggregate and not through any cement and sand particles. Consequently the creep associated with such concrete is low. A coarse aggregate as used in a conventional mix experiences a fair amount of segregation during transportation, and pouring into and out of lorries, etc. Rain also helps segregation in stockpiles. Gap grading avoids these disadvantages by requiring only single sizes of coarse aggregate.

Some advocate two different single sizes of coarse aggregates to be used with sand and cement in a mix. Gap graded concretes as lean as 1(cement):2.45(sand):6.59(gravel), with a water-to-cement ratio of 0.51, increase in strength with age in a similar fashion to conventional concretes.[6] Because of the packing of the aggregate of a gap graded concrete, vertical shutters can often be removed immediately after casting. Walls and columns can then be trowelled if desired or sprayed with a light water jet to expose the aggregate, or if there are surface voids these can be made good.

One disadvantage of gap grading is that if the single-size aggregates supplied contain over 2.5% by weight of undesirable particles, this upsets the grading which is very sensitive to such intrusions. If however such irregularities are to be expected in the supply then the mix can be calculated accordingly to be of reduced efficiency.

2.3.7   No fines concrete

Coarse aggregate (gravel) is mixed with cement and the fine aggregate (sand) is omitted. *No fines* concrete is required to contain a multitude of voids to give good thermal insulation, and these voids need to be large enough to prevent the movement of water through the concrete by capillary attraction. *In-situ* no fines concrete walls have been used in the U.K. for housing, the idea being that good thermal insulation is achieved and that rain beating on a wall penetrates only a short horizontal distance before having dropped to the bottom of the wall, there being no capillary paths to conduct the water completely through the wall. It is, however, often desirable to render and paint exposed no fines concrete walls.

2.3.8 Curing of concrete

After setting or solidifying, concrete increases in strength with age (see Figure 2.3). The strength at a particular age can be further increased by suitable curing of the concrete whilst it is maturing. Such curing comprises the application of heat (not if $CaCl_2$ is present or for high alumina cement or mass concrete) and/or the preservation of moisture within the concrete. The application of heat speeds up the chemical reaction and consequently the rate of hardening of the concrete.

It can be imagined that preventing the escape of moisture from the concrete enables previously unwetted minute particles of cement to participate in the cementing action. If heat is applied to accelerate the hardening of the concrete it is therefore important not to expel the water held within the concrete. In other words, if heat is applied a high humidity is also desirable; steam is therefore a most suitable medium for this purpose. Steam curing can be done at atmospheric pressure or under pressure. The latter method is more effective but far more expensive, as pressure chambers are required. For example, the half-hour strength of concrete steam cured under pressure could equal the 28-day strength of an identical concrete maturing in air.

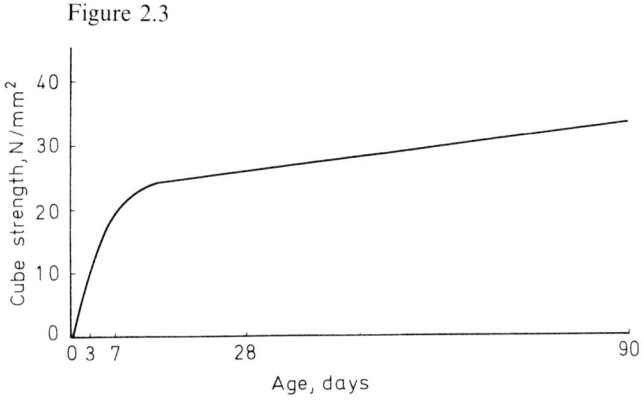

Figure 2.3

Concrete   33

Increasing the strength of concrete by preventing the water used in mixing from escaping is usually done in one of the following ways:

1. *Flooding or submerging the concrete in water.* The floors of basements and reservoirs can fairly easily be flooded with water. Precast concrete units can be immersed in water in special tanks.
2. *Treating the surface of the concrete* so that it cannot dry out. Proprietary products exist for painting, or for applying coverings which adhere to the concrete. Spray guns are often used for applying these surface treatments.
3. *Covering the concrete with damp sand* or *hessian* fabrics, which are kept damp by watering periodically, or with thin polythene sheet. This latter is the more popular these days.

2.3.9   Design of concrete mixes using Road Note No. 4

Most commonly a concrete mix is designed to give the specified strength at the minimum cost. The cost depends upon the value of the materials, the labour required for batching, mixing, transporting, placing and trowelling, and the method of curing adopted.

Mix designs are fairly inaccurate due to the number of possible variables. The D.S.I.R. *Road Note No. 4*[8] of 1950 based a mix design method on the aggregate gradings shown in Figure 2.1. This method is simple and can be used by mixing one's sand and gravel in such proportions as to correspond to one or other of these grading zones. As the method is even then not very accurate, it can be improved upon by casting trial mixes, measuring their workabilities and cube crushing strengths, and then adjusting the mix accordingly. Much of this work can easily be performed in the laboratory. The part of the method with which Table 2.2 is concerned has of course to be established by co-operation with the site. Considerable creditable

Table 2.2

Conditions	Minimum strength as percentage of average strength
Very good control with weight batching, moisture determinations on aggregates, etc.; constant supervision	75
Fair control with weight batching	60
Poor control; volume batching of aggregates	40

research since *Road Note No. 4*[8] has been faced with the inherent complexity of the problem and has seemingly not made this method obsolete as a very simple method of designing a mix. All other methods can still be improved by studying trial mixes, as mentioned previously.

CP 114 used to specify concretes according to their minimum cube crushing strength at 7 and 28 days, and it is still possible for a designer to do this, but CP 110, BS 8110 and BS 5328 have a more scientific approach – unfortunately more complicated. The mix design method presented in this book is based on the required average crushing strength. To design a mix with a certain specified minimum crushing strength, as for CP 114, we use Table 2.2 to obtain the requisite average crushing strength, and then design a mix for this average crushing strength.

BS 8110 and BS 5321 specify a concrete with a *characteristic strength*. For example, they define Grades C25, C30 and C35 of concrete as having

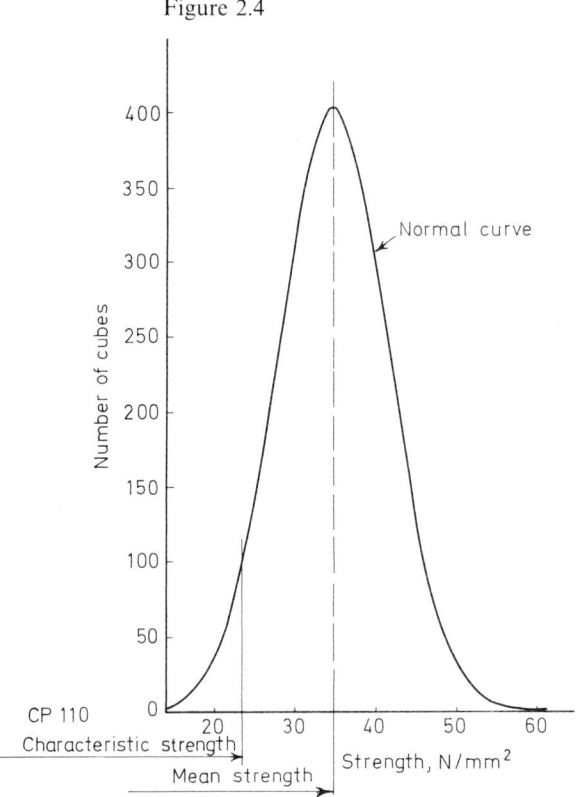

Figure 2.4

characteristic strengths of 25, 30 and 35 N/mm². If a large number of cubes of the same concrete (same age, curing, etc.) are tested the results can be plotted as shown in Figure 2.4. In statistics this figure is known as a *histogram*, and its shape is well known as the normal (Gaussian) distribution. The average (or mean) cube strength is the value of cube strength corresponding to the centroid of this shape. As in this case we assume it to be a normal distribution, not skew[7] (though in fact it is slightly skew), the average (or mean) strength correponds to the centre line of the shape, as shown. Statistical theory gives the formula:

$$\text{Characteristic strength} = \text{Mean strength} - 1.64 \times \text{Standard deviation} \qquad (2.1)$$

The number 1.64 is derived from the fact that BS 8110 and BS 5328 choose characteristic strength as the value below which we can expect 5% of the cubes to fail (see Figure 2.4).

The breadth of the shape of Figure 2.4 gives an indication of the scatter of the results. For statistical purposes this is expressed as *standard deviation*, which can be obtained thus: if we make n cube tests and their crushing strengths are $f_{cu1}, f_{cu2}, \ldots, f_{cun}$ then the mean crushing strength is $f_{cum} = (\Sigma f_{cu1})/n$ and the standard deviation is $\sqrt{[\Sigma(f_{cu1} - f_{cum})^2/(n-1)]}$.

If we are to design a concrete to a particular BS 8110 characteristic strength then we must obtain the mean strength, that is Characteristic strength $+ 1.64 \times$ Standard deviation. Hence, we need to know the standard deviation. Equation (2.1) can be expressed as:

$$\text{Characteristic strength} = \text{Mean strength} - \text{Margin} \qquad (2.2)$$

where

$$\text{Margin} = 1.64 \times \text{Standard deviation} \qquad (2.3)$$

and thus

$$\frac{\text{Characteristic strength}}{\text{Mean strength}} = 1 - 1.64 \times (\text{Coefficient of variation}) \qquad (2.4)$$

where

$$\text{Coefficient of variation} = \frac{\text{Standard deviation}}{\text{Mean strength}} \qquad (2.5)$$

36   *Properties of materials and mix design*

2.3.10   Design of concrete mix of given mean (or average) strength using Road Note No. 4

To design a concrete mix for a site the mean strength has first to be established as in Section 2.3.9. If however the mix is for a laboratory experiment then we design for the mean strength. The required water-to-cement ratio for the mean strength required is obtained from Figure 2.2, which assumes that the concrete is cured in air. Then a decision is made on the degree of workability, using Table 2.1 as a guide. Then the most suitable aggregate-to-cement ratio can be chosen from Table 2.3. This table gives such ratios for different gradings (as given in Figure 2.1), workabilities, water-to-cement ratios, and types of aggregates.

Durability is important and BS 8110 relates this to a minimum cement content according to the condition of exposure and the concrete cover to the reinforcement. Therefore Clause 3.3.3 of BS 8110 should be consulted to see if we have sufficient cement in our mix. If not, we decrease the aggregate-to-cement ratio accordingly. If this has to be done we might perhaps then repeat our design, taking advantage of, say, an increased workability to assist compaction and ease, and therefore cost, of concreting.

Example 2.1. To design a concrete mix for a pretensioned prestressed beam to have a mean (or average) crushing strength of 47 N/mm^2 at an age of 7 days.

The coarse aggregate to be used is a 19.05 mm ($\frac{3}{4}$ in) and down, rounded aggregate with a grading curve approximating to Curve 2 on Figure 2.1. Vibration is to be employed and the prestressing wires cause little obstruction to the placing of the concrete. We shall assume however that the beam is of I-section with narrow flanges and web. Hence it is decided that a *medium* workability is desirable (see Table 2.1).

Using rapid hardening Portland cement and consulting Figure 2.2, the necessary water-to-cement ratio is 0.35.

From Table 2.3, the aggregate-to-cement ratio is therefore 3.

To check that the cement content is adequate for durability, Clause 3.3.3 of BS 8110 gives the minimum mass of cement per m^3 of the concrete. Hence it is necessary to calculate this quantity from (say) Section 2.3.12.

Table 2.3. (1) 19.05 mm (¾ in) rounded aggregate

Degrees of workability	Very low				Low				Medium				High			
Grading of aggregate*	1	2	3	4	1	2	3	4	1	2	3	4	1	2	3	4
Water-to-cement ratio by weight 0.35	4.5	4.5	3.5	3.2	3.8	3.6	3.2	3.1	3.1	3.0	2.8	2.7	2.8	2.8	2.6	2.5
0.40	6.6	6.3	5.3	4.5	5.3	5.1	4.5	4.1	4.2	4.2	3.9	3.7	3.6	3.7	3.5	3.3
0.45	8.0	7.7	6.7	5.8	6.9	6.6	5.9	5.1	5.3	5.3	5.0	4.5	4.6	4.8	4.5	4.1
0.50	—	—	8.0	7.0	8.2	8.0	7.0	6.0	6.3	6.3	5.9	5.4	5.5	5.7	5.3	4.8
0.55	—	—	—	8.1	—	—	8.2	6.9	7.3	7.3	7.4	6.4	6.3	6.5	6.1	5.5
0.60	—	—	—	—	—	—	—	7.7	—	—	8.0	7.2	×	7.2	6.8	6.1
0.65	—	—	—	—	—	—	—	8.5	—	—	—	7.8	×	7.7	7.4	6.6
0.70	—	—	—	—	—	—	—	—	—	—	—	—	×	—	7.9	7.2

(2) 19.05 mm (¾ in) irregular gravel aggregate

Degrees of workability	Very low				Low				Medium				High			
Grading of aggregate*	1	2	3	4	1	2	3	4	1	2	3	4	1	2	3	4
Water-to-cement ratio by weight 0.35	3.7	3.7	3.5	3.0	3.0	3.0	3.0	2.7	2.6	2.6	2.7	2.4	2.4	2.5	2.5	2.2
0.40	4.8	4.7	4.7	4.0	3.9	3.9	3.8	3.5	3.3	3.4	3.5	3.2	3.1	3.2	3.2	2.9
0.45	6.0	5.8	5.7	5.0	4.8	4.8	4.6	4.3	4.0	4.1	4.2	3.9	×	3.9	3.9	3.5
0.50	7.2	6.8	6.5	5.9	5.5	5.5	5.4	5.0	4.6	4.8	4.8	4.5	×	4.4	4.4	4.1
0.55	8.3	7.8	7.3	6.7	6.2	6.2	6.0	5.7	×	5.4	5.4	5.1	×	4.8	4.9	4.7
0.60	9.4	8.6	8.0	7.4	6.8	6.9	6.7	6.2	×	6.0	6.0	5.6	×	×	5.4	5.2
0.65	—	—	—	8.0	7.4	7.5	7.3	6.8	×	×	6.4	6.1	×	×	5.8	5.6
0.70	—	—	—	—	8.0	8.0	7.7	7.4	×	×	6.8	6.6	×	×	6.2	6.1

Table 2.3 (cont.)

(3) 19.05 mm (¾ in) crushed rock aggregate

Degrees of workability	Very low				Low				Medium				High			
Grading of aggregate*	1	2	3	4	1	2	3	4	1	2	3	4	1	2	3	4
Water-to-cement ratio by weight 0.35	3.2	3.0	2.9	2.7	2.7	2.7	2.5	2.4	2.4	2.4	2.3	2.2	2.2	2.3	2.1	2.1
0.40	4.5	4.2	3.7	3.5	3.5	3.5	3.2	3.0	3.1	3.1	2.9	2.7	2.9	2.9	2.8	2.6
0.45	5.5	5.0	4.6	4.3	4.3	4.2	3.9	3.7	3.7	3.7	3.4	3.3	3.5	3.5	3.2	3.1
0.50	6.5	5.8	5.4	5.0	5.0	4.9	4.5	4.3	4.2	4.2	3.9	3.8	×	3.9	3.8	3.5
0.55	7.2	6.6	6.0	5.6	5.7	5.4	5.0	4.8	4.7	4.7	4.5	4.3	×	×	4.3	4.0
0.60	7.8	7.2	6.6	6.3	6.3	6.0	5.6	5.3	×	5.2	4.9	4.8	×	×	4.7	4.4
0.65	8.3	7.8	7.2	6.9	6.9	6.5	6.1	5.8	×	5.7	5.4	5.2	×	×	5.1	4.9
0.70	8.7	8.3	7.7	7.5	7.4	7.0	6.5	6.3	×	6.2	5.8	5.7	×	×	5.5	5.3

* Curve No. on Figure 2.1.
— Indicates that the mix was outside the range tested.
× Indicates that the mix would segregate.

Concrete

2.3.11 Combining aggregates to obtain a grading for the Road Note No. 4 mix design method

Available sands and gravels need to be combined in suitable proportions so that the resultant grading approximates to one of the curves of Figure 2.1, so that the method of mix design of Section 2.3.10 can be used. A graphical method is given for obtaining these proportions in *Road Note No. 4*[8], but the method of calculation, illustrated by the following example, is simpler to explain and understand and the calculations are trivial.

Example 2.2. The gradings of sand and two coarse aggregates kept in stock in the concrete laboratory at Bradford University are given in Columns (a), (b) and (c) respectively of Table 2.4. Suppose that these are combined to approximate to Curve 1 of Figure 2.1, whose grading is listed in Column (i) of Table 2.4.

To 1 kg of sand we can only decide how many kg x of 9.52 mm ($\frac{3}{8}$ in) gravel and how many kg y of 19.05 mm ($\frac{3}{4}$ in) gravel to mix with it to obtain the grading of Curve 1. Two unknowns only need two equations. Hence we can only make Curve 1 correct for the percentages passing two chosen sieve sizes. Suppose we choose the percentages passing apertures 9.52 mm ($\frac{3}{8}$ in) and 4.76 mm ($\frac{3}{16}$ in).

According to Curve 1, the percentage passing 9.52 mm ($\frac{3}{8}$ in) aperture is 45%, hence using Table 2.4:

$$100 \times 1 + 96x + 19y = 45(1 + x + y)$$

According to Curve 1, the percentage passing 4.76 mm ($\frac{3}{16}$ in) aperture is 30%; hence using Table 2.4:

$$100 \times 1 + 13x + y = 30(1 + x + y)$$

Solving these two equations, $x = 0.1172$, and $y = 2.345$. Thus the sand, 9.52 mm ($\frac{3}{8}$ in) gravel and 19.05 mm ($\frac{3}{4}$ in) gravel must be combined in the proportions 1:0.1172:2.345, respectively.

The grading of the combined aggregate is obtained by multiplying Columns (a), (b) and (c) of Table 2.4 by 1, 0.1172 and 2.345, respectively, the products being shown in Columns (d), (e) and (f), respectively. The values in these columns are added together to give the values in Column (g) and then divided by $1 + 0.1172 + 2.345 = 3.462$ to give the values in Column (h), and this is the grading of the combined aggregate. Comparing this with Column (i) we have achieved the same percentages passing

Table 2.4

Aperture size	BS Sieve	Percentage passing			(d) (a)×1	(e) (b)×0.1172	(f) (c)×2.345	(g) (d)+(e)+(f)	(h) (g)÷3.462	(i) Curve 1
		(a) Sand	(b) 9.52 mm gravel	(c) 19.05 mm gravel						
19.05 mm	3/4 in	100	100	97	100	11.72	227.5	339.2	98.0	100
9.52 mm	3/8 in	100	96	19	100	11.25	44.56	155.8	45.0	45
4.76 mm	3/16 in	100	13	1	100	1.524	2.345	103.9	30.0	30
2.40 mm	No. 7	85	1	0	85	0.1172	0	85.12	24.6	23
1.20 mm	No. 14	72	0	0	72	0	0	72	20.8	16
600 μm	No. 25	53	0	0	53	0	0	53	15.3	9
300 μm	No. 52	10	0	0	10	0	0	10	2.9	2
150 μm	No. 100	1	0	0	1	0	0	1	0.3	0

Concrete   41

9.52 mm and 4.76 mm apertures, as calculated. Our error is mainly for percentages passing 1.20 mm and 600 μm apertures. We could repeat the calculation say making the percentages passing apertures 9.52 mm and 1.20 mm equate in Columns (h) and (i). Mix design is not a very accurate science and this is probably not worth the trouble and its result would not really be known to be any better. Various sets of two percentages passing certain sizes could be made equal in Columns (h) and (i) by calculation and all the various results plotted on a graph such as Figure 2.1, and one could choose the combined grading which looks generally closest to the graph of Curve 1. Again it is extremely doubtful if this is worth doing.

2.3.12   Design of concrete mixes and the D.O.E. method

The *Road Note No. 4*[8] method described in Sections 2.3.9 to 2.3.11 inclusive has been followed by a Department of the Environment (D.O.E.) publication[9] which is more comprehensive, more complicated to describe and requires more tables and figures than *Road Note No. 4*. Hughes[7] presents an even more comprehensive and complex method than the two methods just mentioned. In the U.S.A. the A.C.I.[10] recommends a different method.

As explained in Section 2.3.9 mix designs are not very accurate and need to be adjusted by making tests of trial mixes, but are naturally useful for determining the first trial mix. The *Road Note No. 4* method described previously is the simplest of the above-mentioned methods to describe to students and to understand. It has been used for many years and is still used although the D.O.E. presumably hope that their method will replace it. The other methods are easy to use without much understanding by following through the examples and using the tables and graphs given in the publications already cited.

2.3.13   D.O.E. mix design method

This method[9] is simply explained in the following example.

Example 2.3. Design a concrete mix as follows (the item numbers refer to where this information is entered in Table 2.5):

1. Characteristic compressive strength at 28 days = 30 N/mm². (Item 1.1)
2. Referring to Figure 2.4, equation (2.1) and the paragraph following it, standard deviation = k = 1.64 and the 'defective rate' = 5%. (Item 1.1)

Table 2.5. *Completed concrete mix design form for unrestricted design*

Stage	Item		Reference or calculation	Values
1	1.1	Characteristic strength	Specified	Compressive 30 N/mm² at 28 days Proportion defective 5 per cent
	1.2	Standard deviation	Fig. 2.5	— N/mm² or no data 8 N/mm²
	1.3	Margin	C1	(k = 1.64) 1.64 × 8 = 13 N/mm²
	1.4	Target mean strength	C2	30 + 13 = 43 N/mm²
	1.5	Cement type	Specified	OPC/~~SRPC/RHPC~~
	1.6	Aggregate type: coarse		Uncrushed
		Aggregate type: fine		Uncrushed
	1.7	Free-water/cement ratio	Table 2.7. Fig. 2.6.	0.494 ⎱ Use the lower value
	1.8	*Maximum free-water/cement ratio*	*Specified*	0.55 ⎰
2	2.1	Slump or V-B	Specified	Slump 10 to 30 mm or V-B — s
	2.2	Maximum aggregate size	Specified	20 mm
	2.3	Free-water content	Table 2.8	160 kg/m³

Concrete

3	3.1	Cement content	C3	$160 \div 0.494 = 324$ kg/m³
	3.2	*Maximum cement content*	*Specified*	550 kg/m³
	3.3	*Minimum cement content*	*Specified*	290 kg/m³ — use 3.1 if ⩽ 3.2 use 3.3 if > 3.1
	3.4	Modified free-water/cement ratio		—
4	4.1	Relative density of aggregate (SSD)		2.6 known/assumed
	4.2	Concrete density	Fig. 2.7	2400 kg/m³
	4.3	Total aggregate content	C4	$2400 - 324 - 160 = 1916$ kg/m³
5	5.1	Grading of fine aggregate	Percentage passing 600 μm sieve	70 per cent
	5.2	Proportion of fine aggregate	Fig. 2.8	27 per cent
	5.3	Fine aggregate content	⎫ C5	$1916 \times 0.27 = 517$ kg/m³
	5.4	Coarse aggregate content	⎭	$1916 - 517 = 1399$ kg/m³

Quantities	Cement (kg)	Water (kg or l)	Fine aggregate (kg)	Coarse aggregate (kg)
per m³ (to nearest 5 kg)	324	160	517	1399
per trial mix of 0.05 m³	16.2	8.0	25.9	70.0

Items in italics are optional limiting values that may be specified.

OPC = ordinary Portland cement; SRPC = sulphate-resisting Portland cement; RHPC = rapid-hardening Portland cement.
Relative density = specific gravity.
SSD = based on a saturated surface-dry basis.

3. Ordinary Portland cement. (Item 1.5)
4. Slump = 10 to 30 mm. (Item 2.1)
5. Maximum aggregate size = 20 mm. (Item 2.2)
6. Maximum 'free-water'-to-cement ratio = 0.55. This does not need to be specified as it is calculated in item 16. Free-water is the water available for chemical action with the cement. That is, it includes surface water on, but excludes water which has been absorbed by, the aggregates. If this amount of free water is added to the mix the aggregates need to be in a saturated surface-dry condition. (Item 1.8) (More information on this subject is given in (24) following.)
7. Minimum cement content = 290 kg/m^3. (Item 3.3) A minimum is specified in connection with durability.
8. Maximum cement content = 550 kg/m^3 as specified in BS 8110 clause 6.2.4.1. (Item 3.2) A maximum is sometimes specified to avoid the risk of cracking due to drying shrinkage in thin sections and thermal stresses in thicker sections and to reduce shrinkage for example in mass concrete dams, etc.
9. No previous control data. Therefore, from Figure 2.5 the standard deviation = 8 N/mm^2. (Item 1.2)
10. Fine and coarse aggregates are uncrushed. (Item 1.6) As the relative density is unknown, D.O.E. recommend taking it as 2.6 (for crushed aggregates they recommend 2.7). (Item 4.1)
11. The 1975 edition of reference 9 used fine aggregates complying with Table 2.6 which was published in BS 882:1973. This latter

Figure 2.5 Relationship between standard deviation, s, and characteristic strength.

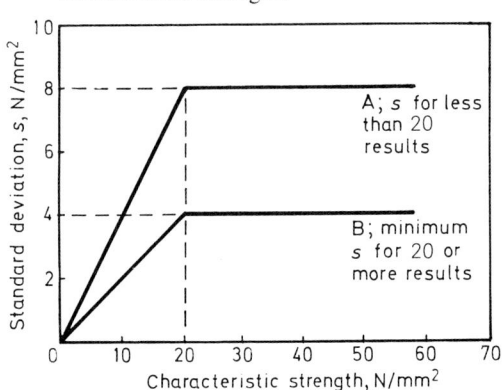

was revised in 1983 and reference 9 makes use of this revised edition. The grading zones of Table 2.6 used by B.R.E were based on materials available to B.R.E. in the London area. They were used to define materials for their work on the development of mix design methods. However, there were fine aggregates, for example at Finningley in Yorkshire, which made very good concrete but which had quite different gradings. To use the mix design methods this meant uneconomic altering of gradings in various localities. To quote from reference 9 'In many parts of the country fine aggregates having such restricted gradings are not available, although the gradings of the available fine aggregates are still suitable for making good quality concrete.' Therefore reference 9 refers instead to the percentage of fine aggregate passing the 600 μm test sieve. The higher the percentage passing the 600 μm test sieve, the finer the fine aggregate. Fine aggregates should comply with the C, M, or F grading requirements of BS 882: 1983, but these limits overlap and are too wide for mix design purposes. The method for deriving a suitable fines content takes into account the many relevant factors, i.e. the type and maximum size of coarse aggregate, the grading of the fine aggregate, characterised by the percentage passing the 600 μm test sieve, and the cement content and workability of the concrete.

Table 2.6. *Fine aggregate, BS 882: Part 2: 1973*

BS 410 test sieve	Percentage by weight passing BS sieves			
	Grading Zone 1	Grading Zone 2	Grading Zone 3	Grading Zone 4
mm				
10.0	100	100	100	100
5.00	90–100	90–100	90–100	95–100
2.36	60–95	75–100	85–100	95–100
1.18	30–70	55–90	75–100	90–100
μm				
600	15–34	35–59	60–79	80–100
300	5–20	8–30	12–40	15–50
150	0–10*	0–10*	0–10*	0–15*

* For crushed stone sands, the permissible limit is increased to 20%. The 5% tolerance permitted by Item 5.2 may, in addition, be applied to the percentage in light type.

46   *Properties of materials and mix design*

12. In Table 2.5 calculations are performed and referenced C1 to C5.
13. Calculation C1 (Item 1.3) uses equation (2.3) to obtain the 'margin', see Section 2.3.9.
14. Calculation C2 (Item 1.4) obtains the 'target mean strength' (this is just the mean strength, see Section 2.3.9, we are trying to achieve with our design) by using equation 2.2 to obtain mean strength.
15. From Table 2.7 for the materials being used and a free-water-to-cement ratio of 0.5 an estimate of the compressive strength at 28 days would be 42 N/mm^2.
16. On Figure 2.6 the line for values of free-water-to-cement ratio of 0.5 is referred to and point A is located on this line corresponding to a compressive strength of 42 N/mm^2, both values from (15) above. Then the curve upon which point A lies is followed to the point B corresponding to a compressive strength of 43 N/mm^2 (the target mean strength from Item 1.4). This point B is then seen to correspond to a free-water-to-cement ratio of 0.494 (Item 1.7). This is satisfactory as it is less than the specified maximum value of 0.55. (Item 1.8)
17. From Table 2.8 the free-water content = 160 kg/m^3. (Item 2.3)
18. As the free-water/cement ratio was 0.494 (Item 1.7), from (17), the cement content = 160/0.494 = 0.324 kg/m^3 (Item 3.1 and calculation C3). This is satisfactory as it is greater than the specified minimum of 290 kg/m^3. (Item 3.3)

Table 2.7. *Approximate compressive strengths (N/mm^2) of concrete mixes made with a free-water-to-cement ratio of 0.5*

Type of cement	Type of coarse aggregate	Compressive strengths (N/mm^2) Age (days)			
		3	7	28	91
Ordinary Portland (O.P.C.) or sulphate-resisting Portland (S.R.P.C.)	Uncrushed	22	30	42	49
	Crushed	27	36	49	56
Rapid-hardening Portland (R.H.P.C.)	Uncrushed	29	37	48	54
	Crushed	34	43	55	61

Concrete   47

19. From Figure 2.7 and Items 2.3 to 4.1 the wet density of the concrete = 2400 kg/m³. (Item 4.2)
20. Calculation C4 (Item 4.3) gives the total aggregate content by subtracting from the weight of 1 m³ of wet concrete (Item 4.2) the weights of cement (Item 3.1) and free-water (Item 2.3) in the 1 m³.
21. Figure 2.8 refers to aggregate of 20 mm maximum size. (The D.O.E. booklet also gives figures for aggregate of maximum sizes 10 and 40 mm.) From Figure 2.8 and Items 2.1, 1.7 and 5.1 the proportion of fine aggregate is obtained. (Item 5.2)
22. Calculation C5 (Items 5.3 and 5.4) obtains the fine aggregate content by multiplying Item 5.2 by the total aggregate content (Item 4.3) and it obtains the coarse aggregate content by subtracting this fine aggregate content from the total aggregate content.

Figure 2.6 Relationship between compressive strength and free-water/cement ratio.

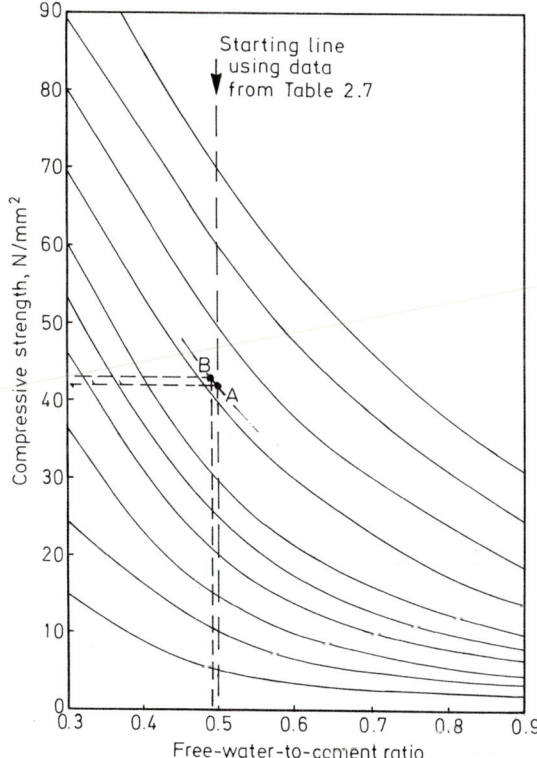

48 Properties of materials and mix design

23. The quantities of constituent materials are given at the bottom of Table 2.5 for a mix of 1 m³ and for a trial mix of 0.05 m³.
24. To obtain the weight of the oven-dry aggregates when aggregates are to be batched in an oven-dry condition for a trial mix, the weights of the saturated surface-dry aggregates derived from calculations C5 are multiplied by $100/(100+A)$ where A is the

Table 2.8. *Approximate free-water contents (kg/m^3) required to give various levels of workability*

Slump (mm)		0–10	10–30	30–60	60–180
V-B (s)		>12	6–12	3–6	0–3
Maximum size of aggregate (mm)	Type of aggregate				
10	Uncrushed	150	180	205	225
	Crushed	180	205	230	250
20	Uncrushed	135	160	180	195
	Crushed	170	190	210	225
40	Uncrushed	115	140	160	175
	Crushed	155	175	190	205

Note: When coarse and fine aggregates of different types are used, the free-water content is estimated by the expression $\frac{2}{3}W_f + \frac{1}{3}W_c$ where W_f = free-water content appropriate to type of fine aggregate and W_c = free-water content appropriate to type of coarse aggregate.

Figure 2.7 Estimated wet density of fully compacted concrete.

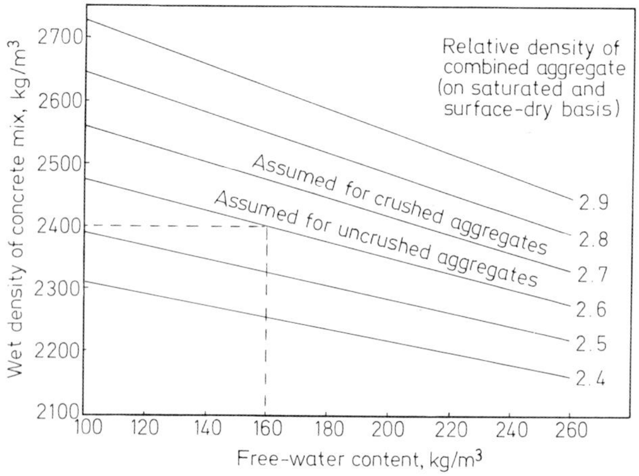

percentage by weight of water needed to bring the dry aggregates to a saturated surface-dry condition. The amount of mixing water should be increased by the weight of water absorbed by the aggregates to reach the saturated surface-dry condition. For example, if the absorption of the fine aggregate is 2% and of the coarse aggregate is 1%, then in the above trial mix:

Weight of oven-dry fine aggregate $= 25.9 \times 100/102 = 25.4$ kg

Weight of oven-dry coarse aggregate $= 70.0 \times 100/101 = 69.3$

$$\text{Water required for absorption} = (25.9 - 25.4) + (70.0 - 69.3)$$
$$= 0.5 + 0.7 = 1.2 \text{ kg}$$

Thus, the quantities for the trial mix are: cement 16.2 kg, water 9.2 kg, fine aggregate 25.4 kg (oven-dry) and coarse aggregate 69.3 kg (oven-dry).

2.3.14 Quantities of materials required to make 1 m³ of concrete

A very simple method is illustrated in Example 2.4. This is useful for individual beams. If one needed considerable accuracy for a large quantity such as a dam, this can easily, and best, be established experimentally in the laboratory.

Example 2.4. Calculate the quantities of ingredients required for casting a beam and cubes in the laboratory having a total volume of 0.4 m³. The mix is to be in the proportions of 1 part cement to 0.87 parts sand to 0.10

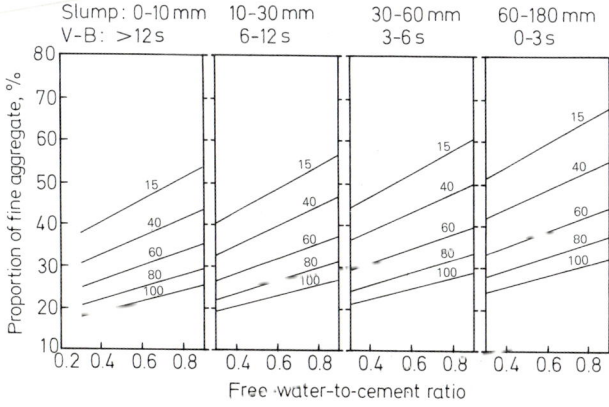

Figure 2.8 Recommended proportions of fine aggregate according to percentage passing a 600 μm sieve.

50 *Properties of materials and mix design*

parts 9.52 mm gravel to 2.03 parts 19.05 mm gravel by dry volumes with a water-to-cement ratio of 0.35 by masses.

Assume the bulk density of cement, sand and gravel to be 1440 kg/m^3 (reasonably true if not using lightweight aggregates). Assume the density of the matured concrete to be 2400 kg/m^3 (again reasonably true). The mass of the concrete is equal to the mass of its ingredients, except that much of the water will evaporate. Assume all the water vanishes – this will very slightly underestimate the cement, sand and gravel. Therefore:

$$1 \text{ kg} + 0.87 \text{ kg} + 0.10 \text{ kg} + 2.03 \text{ kg} = 4 \text{ kg}$$
cement   sand   small gravel   large gravel   concrete

Mass of mature concrete = $0.4 \times 2400 = 960$ kg. Therefore requirements are:

$1 \times 960/4 = 240$ kg cement

$0.87 \times 960/4 = 208.8$ kg sand

$0.10 \times 960/4 = 24$ kg small aggregate

$2.03 \times 960/4 = 487.2$ kg large aggregate

$240 \times 0.35 = 84$ kg water

Then add 10% to these figures to allow for small underestimation and waste. This figure may need small adjustment according to experience of the concreting conditions, the particular mix and type of aggregates, etc.

2.3.15 Prescribed mixes

CP 110 gives prescribed mixes in Table 50 to replace the nominal mixes of CP 114. Generally these will give uneconomic concretes stronger than required, but they have the advantage that proper mix design procedures do not need to be established for the concreting plant. BS 8110 refers to BS 5328:1981 which recommends six grades of concrete for prescribed mixes each with a medium and high workability. In BS 5328 each mix is defined in Table 1 as a ratio by mass of aggregate to cement and then Table 2 recommends the ratio of fine to total aggregate for each mix.

2.3.16 Shrinkage

When cement, sand, gravel and water are mixed together the gross volume decreases as the finer particles arrange themselves in the interstices of the larger particles. This shrinkage continues as the concrete

Concrete

is being worked into place. Evaporation of water in the mix also decreases the volume of such concrete. It is possible to fill a mould, for example of a 150 mm cube, and observe the concrete retract into the mould. Shrinkage, when the concrete is in a fluid state, does not matter structurally because no internal stresses can be instigated. There is an inaccurately known point at which the concrete changes from a fluid to a solid and immature fragile material. The exact time when this occurs depends upon the water-to-cement ratio, the type of cement, and the ambient humidity and temperature. After this time, further shrinkage of the concrete will cause internal stresses and even cracks to occur. The time when the transition occurs from liquid to solid is not precisely determinable, and it is difficult to know exactly when to commence measuring the shrinkage of the concrete in its solid state.

Measurements of the coefficient of shrinkage are possibly commenced too late to be of real mathematical value in research, because such readings are often commenced just when the specimen is hard enough to strip and handle for the purposes of the test. On such a basis the shrinkage coefficient can be of the order of 0.0005 at an age of 12 months and a typical relationship between shrinkage and age is illustrated in Figure 2.9. Initially the rate of shrinkage is high so that the error in not knowing the precise time to start measurements is quite appreciable. With the above coefficient, and supposing, for simplicity, the modulus of elasticity of concrete is $28\,000$ N/mm^2, then if the concrete were restricted from

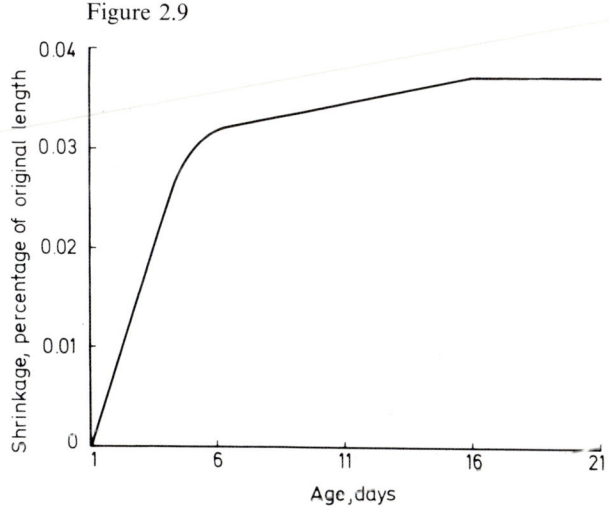

Figure 2.9

shrinking the tensile stress induced in the concrete would be $0.0005 \times 28\,000 = 14\ N/mm^2$. The concrete would certainly crack as its ultimate tensile strength would generally only be about $2.8\ N/mm^2$. The coefficient of shrinkage is less for a lean mix than for a rich mix (in cement content). It is less for a low water-to-cement ratio than for a high one, is very sensitive to the method of curing, and is influenced to a lesser extent by all the other possible variables.

Shrinkage after the concrete has solidified continues as and when further water evaporates. The chemical reaction of cement with water, and thus the shrinkage, continues in the concrete seemingly indefinitely. A gel is formed which contracts upon desiccation and becomes very hard (see Section 2.1). If concrete is submerged in water this cement gel expands with considerable force, so that the whole mass of concrete expands. This expansion, however, can never equal the shrinkage which has already taken place. On drying the concrete in air shrinkage again occurs. Therefore, when concrete is subjected to continual wetting and drying, as for example due to tidal action, it experiences corresponding expansions and contractions. If concrete is cast beneath water then it does not shrink at all but expands, owing to the cement gel absorbing water. This is sometimes done with dock work when concrete is placed or pumped through a tube into its own mass beneath the sea.

If a mass of concrete shrinks (or expands) uniformly and its movement is not restricted by any external forces, then no internal stresses can be induced in the concrete. This seldom happens in practice; usually any movement of the concrete is restricted internally by reinforcement embedded in the concrete, and often externally by its surroundings. Also, the surface of concrete will often dry out (and therefore shrink) faster than the internal particles of concrete. When the concrete of a reinforced beam is in the solid state, as it shrinks it also bonds to the reinforcement. The resistance of the reinforcement to contraction opposes the shrinkage of the concrete. Thus the concrete near to the reinforcement is in tension, a bond stress is developed between the two, and the reinforcement is in compression. Shrinkage cracks often exist in reinforced concrete beams at intervals along the length of the reinforcement. These are sometimes too small to be observed with the instruments normally available. When a reinforced concrete beam is tested, cracks can usually be observed at a lighter loading than predicted from the modulus of rupture of the concrete, indicating that cracks or tensile stresses are already present due to shrinkage.

Designs concerning conventional reinforced concrete work do not usually attempt to estimate the quantitative effect of shrinkage, because such calculations cannot be made with any degree of confidence and the basic assumptions of any mathematical analysis can be adversely criticised. Prestressed concrete designers simply treat shrinkage as a 'loss' reducing the prestressing force. The ultimate strength of a beam is not altered by shrinkage because when cracks occur the initial internal stress systems are released, yet shrinkage affects the size of cracks and deflections at working loads.

The particles at the surface usually experience different conditions of curing to internal particles. Their rates of shrinkage thus differ and this 'differential shrinkage' can cause troublesome stresses, cracks and movements, for example the surface crazing of artificial stones and the warping of ground floor and road slabs. This effect can be reduced by endeavouring to cure the surfaces similarly to the internal fibres. The latter are fairly well sealed from the atmosphere so that to reduce differential shrinkage it is therefore desirable to seal the surfaces from the atmosphere. One way of achieving this is to immerse the concrete member in water for as long as possible. It is often more economical to cover with damp hessian sacks, sand, or water-proof sheets, or to spray periodically with water. A granolithic topping on a floor is very vulnerable to the detrimental effects of differential shrinkage and is usually kept damp for as long as practicable, and for at least seven days.

Shrinkage must always be borne in mind in the design and construction of structures. Whenever possible, concreting programmes aim at minimising the detrimental effects of shrinkage. For example, ground floor *slabs on solid* (placed over either suitable subsoil or suitably consolidated blinded hardcore), are often concreted in numerous independent portions each of about 4.5 m square, which are able to shrink before being joined together. Plain concrete roads are similarly constructed. This is not considered necessary when reinforcement is present. Numerous minute cracks are formed, but as the reinforcement resists shrinkage the overall contraction is negligible. Some engineers will attribute almost any serious crack in a structure solely to shrinkage. This is often a fallacy because the reinforcement of most structures has a considerable resistance to the forces exerted by the shrinkage of the concrete, so that shrinkage cracks in a long structure will take the form of very small cracks fairly regularly spaced throughout the length of the structure. A serious crack is more often caused by thermal expansion and contraction, and settlement.

Properties of materials and mix design

2.3.17 Relationship between stress and strain for concrete

If a graph is plotted relating stress and strain, the shape of the curve obtained is very much influenced by the rate at which the stress is applied. It is also dependent upon the strength of the concrete under question and indeed to some degree upon all the other possible variables. Figure 2.10 shows a relationship OAF which is typical of a concrete specimen loaded at a uniform rate. If the stressing had been held at the point A the concrete would have continued to strain under this particular constant stress. After a certain lapse of time, when the strain had reached the point B, had the stressing been recommenced at the previous rate, the relationship would have been the curve BC. Had the stressing been stopped at C, the same phenomenon of *creep* would have occurred on C to D as occurred on A to B, that is the specimen strained or crept under constant stress until the stressing was recommenced at the point D, and the relationship then took the form represented by DE.

This phenomenon of creep (known in the U.S.A. as *plastic strain* or *time flow*) has been the subject of many investigations. Figure 2.11 shows a curve CD which relates the creep (or strain) to time when the specimen is subjected to a constant stress. In this instance it took 5 s to apply the stress, so that the readings commenced from this time. It was once imagined that if this loading had been instantaneous and the observations of creep had been commenced immediately then this curve would have taken the form BCD. This not so: the relationship is as ACD. Evans[11] constructed an apparatus which could load a specimen and record the

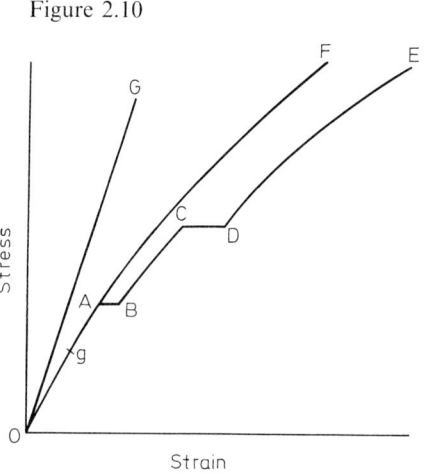

Figure 2.10

strains at an extremely high speed. This enabled him to obtain readings of creep after an instantaneous loading to the stress in question, and enabled him to plot the curve AC in Figure 2.11. The same apparatus enabled him to discover an interesting relationship between stress and strain. At any particular stress an instantaneous increase in stress always gave a directly proportional increase in strain. Thus he obtained the linear relationship OG shown in Figure 2.10. This was an attempt to find a modulus of linear elasticity (Young's modulus) for concrete and thus to divorce the elastic from the plastic action, as in the early days attempts were made to use the elastic theories of design, which had been developed for steelwork, for reinforced concrete. This endeavour to separate elastic and plastic action was not subsequently favoured and creep cannot exactly be divorced from elasticity, shrinkage and other possible variables. Investigators generally agree that creep is mainly directly proportional to the constant stress causing it and proportional to a function of time. Various functions have been recommended for this.

It is thus distinctly noticeable that with regard to the relationship between stress and strain, concrete is comparable in behaviour to natural stones and timber, but certainly not to mild steel, because there is no period of proportionality, no marked elastic limit and no yield point. Apologies must therefore be made for using the term 'modulus of elasticity' for concrete. However, from the early days this has been done

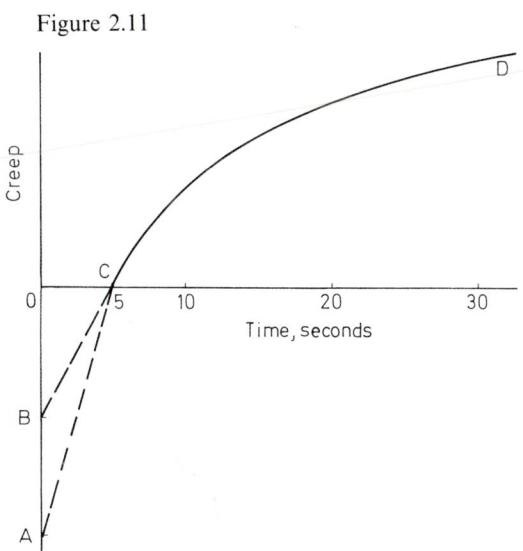

Figure 2.11

56   *Properties of materials and mix design*

in connection with the elastic theory which still has its uses. Therefore some value or values must be attributed to a rather mythical modulus of elasticity. Figure 2.12 illustrates a typical stress–strain diagram for a concrete specimen and shows various ideas which have been propounded for the modulus of elasticity. OT_0 is tangential to the function at the origin and is called the *initial tangent modulus*. TPT' is tangent at the point P and is known as the *tangent modulus* at this point. Similarly T_1QT_1' is the tangent modulus at point Q. The straight line PQ is called the *chord modulus* for the range P to Q. OP is the *secant modulus* for point P, and similarly OQ is the secant modulus for point Q. In Figure 2.10 the slope of the curve OG is Evans' *short range* or *instantaneous modulus* of elasticity. This modulus is suitable for use in predicting the stresses caused in concrete structures by shocks from bombing or earthquakes.

The maximum permissible compressive stress in bending at working loads is often specified, for designs based on elastic theory, to be about one third of the crushing strength. Up to such working stresses the relationship between stress and strain approximates with reasonable accuracy to a straight line and most engineers utilise a secant modulus of elasticity corresponding to the maximum allowable working stress. This is the modulus of elasticity implied when reference is subsequently made to the modulus of elasticity of concrete, unless stated otherwise.

If points A and B in Figure 2.10 were at the allowable working stress of the concrete under investigation, then the moduli of elasticity at points A and B are obviously different. One can take the modulus of elasticity for

Figure 2.12

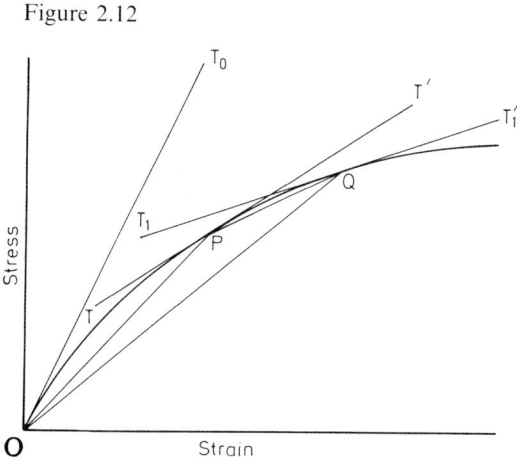

A and then make a separate calculation for creep. The former depends on the speed of loading to A, and the latter relies on debatable methods. It is usual, and simpler, to take the secant modulus of elasticity of point B, or whatever point on A to B one considered relevant to the time creep has been occurring. For example, concrete at the age of one year can have a modulus of elasticity of about two thirds of its value at the age of one month. When creep tests are made, specimens are cast out of the same mix, for the purpose of measuring the shrinkage which occurs. The shortening due to shrinkage can then be deducted, to give the true creep over the period independently of the effect of shrinkage.

Concrete made with certain popular lightweight aggregates can have a modulus of elasticity of only two thirds of the value of a conventional type of concrete of the same ultimate compressive strength. For elastic design the modulus of elasticity has been related to the concrete strength, but then for simplicity CP 114 adopted a constant value of 14000 N/mm². For example, the modular ratio α_e used by CP 114, is Young's modulus for steel 210000 N/mm² divided by 14000, which equals 15.

If a reinforced concrete beam is subjected to a loading test α_e could be about 9 for use in calculations predicting deflection or stresses. If the load were maintained for say one year then α_e would be about 15.

With time, creep causes beams to deflect more, causes compression steel to be more highly stressed, and causes long slender columns to increase their lateral deflection. This causes the bending moments to be higher and research by the author shows this to be a very important effect.

With regard to prestressed concrete, creep of steel (relaxation) and of concrete are calculated as losses of prestressing force.

2.3.18 Tensile strength of concrete and granolithic toppings

For R.C. design and construction it has traditionally been the case that the tensile strength of the concrete has been abandoned in favour of the reinforcement taking the tension, so the technology of mix design for R.C. has been to make concrete as economically strong in compression as possible. Compressive strength design has been the main objective and whatever the associated tensile strength turned out to be, was assessed from tests which gave very low strengths and erratic results. The relationship between compressive and tensile strength for a concrete designed for R.C. work and one designed primarily for tension are different.

Granolithic concrete was pioneered by Stuart's Granolithic Co. Ltd. of the U.K. (initially Edinburgh and London) from 1840. They constructed suspended floors of notable Victorian and subsequent buildings in the U.K., a substantial number of which still exist, of quite thin plain (i.e. unreinforced) concrete before the late nineteenth century/early twentieth century pioneers of reinforced concrete (e.g. Wilkinson in the U.K. and various names in France such as Hennebique – who had great influence in the U.K. e.g. the Yorkshire Hennebique Contracting Co. Ltd, of Leeds, established 1904). The great selling point was fire resistance compared to the wooden floors of those times. For example, the following were all built before 1901:

1. Cranston's 'Waverley,' Temperence Hotel, Southampton Row, London. 'The fireproof floors are constructed of Stuart's Granolithic. There are 31,500 ft.super (2926 m^2) of paving, 1,650 ft.super (153 m^2) of Landings and 1,700 ft.lineal (518 m) steps'.
2. New City Hall, Philadelphia, U.S.A. '79,110 ft.super (7350 m^2) of Stuart's Patent Granolithic in this building. The highest tower in the world'.
3. Jenner's Warehouse and Showrooms, Princes Street, Edinburgh. 'The six floors and roof contain 108,900 ft.super (10117 m^2) of Stuart's Fireproof Granolithic. The floors are from 2.5 to 3 inches (64 mm to 76 mm) thick, and can carry a weight of 10 cwt. per foot super. (53.7 kN/m^2)'
4. New Municipal Technical Schools, Birmingham (Suffolk Street). 'There are over 103,000 ft.super (9569 m^2) of Stuart's Patent Granolithic Fireproof Flooring in this building.' The floor of the Examination Hall is 20.2 m by 13.56 m and 82.6 mm thick. It is supported by a grillage of beams so that the slab is two way spanning 2.7 m by 2.7 m. It was designed to carry 'a safe distributed load of' 8.05 kN/m^2.
5. Hollingdrake and Clegg, Miall Street Mills, Battinson Rd, Halifax, Yorkshire, Warehouse (5 storey) – the flat floors are 'formed of Granolithic Arches 3 ins. (76 mm) thick at the crown and 10 ft. 6 in. (3.2 m) span between joists. It was specially designed to resist the great vibration caused by quick running machinery in the Woollen Mill. It has proved itself after a lengthened period of trial in every respect satisfactory, and other floors at the Clegg St. Mills have since been laid for the same purpose. Safe load on floor 3 cwts per sq. ft. (16.1 kN/m^2).'

Granolithic concrete has been used for high quality pavements and paving slabs and the Company introduced granolithic for this purpose into Washington, U.S.A. in 1882 and France and Duisburg a/Rhine, Germany, in 1883. The company used to pave the holds of cargo and other ships with *in-situ* granolithic concrete e.g. the S.S. *Spartan* Hospital ship for the S. African War in 1899. With regard to weathering properties the paving flags of Stuart's granolithic on the promenade at Eastbourne, England, have been in good condition for more than half a century.

Good quality granolithic concrete, e.g. made with Shap granite from the Lake District of England, was expensive and the very arduous and skilled laying of top quality granolithic concrete was very expensive, so it often became used as a topping to industrial R.C. suspended floors and slabs on solid. Toppings used when the author was 'the Company's Development Engineer' have been 76, 64, 51, 38, 25, 13 and 3 mm thick. These have considerable resistance to wear from heavy steel wheeled trolleys and the like and a good resistance to industrial fluids which attack concrete such as lactic acid from milk in dairies, etc.

To design concrete for compressive strength see earlier in this Chapter. One feature of this which soon became apparent was to maximise the use of large rounded aggregate and minimise the use of small to dust aggregate in a mix, so that the water/cement slurry required for coating the large aggregate for the workability required for good compaction is a minimum and therefore the water to cement ratio of the whole mix can be kept low giving high strength (see Figure 2.2).

The author has studied tests made in the nineteenth century by Stuart's Granolithic Co. Ltd and analysed these and the plain suspended slabs they constructed to certain buildings. Undoubtedly the indication is that the granolithic concrete was capable of very high flexural tensile strengths, for example:

1. For a two-way spanning slab design, one side simply supported, the others fixed; for the design loads the flexural tensile stress would need to resist *c*. 8 N/mm.
2. For a two-way spanning slab, all sides simply supported, the loading test did not achieve failure even though the flexural tensile stress would be *c*. 35 N/mm.
3. For a two-way spanning slab, two opposite sides simply supported, the others fixed, the test did not achieve failure even though the flexural tensile stress would be *c*. 36 N/mm.
4. For a two-way spanning slab, all sides simply supported, the test

did not achieve failure even though the flexural tensile stress would be $c.$ 30 N/mm.

For their designs of plain suspended concrete floors the prime consideration was tensile strength. If this were satisfactory then the associated compressive strength was more than adequate and the durability against wear, in the case of industrial floors, and weathering was exceedingly good.

To design primarily for this tensile strength, ideally the particles of an angular aggregate with rough fractured surfaces, resulting from crushing, need to be stuck together with the minimum amount of a mixture of cement, very fine sand and water; to coat the pieces of angular aggregate with the minimum thickness of this in effect 'glue' and to fill the small voids between the pieces of angular aggregate with the minimum amount of this adhesive. In effect the particles of granite are stuck together as closely as possible with this cement/fine sand/water adhesive.

For top quality granolithic concrete the Company mentioned used to use Shap granite and would not regard basalt from North Wales or limestone from Derbyshire or Yorkshire as used by many firms competing in this market, as suitable for granolithic toppings. Nevertheless many 'granolithic' toppings not using granite chippings are laid by plasterers as opposed to skilled granolithic layers, and are cheaper than quality granolithic toppings, and significantly inferior.

Mr G. Kirby, formerly M.D. of Stuart's Granolithic Co. Ltd. and who worked a lifetime for the Company, always furiously disagreed with publications/guidance on granolithic flooring, which based their approach on traditional compressive concrete strength rather than granolithic tensile strength and ignored his other knowledge from long experience of specialist granolithic floor laying.

Mr George Brown, a former colleague of Kirby's, and now M.D. of Kontrad Associates Ltd. of Stockport, Cheshire has used the approach of designing concrete primarily for tensile strength for granolithic toppings and applied it to slabs on solid and designed mixes with high flexural concrete tensile strength with crushed angular limestone aggregates. A slab like this is much stronger in flexural tension than one designed traditionally for primary compressive strength and can be thinner and overall more economic (by up to 20% he claims) and better at resisting cracking.

Slabs on solid to this design were tested by Brown and the author[11a] in

Types of reinforcement 61

the presence of representatives from certain important British research organisations and, for example one 75 mm thick slab was considered to have a flexural concrete tensile strength of over 11 N/mm^2 without failing; i.e. more than three times what the author would expect from a traditional concrete designed, for R.C. work, primarily for compressive strength.

2.4   Types of reinforcement

Much reinforced concrete construction employs 'black' mild steel bars of circular cross section. In the early days, engineers often worried that such bars might not grip or 'bond' to the concrete. Consequently, numerous bars were devised with surface deformations. As knowledge advanced, it became accepted that a mild steel bar of circular cross section could grip adequately to the concrete to develop its full tensile strength, surface deformations on the bar being superfluous.

Engineers generally are now happy to use high tensile steel provided the bar mechanically bonds with the concrete. If a mild steel bar of square cross section is twisted, this *cold working* converts it into a high tensile steel bar which can mechanically grip to the concrete. Another type of bar is made by rolling a round mild steel bar with a slight patterning on its surface, then subjecting it to cold working by tensioning and twisting to give a high tensile bar with a mechanical bond. Another type of high tensile bar is a hot rolled high tensile steel bar with a deformed surface. These high tensile reinforcements are called *high yield* by BS 8110 because it is the yield stress which is of interest in our theories for ultimate strength. Cold working, for example, can increase the yield stress of mild steel much more than its ultimate stress. The advantage of using high yield bars is that the mass of steel required is reduced, and even though its cost per kilogram is higher than mild steel the total cost of the reinforcement and its fixing can be reduced. This does not apply in the case of the nominal reinforcement, which is usually more economic in mild steel, in a structure. Square twisted bars are bulkier for detailing, concreting, etc., than round deformed bars of the same strength. This disadvantage is reduced for a square twisted bar with chamfered corners. Sometimes square twisted bars have the advantage of bulk per unit cost for use as 'spacer bars' in cylindrical shells – for keeping the fabrics apart and aiding concreting on the sloping surfaces.[17] The appropriateness of a bar for a purpose and its cost and availability will usually decide which type of reinforcement to use. Mild steel is usually the most universally available and because more is required than high yield steel, for example, as longitudinal tensile reinforcement in a cylindrical shell, then as Young's modulus is the same for both, the moment of inertia will be greater and hence the deflection less for shells with such mild steel. Also, the design has been elastic so that the lower strains of the mild steel do not conflict as much with the assumptions of the design. This also applies to frames which have their bending moments decided on elastic theory.

One should ascertain that any high yield reinforcement to be used bent does not have its strength seriously impaired by 'overstrain'. For example, the cold working of a bar introduces internal stresses in the bar. If the bar is then bent, further high stresses are superimposed on these stresses. It has been known for the fibres of steel on the inside of a bend to crush and for this not to be noticed until the bar was accidentally gently knocked, when the bar then came apart at the bend. Reference 12 explains this problem and establishes that for two particular high yield bars, at the time, overstrain was not a practical worry. One of these bars had less cold working than the same make of bar at an earlier time. The amount of cold working is very important and a certain bar can have this altered for policy reasons from time to time without the designer necessarily realising that this has happened. A disadvantage of high yield bars is that the percentage of longitudinal tensile reinforcement is reduced, and it has been proved by many that this reduces the strength of a beam in shear. Research shows that at a given stress in the reinforcement the cracks will be more numerous and smaller for a mechanically bonded bar than for a plain bar. Certain recommendations for the design of structures to resist bombing do not allow high yield steels to be stressed as highly as mild steel reinforcements, because they are more brittle than mild steel, so that their strength can be impaired by sudden shocks.

High yield wires are used to make fabrics for reinforcing slabs. Cross wires are welded to the main wires and enable the main high yield wires to be mechanically bonded to the concrete. The chief advantage of such fabric reinforcements is the speed and low cost of fixing. A disadvantage is the high cost of fabrics. Also, fabrics do not commonly allow comparable economies to those effected by bending up or curtailing alternate bars in slabs. The steel over the supports of continuous slabs is far more rigid for concreting purposes when bars are used as opposed to fabrics. The main steel in a slab is sometimes inadequately anchored into the supporting beams when fabrics are used. The cross wires of rectangular BS fabrics do not normally satisfy the recommendation of BS 8110, to the effect that the high yield reinforcement in any direction should be not less than 0.13% of the gross cross-sectional area. Sometimes additional bars are laid on the fabric to supplement the area of the cross wires to comply with the recommendation, but quite often this has not been done. Such steel is important, however, when temperature stresses are liable to occur or when the slab is of a substantial length (or width) in the direction of the cross wires.

In the U.K., wires commonly used for prestressed concrete are of 2, 5 and 7 mm diameter. Some are also available crimped or with indented surfaces. The wires usually need to be degreased before use either with carbon tetrachloride or by allowing them to rust very slightly and then removing any loose rust. Some favour the latter with plain wires (ones not provided with a mechanical bond) as the rust pitting can increase bond. The author consistently found both methods unsatisfactory for certain laboratory tests of beams with 2 mm diameter plain wires and reliably cured this trouble by using crimped wires. Strand is also very popular in the U.K. – this is essentially a wire rope. When stretched the wires tend to pull in laterally, resulting in a lower modulus of elasticity and also greater relaxation (creep) losses than with straight wires or bars. To reduce these disadvantages strand can be cold-drawn, which also makes it less bulky and stronger. Much work has also been done in the U.K. with high tensile steel bars having rolled-on threads. These threads do not weaken the bar like cut threads.

2.5 Practical use, creation and economics of structural concrete

Concrete is a heavy structural material. The largest spans of bridges are steel suspension bridges, next largest are the steel trusses, then steel girders, reinforced concrete arches, prestressed concrete girders and then reinforced concrete girders. The material cost of concrete is very low per unit of compressive strength. This strength is weak relative to steel, so that in compression it has larger sections and does not have buckling problems as limiting as do steel columns and beams. This explains its economy for columns, arches and prestressed concrete, all essentially concrete in compression. Also many columns, say in a building, are within reason more economic than few, as the columns are more economic than longer span reinforced concrete beams.

The large sections cause members to be heavy. It is important for economy to minimise the weight of suspended floors and roofs. Slabs cannot be too thin because of cracks due to shrinkage and temperature and thus the danger of a miscellaneous point load punching through. A minimum reinforced concrete floor or roof thickness is about 125 mm.

Properties of materials and mix design

For lightness and economy for *in-situ* reinforced concrete construction a slab 125 mm thick can be spanned continuously as far as possible and supported by T-beams which use the slab as their flanges. If the spans required are greater, then this system of beams can be supported by main T-beams. With this system, for economy, the length-to-breadth of the slab panels should be $\geqslant 2:1$. If this ratio $<2:1$ then the slabs should be designed less economically to be two-way spanning. If because of supporting columns the grid of beams is required to be square, then 'two-way spanning slabs' will be useful. If the overall floor or roof thickness needs to be reduced then a 'flat-slab system' may be used. Because of its shallow depth the amount of reinforcement needed is high and it is a heavy construction as none of the concrete not required in flexural tension is eliminated. Economy is improved in this latter respect by having dropped panels, but these can only be used economically for thicker floors or roofs of more than about 220 mm total thickness. Both types of flat slab have inexpensive shuttering but drop panels cause significantly more expense. Sometimes the thinner part of the slab between the dropped panels uses double shuttering or hollow tiles. As the self-weight is high they tend to be less economic for light loadings. 'Waffle floors or roofs' can help the economy of this type of construction, but if the minimum crown thickness is too small and inadequately reinforced they can crack noticeably due to shrinkage, and for some structures this can interfere with serviceability.

The weight of reinforced concrete roofs can be reduced by using shell roofs or folded plate roofs, and weight reduction is more important because the superimposed load is usually very light – even with a shell roof only 63 mm thick the self-weight is often 60% of the total load in the U.K. The minimum thickness of a roof slab would be about 110 mm, and this plus supporting beams is far heavier than a shell roof.

Hollow tile roofs and floors are economic for *in-situ* constructions where floors are required to be say $\geqslant 200$ mm thick, and they can have the advantage of continuity and can provide flanges for T-beams.

The previous remarks apply to *in-situ* concrete. Lightness and economy can be assisted by the use of precast concrete floor and roof units. Generally, they are less expensive than *in-situ* floors and roofs, but the supporting beams lose efficiency and generally the structure is less robust. The great advantage and economy of the continuity of beams and framing action of *in-situ* work, indeed the monolithic advantage of concrete construction, is reduced.

Prestressed concrete tends to be economic mainly when the construction depth allowed is inadequate for reinforced concrete construction.

The heavy weight problem of concrete when overcome in a design automatically gives other advantages in the final structure, such as high natural frequency, easily spread small point loads and damping of small vibrations. Other advantages automatically obtained from R.C. construction are good fire resistance and durability.

The structure is often dictated by client layout requirements. Aesthetics have not been mentioned because there are so many claddings and finishes available, for example a beam and slab floor often has a suspended ceiling to accommodate services so that the final appearance can be the same as a flat slab. Structural concrete usually looks best when the prime aesthetics of the building are based on the structure as opposed to the cladding. Both truism and proportioning according to strength requirements have parts to play, for example a pseudo-reinforced concrete shell roof composed of rolled steel girders and a curved slab can look wrong and unattractive – the girders have constant depth, looking too much in some places and too little in others.

Reference should be made to Section 7.3 and Figures 7.5 to 7.8.

2.6 'Bond' between concrete and steel

This is a most necessary requirement of reinforced concrete construction. If, for example, no 'bond' existed between the tension reinforcement of a beam and the surrounding concrete, then the system would behave in the same way as a carriage spring, having two leaves of different inertias and strengths, namely a relatively large concrete leaf (possibly with a modulus of rupture of only say 3.5 N/mm^2) and a comparatively small steel leaf (relatively strong with a maximum ultimate fibre stress in bending of, say, 520 N/mm^2). Under these conditions the stiffer concrete member would resist most of the superimposed bending moment and its ultimate strength would very soon be realised, at such a load that the assistance of the reinforcement in resisting bending moment could be described as negligible. Thus for the reinforcement to be utilised satisfactorily it has to bond to the concrete so that a reinforced concrete beam bends as though it is a homogeneous member (the strain in the reinforcement being the same as the strain in the surrounding concrete fibres).

Pretensioned tendons must bond to the concrete which is cast around them. Otherwise when released after the concrete has adequately matured, no precompression would be induced in the concrete, the wires just sliding relative to the concrete.

Bond comprises two different actions. Firstly, there is the ability of the concrete to stick to the steel. This is usually referred to as *adhesion*. An illustration of this is the ability of concrete to most forcibly stick to steel shovels, trowels and parts of mixers. Secondly, there is the frictional resistance between the steel and the concrete, often called *grip*. When a bar is tending to pull out of its surrounding concrete the relative movement of the bar to such concrete is known as *slip*. A bond stress cannot exist without its coexistent strain, that is without slip. Adhesion is an initial resistance to bond and occurs when the slip is minute. With a smooth cylindrical bar for example, adhesion is often attributed to micromechanical locking (minute irregularities on the bar mechanically locking to the concrete). As soon as a small amount of slip occurs the adhesion is ruptured and takes no further part in the bond resistance. For such slips a bond resistance is developed by the friction between the bar and the surrounding concrete. This is aided by the shrinkage of the concrete upon setting, as this causes the concrete to exert a radial pressure on the reinforcing bar, thus increasing the frictional resistance between the two materials.

The frictional resistance can be assessed by multiplying such a pressure due to shrinkage by some suitable coefficient. Certain coefficients suggested by Armstrong[13] illustrate the sensitivity of the frictional resistance to the grease and rust on the surface of a bar. *Dilatancy* is a resistance to slip resulting from the wedging action of the small particles of concrete loosened after an initial slip has occurred. This effect constitutes a part of the general frictional resistance mentioned previously. The frictional resistance is enhanced at the locality of a crack where a *tangential friction* occurs because of the slight change in direction of the reinforcement bar.

Another contribution to the frictional resistance can be called *wedge action*. When the stress in a bar changes along its length due to its bond to the surrounding concrete, the effect of Poisson's ratio will cause a corresponding change in its cross-sectional area. Thus, such a reinforcement bar becomes slightly tapered and hence the term *wedge action*. With non-prestressed reinforced concrete this effect is extremely small. For prestressed concrete where steel stresses are much greater the wedge action is a significant asset. To illustrate this point, Figure 2.13 exaggerates the effect; the pretensioned wire is unstressed after release at A and has therefore a larger diameter here than at B where the wire is in its fully stressed condition.

Both the adhesion and the frictional resistance are increased by *mechanical locking*, that is by using reinforcement bars with surface deformations which mechanically lock to the concrete.[14] These have now become very popular in the U.K. whereas they were used by a minority in 1950 to 1960.

Bond therefore consists of firstly an *adhesive resistance* and then a *frictional resistance*. As a simple illustration, Figure 2.14 refers to a *pull-out test* of a steel rod from a concrete block. When the pull in the rod is P the portion of the graph AB represents the way in which the force in the rod is gradually transmitted to the concrete by frictional resistance. At the point B, the force still in the bar is insufficient to overcome the adhesive resistance of the remainder of the bar, and therefore BC represents the way in which the force in the rod is gradually transmitted to the concrete by adhesion. When the load in the bar is increased to P', the length of the bar slipping increases and the curve becomes A'B'C', A'B' being the frictional stage and B'C' the adhesive stage.

Figure 2.13

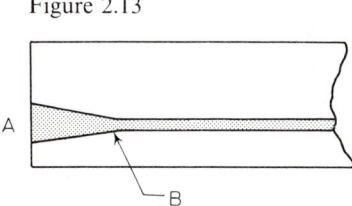

Figure 2.14

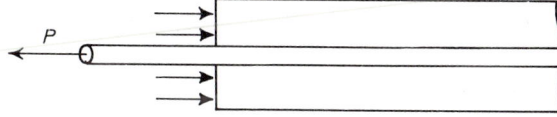

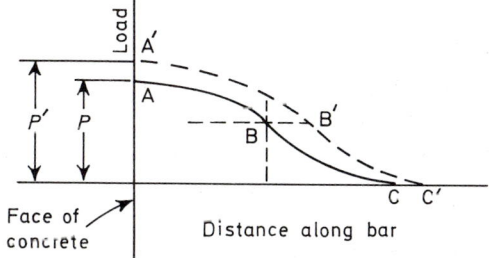

Properties of materials and mix design

When a reinforced concrete beam is subjected to bending, the tension reinforcement which is bonded to the concrete is such that both the steel and the concrete are in tension. This is a criticism of the above mentioned pull-out test in which the steel is in tension and its surrounding concrete is in compression. Tests[15] of bond stress are therefore made by measuring the strain in the steel, and the strain in the concrete touching such steel, along the lengths of the bars provided as tension reinforcement in beams.

2.6.1 Anchorage or bond length

Figure 2.15 shows a bar anchored into a block of concrete. The necessary bond length l_b is to be determined so that the bar can develop a tensile stress of f_s at section B. If the bar has a cross-sectional area A_s and perimeter u, then the force in the bar N_s is given by

$$N_s = A_s f_s \tag{2.6}$$

If f_{mbs} is the average bond stress between the steel and the concrete, this exists over an area of contact equal to ul_b, therefore

$$N_s = f_{mbs} u l_b \tag{2.6a}$$

Eliminating N_s between these two equations

$$l_b = A_s f_s / (f_{mbs} u) \tag{2.7}$$

If diameter of bar $= d_b$, then from equation 2.7

$$l_b / d_b = f_s / (4 f_{mbs}) \tag{2.8}$$

BS 8110 expresses equation 2.7 as

$$l = \frac{A_s f_s}{f_b \pi \varphi_e} \tag{2.7a}$$

Figure 2.15

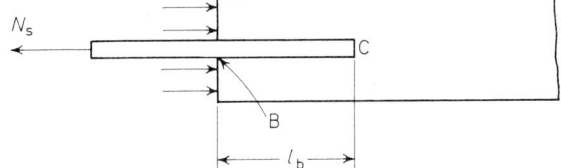

'Bond' between concrete and steel

where $l = l_b$, $f_b = f_{mbs}$ and $\varphi_e = d_b$. φ_e is the equivalent diameter in the case of deformed bars.

$$\therefore l = \frac{f_s \varphi_e}{4f_b} \qquad (2.8a)$$

BS 8110 gives the design ultimate value of f_b as $f_{bu} = \beta\sqrt{(f_{cu})}$ and values of β (see Table 2.10) are given in its Table 3.28 and depend upon whether the bars are plain or deformed or in fabric form. It also gives the design ultimate value of $f_s = \gamma_m f_y = 0.87$. Therefore from equation 2.8a

$$l = \frac{0.87 f_y}{4\beta\sqrt{f_{cu}}} \varphi_e \qquad (2.8b)$$

Table 2.9 enables anchorage lengths to be easily determined for bars in tension; values of the ratio l to φ_e are read off for values of f_{cu} (= concrete grade or characteristic strength) and f_y (characteristic strength of steel). Values of β and f_y are shown in Table 2.10.

Table 2.9. *Tension anchorage lengths (mm)*

f_{cu}	25	30	35	40 or more
f_y				
250 (plain)	39	36	33	31
460 (deformed type 1)	51	46	43	40
460 (deformed type 2)	41	37	34	32
Stresses, N/mm²	Ratios l to φ_e			

Table 2.10

		β	
Designation	f_y	Tension	Compression
Plain hot rolled mild steel	250	0.28	0.35
Square-twisted bars (i.e. cold worked) (BS 8110 deformed type 1)	460	0.40	0.50
High-yield rolled ribbed bars (BS 8110 deformed type 2)	460	0.50	0.63

When bars in compression are anchored, the compression on a bar is also resisted by the pressure on its end (e.g. end C in Figure 2.15). To allow for this it is simple to add a suitable amount to equation 2.6a, namely A_s times the compressive stress on the concrete. This was once done, but BS 8110 prefers simply, but less logically and precisely, to increase the values of β for bars in compression. On this basis Table 2.11 enables anchorage lengths to be easily determined for bars in compression, similarly to Table 2.9.

2.6.2 End anchorages

In practice reinforcement is seldom, if ever, perfectly clean of rust and/or mill scale and/or grease. This can have a more disastrous effect upon the anchorage in tension of plain than of deformed bars. Hence it is a good practice always to provide plain bars, when used in tension, with end anchorages such as hooks or nibs. These end anchorages are disadvantageous for deformed high yield steel because of cost, efficiency

Table 2.11. *Compression anchorage lengths (mm)*

f_{cu}	25	30	35	40 or more
f_y				
250 (plain)	32	29	27	25
460 (deformed type 1)	41	37	34	32
460 (deformed type 2)	32	29	27	26
Stresses, N/mm²	Ratios 1 to φ_e			

Table 2.12. *Anchorage values of hooks and nibs*

mm	d_b	6	8	10	12	16	20	25	32
Mild steel									
Hook	$16d_b$	96	128	160	192	256	320	400	512
	$l_h(9d_b)$	54	72	90	108	144	180	225	288
Nib	$8d_b$	48	64	80	96	128	160	200	256
	$l_n(5d_b)$	30	40	50	60	80	100	125	160
High yield									
Hook	$24d_b$	144	192	240	288	384	480	600	768
	$l_h(11d_b)$	66	88	110	132	176	220	275	352
Nib	$12d_b$	72	96	120	144	192	240	300	384
	$l_n(5.5d_b)$	33	44	55	66	88	110	138	176

when stopping off bars in beams, and overstrain,[12] but can be used if essential (for example lack of space in which to anchor at end of beam in some instances). Similarly it is disadvantageous in cost and efficiency to use end anchorages on bars in compression, but they can be used if essential. End anchorages are commonly hooks and nibs as shown in Figure 2.16 (a) and (b) respectively. To anchor a bar, the overall length a required is the value of l_b from the tables of Section 2.6.1, less $16d_b$ and $8d_b$ for a mild steel hook and nib, respectively, and $24d_b$ and $12d_b$ for a high yield steel hook and nib, respectively. After determining a for a bar we need to determine its total length. The total lengths of bars with hooks and nibs are $a_h + l_h$ and $a_n + l_n$, respectively. All these values are given in

Figure 2.16

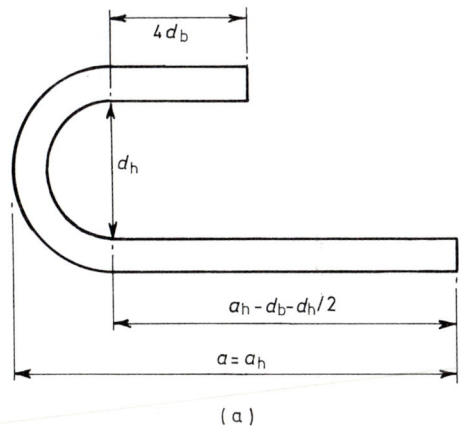

(a)

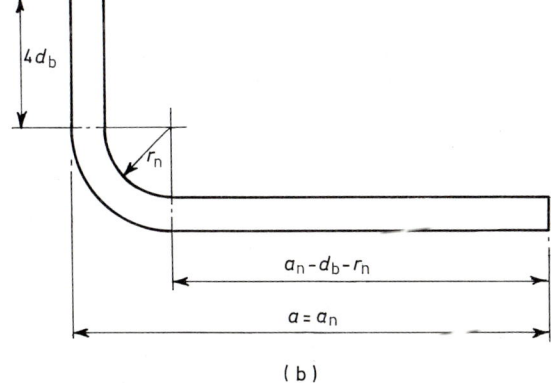

(b)

Table 2.12 to aid designers. From the geometry of Figure 2.16(a), the total length of the bar

$$= a_h + l_h = (a_h - d_b - 0.5d_h) + 0.5\pi(d_h + d_b) + 4d_b$$

$$\therefore l_h = 3d_b - 0.5d_h + 0.5\pi(d_h + d_b) \tag{2.9}$$

From Figure 2.16(b), the total length of the bar

$$= a_n + l_n = (a_n - d_b - r_n) + 0.5\pi(r_n + 0.5d_b) + 4d_b$$

$$\therefore l_n = 3d_b - r_n + 0.5\pi(r_n + 0.5d_b) \tag{2.10}$$

From these equations: for mild steel $d_h = 4d_b$ and $r_n = 2d_b$, thus $l_h = 8.85d_b$, say $9d_b$, and $l_n = 4.93d_b$, say $5d_b$; for high yield steel $d_h = 6d_b$ and $r_n = 3d_b$, thus $l_h = 11d_b$ and $l_n = 5.5d_b$.

Hooks are worth much more as an anchorage per unit length of material than nibs and cost little more to produce.

Tables 2.13 and 2.14 are based on Table 2.10 and $f_{cu} = 25$ N/mm² and should be useful for designers of *in-situ* concrete, because the weakest structural concrete is generally used for such work. With regard to Table 2.13, plain mild steel bars are not recommended to be anchored without end anchorages and are therefore excluded from the table.

Table 2.13. *Straight anchorage lengths ($f_{cu} = 25$ N/mm²)*

φ_e, mm	6	8	10	12	16	20	25	32	f_y, N/mm²
	Compression anchorage lengths (*l*), mm								
$32\varphi_e$	192	256	320	384	512	640	800	1024	250
$41\varphi_e$	246	328	410	492	656	820	1025	1312	460 (1)
$32\varphi_e$	192	256	320	384	512	640	800	1024	460 (2)
	Tension anchorage lengths (*l*), mm								
$51\varphi_e$	306	408	510	612	816	1020	1275	1632	460 (1)
$41\varphi_e$	246	328	410	492	656	820	1025	1312	460 (2)

Table 2.14. *Overall tension anchorage lengths (mm) for hooks and nibs ($f_y = 250$ N/mm², $f_{cu} = 25$ N/mm²)*

φ_e, mm	6	8	10	12	16	20	25	32
a_h ($23\varphi_e$)	138	184	230	276	368	460	575	736
a_n ($31\varphi_e$)	186	248	310	372	496	620	775	992

Example 2.5. A plain mild steel bar of 12 mm diameter is to be anchored with a hook. The characteristic strength of the concrete is 25 N/mm². Determine the overall length of the anchorage and the total length of the bar required for this anchorage.

From Table 2.10, $f_y = 250$ N/mm².

From Table 2.9, $l_b/d_b = 39$, ∴ $l_b = 468$ mm.

From Table 2.12, $16d_b = 192$ mm, $l_h = 108$ mm

$$\therefore a_h = 468 - 192 = 276 \text{ mm}$$

$$\text{Total length} = 276 + 108 = 384 \text{ mm}$$

Alternatively, for these particular stresses, using Table 2.14, $a_h = 276$ mm, and from Table 2.12, $l_h = 108$ mm, therefore total length $= 276 + 108 = 384$ mm.

2.6.3 Laps in reinforcement

To lap bars in compression, for example in columns, walls and sometimes over the supports of continuous T-beams, normally straight lengths are lapped the distance of the compression anchorage length (see Sections 2.6.1 and 2.6.5, and Figure 2.17). There is rarely any advantage in using hooks or nibs and so reducing the lap length to the overall length of anchorage (see Section 2.6.2).

Lapping bars in tension is to be avoided. Plain bars without end anchorages should not be lapped in tension. When bars have to be lapped (see Sections 2.6.1, 2.6.2 and 2.6.5) in tension one should try to make laps, which need to be the distance of the tension anchorage length, as far from the places of maximum stress as possible and to stagger laps so that they do not overlap one another. For example, for a particular folded plate[16] about 26 m long the tension steel to be used was in 12 m lengths. The number of bars of the same diameter which needed to be provided for the full length was increased by one; then each plain bar could be discontinued

Figure 2.17

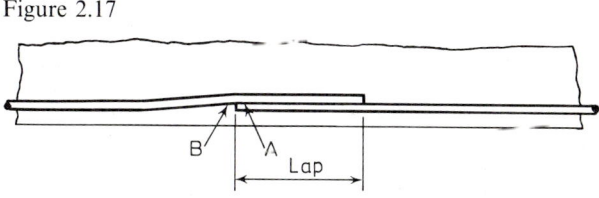

74 *Properties of materials and mix design*

at any position. The system is indicated in Figure 2.18, bars A being of the maximum length possible and lengths a_h being the overall length of the hook anchorage. Adjacent hooks had a clear distance between them of about 75 mm to give a tolerance to the bar bender and fixer and to aid concreting.

The compression lap shown in Figure 2.17 should not be used in tension as the bars try and pull into line and thus outwards at A and B, trying to split off the concrete cover. If one is desperate to use this type of lap in tension, then the only chance of success is to use deformed bars and a stirrup at A, designed to resist the splitting force. The effective depth of reinforcement is reduced at B – to avoid this, the lap shown can be rotated through a right-angle if detailing permits.

It is good practice to have a gap of about 15 mm between the lapped bars (Figure 2.17), to avoid voids in the concrete between the bars.

2.6.4 Curtailment of reinforcement in beams

Table 2.15 is useful for designers giving the points B where bars are no longer required for resisting bending moment in a beam of span l. It is based on uniformly distributed loads. For continuous spans it assumes that the bending moments at mid span and support are equal. Column β gives the number of bars at the position of maximum sagging bending moment at or near to mid span. The coefficients α are given for the order in which these bars are no longer required for considerations of bending moment, counting from the position of maximum sagging bending moment. The bending moment diagrams to which the coefficients relate are shown below the table.

Strictly speaking, at the point when a bar is no longer required, if it is not immediately bent-up for shear it can be just terminated, but it must be checked that it has sufficient anchorage length to develop its full tensile strength from the point where this is needed. However, for plain bars a mechanical end anchorage is desirable (Section 2.6.2) so the curve of the

Figure 2.18

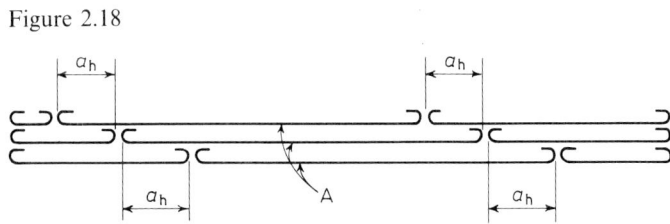

'Bond' between concrete and steel

Table 2.15

	α_1						α_2						α_3						α_4					
	Order of stopping-off or bending-up bars																							
β	1st	2nd	3rd	4th	5th	6th	1st	2nd	3rd	4th	5th	6th	1st	2nd	3rd	4th	5th	6th	1st	2nd	3rd	4th	5th	6th
1	0	—	—	—	—	—	—	—	—	—	—	—	—	—	—	—	—	—	—	—	—	—	—	—
2	.15	0	—	—	—	—	.11	—	—	—	—	—	0	—	—	—	—	—	.09	—	—	—	—	—
3	.21	.09	0	—	—	—	.24	.11	—	—	—	—	.13	0	—	—	—	—	.21	.09	—	—	—	—
4	.25	.15	.07	0	—	—	.30	.19	.11	—	—	—	.18	.08	0	—	—	—	.27	.16	.09	—	—	—
5	.28	.19	.12	.05	0	—	.33	.24	.17	.11	—	—	.22	.13	.05	0	—	—	.30	.21	.15	.09	—	—
6	.30	.21	.15	.09	.04	0	.35	.28	.21	.16	.11	—	.25	.15	.09	.04	—	—	.31	.24	.18	.13	.09	—
7	.31	.23	.17	.12	.08	.04	.37	.30	.24	.19	.14	.11	.27	.18	.13	.08	.03	0	.33	.27	.21	.16	.12	.09
8	.32	.25	.19	.15	.10	.07	.39	.32	.26	.22	.17	.14	.29	.20	.15	.11	.07	.03	.34	.29	.23	.19	.15	.12
							.40	.33	.27	.24	.20	.17	.30	.22	.17	.13	.09	.05	.35	.30	.25	.21	.18	.15

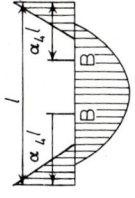

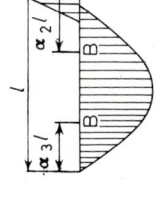

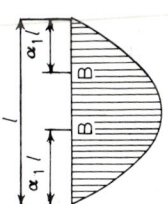

76 *Properties of materials and mix design*

hook or nib can be commenced at this point where the bar is no longer required.

Example 2.6. A continuous beam carries a uniformly distributed load over an internal span of 8 m and the design for ultimate limit state of bending requires five 25 mm diameter deformed bars of hot rolled high yield steel in tension at mid span. One of these bars is to be curtailed; determine the length of this bar from mid span, assuming $f_{cu} = 25$ N/mm².

From Table 2.15, $\alpha_4 = 0.31$, $\therefore \alpha_1 l = 0.31 \times 8 = 2.48$ m.

Allow no extra anchorage length (see later) but check that this bar has sufficient anchorage length from mid span where it is fully stressed.

From Table 2.10, $f_y = 460$ N/mm².

From Tables 2.9 and 2.13, $l_b = 1025$ mm.

Length of bar from mid span $= 4 - 2.48 = 1.52$ m and this is all right, as it is greater than 1025 mm. (Also see remainder of this section.)

But then BS 8110 expresses concern that in practice the distribution of live loading may not be as assumed and this would make errors in the values of α. Hence it recommends that an extra anchorage length be added to each curtailed bar of $12\varphi_e$ or its effective depth. Against this is the fact that design loadings are sometimes very conservative and when the distribution is wrong the total is usually less.

BS 8110 also expresses concern about anchoring bars in tension zones and recommends bars extending 'an anchorage length appropriate to their design strength $(0.87f_y)$ from the point where they are no longer required to resist bending'. This seems very conservative relative to past practice and experience. In addition it recommends that a bar should not be stopped in a tension zone unless the shear capacity is twice the actual shear present; or the continuing bars have twice the area required to resist the moment at that section.

A method used successfully over many years by the author is simpler than the requirements of the preceding two paragraphs. It is based on the idea that the bar to be curtailed will be continued to some extent beyond the point where it is no longer required. There will thus be no sudden change in total tensile force on either side of this point, because the beam curvature and bending moment do not suddenly alter. Hence it is good practice to assume that all the bars have the same strain and stress at this point. Thus the bar to be curtailed is anchored for this stress, whether in a zone of tension or compression.

'Bond' between concrete and steel

Example 2.7. If the 20 mm diameter bar is to be curtailed out of a group of two 25 mm diameter and one 20 mm diameter deformed bars, determine the length of this bar which must be continued past the point P where it is no longer required. Suppose for its design stress the 20 mm diameter bar needs an anchorage length of 820 mm.

Tensile force required at point P $= 2 \times (\pi/4) \times 25^2$
$\times$ Design stress

Stress in all bars at this point $= \dfrac{2 \times (\pi/4) \times 25^2 \times \text{Design stress}}{2 \times (\pi/4) \times 25^2 + (\pi/4) \times 20^2}$

$= 0.7576 \times$ Design stress

Anchorage length required $= 0.7576 \times 0.82 = 0.621$ m

Example 2.8. Repeat Example 2.6 with this alternative method.
As before $\alpha_1 l = 2.16$ m, $f_y = 460$ N/mm², $l_b = 1025$ m.
The anchorage length required from point P $= (4/5) \times 1025 = 820$ mm.
Length of bar from mid span $= 4 - 2.16 + 1.2 = 3.04$ m, and this is all right as it is greater than 1.5 m.

2.6.5 Anchorage of bent-up shear bars

Bars bent up as shown in Figure 3.5(b) can be used as shear reinforcement. The anchorage length NBH is that required for the bar to be able to develop its design strength at the neutral axis N.

Example 2.9. A 25 mm diameter mild steel bar is bent up at 45° to resist shear. Its design strength is $f_y/\gamma_m = 250/1.15 = 217.4$ N/mm², and $f_{cu} = 25$ N/mm². The effective depth of the bottom tensile reinforcement $= 450$ mm, and the cover to the top steel $= 25$ mm. Determine the length BH.

In calculating BN we should use the depth of the neutral axis but for simplicity and slight extra safety we will use $0.5 \times 450 = 225$ mm. Then BN $= [225 - 25 \text{ (cover)} - 12 \text{ (half dia. bar)}]\sqrt{2} = 266$ mm. The total anchorage length from Table 2.9 is $39 \times 25 = 975$ mm. For detailing such a shear reinforcement system, usually the shorter the length BH the better so economising further it has been common past practice to allow half the value of a nib for the anchorage effect of the bar deviating through 45° at B. CP 114 used to allow this, and it has some value, but it does not seem to be mentioned in BS 8110. From Table 2.12 this reduction in anchorage length is $200/2 = 100$ mm. From the same table the hook at H reduces the overall anchorage length by 400 mm. Hence $a_h = 975 - 100 - 400 = 475$ mm. Hence BH $= 475 - BN = 475 - 266 = 209$ mm.

2.6.6 Bearing stresses inside bends

Figure 2.19 shows a reinforcement bar of diameter d in tension bent to any shape. At point P the tensile force in the bar is F_b and at point P' this force has become $F_b - \delta F_b$. This change δF_b is due to the bond stress over the length PP' shown on Figure 2.19 as a force δF_b (this acts all around the perimeter of the bar). Because of the change of direction of the bar, and thus of the axial force in it, there is a bearing stress f inside the bend. Resolving forces perpendicular to PP'

$$f dr \delta\alpha = F_b \sin(\delta\alpha/2) + (F_b - \delta F_b)\sin(\delta\alpha/2)$$

In the limit when $\delta\alpha \to 0$, $\sin(\delta\alpha/2) \to \delta\alpha/2$, and $\delta F_b \to dF_b$.

$$\therefore 2fr d = 2F_b - dF_b$$

Now dF_b is negligible ($\to 0$) in comparison to the size of the quantities $2frd$ and $2F_b$.

$$\therefore f = F_b/(rd) \tag{2.11}$$

This stress at P does not need to be checked for the standard anchorage hooks and nibs of Section 2.6.2. BS 8110 requires f to be checked when the bar continues more than $4d$ after the bend and is still required for bond resistance – for example the bend at B in Figure 3.5(b) and at b' in Figure 2.21.

Figure 2.19

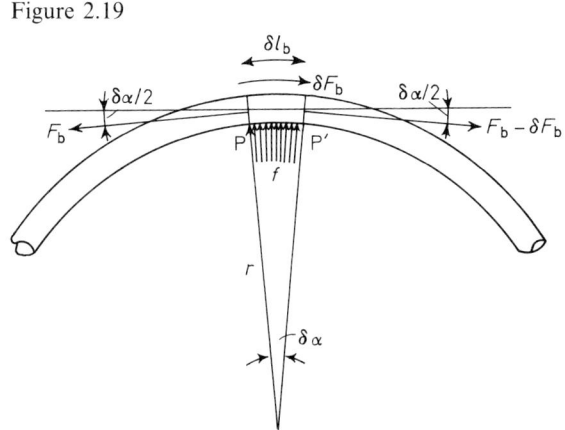

'Bond' between concrete and steel

The same theory and equation (2.11) apply for bars in compression. In Figure 2.19 F_b would be in the opposite direction and f would be at the opposite side of the bar.

A dowel bar under the bend just transmits and concentrates the bearing stress to immediately below it, though it can help to spread this pressure transversely. BS 8110 limits f from equation (2.11) in its equation 50 and so such dowel bars are not considered helpful in reducing the bearing stresses inside bends.

Example 2.10. For example 2.10 determine the minimum radius of curvature allowed at B. The beam is T-shaped and the bend B is in the wide flange.

Stress in bar at B = $217.4 \times (975-266)/975 = 158.1$ N/mm².

$d = 25$ mm, $F_b = 158.1 \times (\pi/4) \times 25^2 = 77612$ N.

The permissible f is, from BS 8110 (formula 3.11.6.8), $a_b = \infty$,

$$= (2 \times 25)/(1 + 2 \times 25/\infty) = 50 \text{ N/mm}^2$$

Hence from equation 2.11

$$r = F_b/(fd) = 77612/(50 \times 25) = 62.1 \text{ mm}$$

2.6.7 Anchorage of stirrups (or links)

BS 8110 recommendation 3.12.8.6 (Anchorage of links) conflicts with the BS 8110 recommendations already referred to in this chapter regarding anchorage length and bearing stress inside bends. Its inadequacy in this respect might be justified on the basis that the design of stirrups for shear is still conservative, but this is not indicated, and is not a sound approach. Against this, the anchorages are sometimes in tension zones and links are sometimes required to resist torsion, for example a beam of square cross section would experience maximum shear stress due to torsion not only at the neutral axis but at the centres of the top and bottom peripheries of the beam – where the links might have inadequate tension anchorage if in accordance with BS 8110. Multitudes of beams in practice, designed with links to resist shear and not designed to resist torsion, do indeed have to resist varying amounts of torsion.

It would seem desirable[17] for the anchorages of links designed to resist shear and/or torsion to be in accordance with the previous sections of this chapter. In addition tests[12] show that deformed bars are only 10% more

80 *Properties of materials and mix design*

effective in shear than plain bars, and that if the deformed bars are high yield then the failure is unexpected and violent. Deformed high yield stirrups should not be stressed any higher than mild steel links in shear.[12] This unfortunately disagrees with BS 8110 but agrees with CP 114 (1957).

Example 2.11. Design the anchorage of an 8 mm diameter mild steel link of design strength $f_y/\gamma_m = 250/1.15 = 217.4 \text{ N/mm}^2$, and $f_{cu} = 25 \text{ N/mm}^2$. The internal dimensions of the link are 175×400 mm.

The tension lap from Table 2.9 is $39d_b$. Two right-angle bends, using say the shape of link of Figure 2.20, are worth $8d_b$ each as anchorage (see Section 2.6.2). Hence tension lap required is $39d_b - 16d_b = 23d_b = 23 \times 8 = 184$ mm. The length of each vertical end is approximately $(184 - 175) \times 0.5 = 4.5$ mm. This should be at least l_n of Table 2.12, i.e. $5d_b = 5 \times 8 = 40$ mm. The link is shown in Figure 2.20, the lap being along the top and down each side a length of 40 mm.

2.6.8   Splitting effects of bar anchorages

Anchoring a bar abcd from a beam into a column as shown in Figure 2.21 is bad practice, causing splitting of the column along bcd. Even if the bearing stress is in order at b, increasing the length bcd does not add useful anchorage length, because of the splitting weakness. The bar should be taken as far across the column as possible, that is ab'c'd'. Designs are made for bending moments and shear forces assuming members to be concentrated at their centre lines. The true internal stress system at a practical junction is difficult to assess; hence the junction should be detailed as conservatively as possible, that is, bb' should be as

Figure 2.20

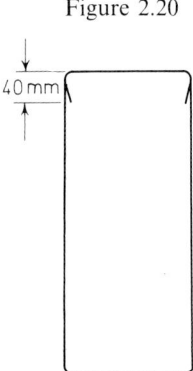

great as possible. In calculating the anchorage length, the bend at b' is worth the values $8d_b$ and $12d_b$ of nibs in Table 2.12.

2.6.9 Anchorage lengths based on elastic analysis

Equation 2.8 can be used provided f_s is taken as the permissible stress for a bar in tension and f_{mbs} as the permissible bond stress. Permissible stresses are stresses at working loads and are given in CP 114 and BS 5337:1976. In a similar way Sections 2.6.2–2.6.9 apply.

Thus BS 5337:1976 gives $f_s = 85$ N/mm² in tension for plain bars and exposure Class A and $f_{mbs} = 1.0$ N/mm² and 0.9 N/mm² for Grade 30 and 25 concretes, respectively. The respective anchorage lengths are thus $85d_b/(4 \times 1.0) = 21.25d_b$ and $85d_b/(4 \times 0.9) = 23.61d_b$. Table 2.16 is to help designers using elastic theory.

Example 2.12. Determine the overall anchorage length of a 20 mm diameter plain bar of mild steel with an end hook, permissible tensile stress $= 85$ N/mm² and permissible average bond stress $= 0.90$ N/mm².

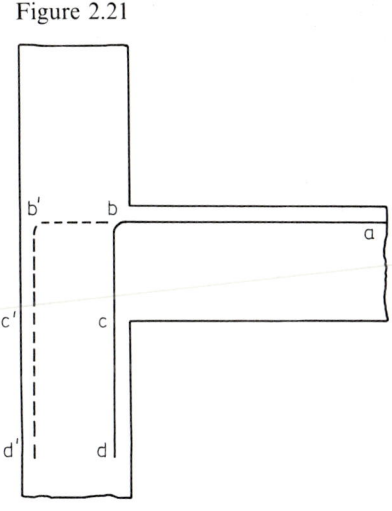

Figure 2.21

Table 2.16

d_b, mm	6	8	10	12	16	20	25	32
$21.25d_b$	127	170	212	255	340	425	531	680
$23.61d_b$	141	188	236	283	377	472	590	755

82 *Properties of materials and mix design*

From Table 2.16, straight anchorage length = 472 mm.

From Table 2.12, hook is worth 320 mm.

Hence (see Figure 2.16(a)) $a_b = 472 - 320 = 152$ mm.

2.7 Corrosion of reinforcement, carbonisation and 'cathodic protection'

Corrosion of reinforcement has been discussed in Section 2.1, item 4, in connection with calcium chloride, and this should be consulted before continuing to read this present Section. Without calcium chloride, corrosion of the reinforcement can take place, more slowly, just the same, by ingress of water through cracks in the concrete to the reinforcement. Oxygen ion must also be present. Rainwater is normally acidic, particularly in industrial areas and contains oxygen ion. If the air is clean it can dissolve carbon dioxide (to give carbonic acid) and oxygen ion. If polluted it may also dissolve CO, SO_2 and SO_3 to give carbonous, sulphurous and sulphuric acids.

It is generally considered that the alkaline nature of the concrete in contact with the reinforcement reduces its tendency to corrode. Over many years if rainwater and/or the gas CO_2 from the atmosphere can penetrate the cracks to the reinforcement or if the concrete is porous enough to allow this to happen, the CaO and $Ca(OH)_2$ in the cement is converted to $Ca(CO_3)_2$ and it loses its alkalinity which was resisting corrosion of the reinforcement. This is known as 'carbonisation'.

Since about 1983, this has been studied considerably in the U.K. because in the terrific building boom of the 1960s many buildings and houses were built with external precast concrete walls exposed to climatic conditions often in industrially polluted cities. In the U.K. prior to *c.* 1940, exposed concrete construction could often be seen to suffer severe corrosion of its reinforcement and subsequently it was usual to clad it with brickwork, glazing, etc. The buildings of the 1960s with concrete exposed often show such troubles again in the 1980s. Carbonisation is one problem but porous concrete, voids in concrete, construction joints opening up due to shrinkage etc. allow the carbonisation to penetrate to the reinforcement and be troublesome there. The 1960s building boom meant that there were staff shortages at all levels and speedy programmes to complete; not conducive to good workmanship well supervised.

To fight against carbonisation there has been much interest in 'cathodic protection'. A covering which conducts electricity (and protects the concrete from the weather) is painted on the surface of the exposed concrete. A simple motor car battery then has one of its terminals connected to this covering and its other terminal connected to the reinforcement. This is then like a simple schoolboy experiment in electrolysis where the covering is the anode and the reinforcement is the cathode. There is considered to be sufficient dampness (i.e. water) in the concrete to enable electrical current to be conducted, that is for electrolysis to take place. Then for example calcium compounds such as calcium carbonate in the concrete will be split so that the calcium ion will be deposited on to the cathode and with the water/dampness there will react chemically with it to form calcium hydroxide on the reinforcement. Thus the chemically neutral calcium carbonate in contact with the reinforcement is replaced with the alkali calcium hydroxide, and an alkaline surrounding for the reinforcement is more resistant to further corrosion than the neutral carbonated compound.

Of course there are problems if the corrosion prevents conduction of current, which is only from a low voltage source, from one piece of reinforcement to another in contact with it.

References

1. Bogue, R. H., *Chemistry of Portland Cement*, Reinhold, New York (1955).
2. Troxell, G. E., Davis, H. E. & Kelly, J. W., *Composition and Properties of Concrete*, McGraw-Hill, U.S.A. (1968).
3. Brunauer, S., Tobermorite gel – the heart of concrete. *American Scientist*, Mar. (1962).
4. Neville, A. M., *Properties of Concrete*, Pitman (1963 and 1973).
5. Robson, T. D., High *alumina* cements and concretes. *Contractors' Record* (1962).
6. Bate, E. E. H. and Stewart, D. A., A survey of modern concrete technique. *Proceedings of the Institution of Civil Engineers, Part* 3, Dec. (1955).
7. Hughes, B. P., *Limit State Theory for Reinforced Concrete*, Pitman (1971 and 1980).
8. *Design of Concrete Mixes*. D.S.I.R. Road Research Laboratory, Road Note No. 4., H.M.S.O., London (1950).
9. Teychenne, D. C., Nicholls, J. C., Franklin, R. E., Hobbs, D. W. and Erntroy, H. C., *Design of Normal Concrete Mixes*, Department of the Environment, H.M.S.O., London (1975).
10. A.C.I. Committee 211, *A.C.I. Standard 211.1–70: Recommended Practice for Selecting Proportions for Concrete*. American Concrete Institute, Detroit (1970).
11. Evans, R. H., Effect of rate of loading on some mechanical properties of concrete, Proceedings of the 1958 London Conference organised by the Mining Research Establishment of the N.C.B. in consultation with the Building Research Establishment, D.S.I.R., pp. 157–175.

11a. Hayward, D., Brown, G. and Wilby, C. B., Slab firm goes for mix not strength. *New Civil Engineer*, 5 Sept. 1985, pp. 26–7. London.
12. Wilby, C. B., Overstrain in high-tensile reinforcing bars at bends and in stirrups, *Indian Concrete Journal*, Jan. (1962).
13. Armstrong, W. E. I., Bond in prestressed concrete. *Journal of the Institution of Civil Engineers*, **33**, Nov. (1949).
14. Regan, P. E. and Yu, C. W., *Limit State Design of Structural Concrete*, Chatto and Windus (1973).
15. Evans, R. H. and Robinson, G. W., Bond stresses in prestressed concrete from X-ray photographs. Paper No. 6025, *Proceedings of the Institution of Civil Engineers*, Part 1, Mar. (1953).
16. Wilby, C. B., *Concrete for Structural Engineers*, Newnes-Butterworths, London–Boston (1977).
17. Evans, R. H. and Wilby, C. B., *Concrete – Plain, Reinforced, Prestressed and Shell*, Art. 5.7, Edward Arnold (1963).

3

Reinforced concrete beams

3.1 Design

To design a reinforced concrete beam a reasonable procedure is as follows:

1. Estimate the dimensions of the beam. The overall depth can be taken as say a proportion of the 'effective span', 1/20 for simply supported, 1/25 for continuous, and 1/10 for cantilever beams. The breadth (or breadth of rib for, say, a T-beam) can be taken as $\frac{1}{3}$ to $\frac{1}{2}$ of this depth.

2. For a rectangular beam, the ratio of the maximum distance between lateral restraints to breadth is ideal if less than 30, reasonable if between 30 and 40, and likely to be impracticable if more than 50. This is because of the possibility of narrow beams buckling sideways.

3. Check the strength in shear, and torsion if present, in the worst case, usually a section adjacent to a support. It may well be that reinforcement is required and this should not normally be greater than say 10 mm diameter stirrups (two, four or six arm according to width of beam) at 80 mm centres. The beam may well eventually be detailed with bent-up bars assisting the stirrups in the localities of maximum shear force.

4. Check the strength in bending. For a rectangular (or simply supported) beam this is best done first at the section subjected to maximum bending moment. For a continuous T-beam, mid span will normally be strong enough in bending if the supports are, because the beam is acting as a rectangular beam at the supports. If compression steel is required at the supports it might be desirable because of detailing to revise the design to eliminate the need for such steel – if this is not done it is useful practically for the area of the compression steel not to exceed 1% of the breadth (of rib for a T-beam) times the overall depth. Then

determine the longitudinal tension reinforcement and see if it can be detailed reasonably in the beam. In the case of simply supported T-beams it is speedier to calculate the reinforcement before the compressive strength of the section.

The beam has now been reasonably well designed and it is now only a matter of checking the limit states of deflection and cracking, and then determining the bending and shear steel at critical sections.

3.2   Elastic analysis for bending moments

At working loads the elastic analysis gives a reasonably accurate assessment of the (longitudinal) stresses in the concrete and reinforcement. It also gives a reasonable assessment of deflections experienced at modest loads in a loading test using a Young's modulus obtained from testing specimens of the concrete. For estimating the deflection of a member in practice, for, say, a BS 8110 Grade 20 or 25 concrete, the elastic analysis is reasonable for design purposes if a Young's modulus of the concrete of say 14 kN/mm^2 is used for continuously sustained loading and 21 kN/mm^2 for loading of short duration.

CP 114 allowed the design of beams to be based on elastic analysis, restricting stresses to within the elastic behaviour of the materials at working loads. Multitudes of structures which have lasted many years illustrate the safety of such designs. BS 8110 mainly does not use this method. With regard to the application of this method refer to Section 1.2.

For shell roofs the analysis for forces and bending moments is elastic, so it would seem logical and safe (because our experience is based on elastic design) to use elastic analysis for designing for these forces and moments. Where experimental evidence has not adequately ratified the methods of predicting ultimate bending moments, members can be designed by elastic theory with confidence, for example shells, or a beam with unsymmetrical section with skew loading. (The author has designed purlins like this.)

For the above and other reasons the elastic analysis will be presented as concisely as possible, using only the moment of inertia of the equivalent concrete section method.

3.2.1   Assumptions made in the elastic design of reinforced concrete

Firstly, it is assumed that *plane sections subjected to bending remain plane after bending* (Bernoulli's theorem). This is found to be reasonably true by experiment, and means that the *distribution of strain is linear* across the section.

It is also assumed that *stress is proportional to strain for both the steel and the concrete*. This is accurately true for the steel up to the limit of proportionality, but only approximately true for the concrete as far as the allowable working stress (permissible stress), and is most inaccurate above this stress towards failure. The elastic method of design endeavours to compute the stresses at working loads, and limits these stresses to amounts below the yield stress of the steel and the crushing stress of the concrete. The respective *factors of safety* are obtained from experience in industry. It can therefore be appreciated that beams designed in such a fashion are safe but are not designed to have specific load factors against their ultimate strengths. Advocates of elastic design feel that stresses and therefore the size of cracks at working loads are controlled. Concerning the design of prestressed concrete beams in bending the 'modulus of elasticity for concrete in tension is assumed to be the same as the value of this modulus for concrete in compression'.

Perfect bond is assumed between the steel and the concrete. The concrete shrinks upon setting and therefore exerts a pressure upon the steel, which assists the resistance to friction between the two materials. This pressure is reduced to some extent when the steel and surrounding concrete are stressed in tension because Poisson's ratio is greater for steel (approximately 0.29) than for concrete (approximately between 0.20 and 0.14). The converse applies when the steel and surrounding concrete are stressed in compression. Irregularities on the surface of the reinforcement lock the steel mechanically to the concrete. Several proprietary high tensile bars and prestressing wires are purposely manufactured to create such an effect.

The depth of the steel reinforcement is considered to be negligible compared with the depth of the beam. This is usually a reasonable assumption.

Normally, temperature and shrinkage stresses are ignored in the design of sections to withstand bending moments, shear forces, and axial forces. It can be mentioned here that fortunately the thermal coefficients of expansion of concrete and steel are sensibly the same. For the design of the structure as a whole, temperature and shrinkage effects must be considered. For example, long buildings need movement joints, temperature stresses are particularly important in the design of chimneys, and losses in prestress due to shrinkage are important.

88   Reinforced concrete beams

Concrete is assumed to be cracked in tension when bending stresses are considered. This is because the tensile strength of concrete is only about one-tenth (and can be as little as one-thirtieth for high strength concretes) of its compressive strength. The same concrete is, however, expected to resist diagonal tensile stresses. If the beam were prestressed it would be permissible for certain small tensile stresses to occur under bending. Concrete has a most unreliable resistance to tension. The ultimate strengths of numerous direct tension specimens made from the same batch of concrete in an exactly similar fashion can vary enormously. The maximum strength can often be as much as twice the minimum strength. The ultimate tensile stress in bending, judged by the extreme fibre stress, using the assumptions of the elastic analysis (and known as the *modulus of rupture*) is higher and more reliable than the direct tensile strength.

3.2.2   Moment of inertia of a reinforced concrete section

Referring to Figure 3.1(a), XX is the neutral axis of any section subjected to bending, δA_{c1} is a small portion of area of the concrete at a distance d_{c1} from the neutral axis, and δA_{s1} is a small portion of area of the steel at a distance d_1 from the neutral axis.

The distribution of strain is assumed linear and is shown in Figure 3.1(b). Let the strain be of magnitude ε_1 at unit distance from the neutral axis. Therefore

Strain for portion $\delta A_{c1} = \varepsilon_1 d_{c1}$

$\therefore$ Stress for portion $\delta A_{c1} = \varepsilon_1 d_{c1} E_c$

If E_c and E_s are the Young's moduli for the concrete and steel respectively, the force for portion $\delta A_{c1} = \varepsilon_1 d_{c1} E_c \delta A_{c1}$ and similarly the force for

Figure 3.1

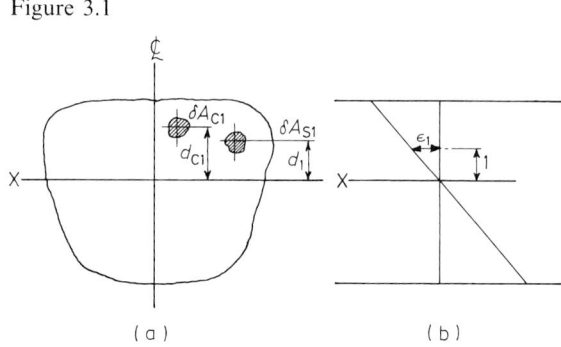

(a)   (b)

portion $\delta A_{s1} = \varepsilon_1 d_1 E_s \delta A_{s1}$. Therefore, the moment of resistance of the section, M, is

$$M = \Sigma(\varepsilon_1 d_{c1} E_c \delta A_{c1}) d_{c1} + \Sigma(\varepsilon_1 d_1 E_s \delta A_{s1}) d_1$$

$$\therefore M = \varepsilon_1 E_c (\Sigma \delta A_{c1} d_{c1}^2 + \Sigma \alpha_e \delta A_{s1} d_1^2) \tag{3.1}$$

where $\alpha_e = E_s/E_c$ is the *modular ratio*.

Comparing equation 3.1 with the classical formula $M = fI/y$, where f is the stress at distance y from the neutral axis, $f/y = \varepsilon_1 E_c y/y = \varepsilon_1 E_c$ if we consider concrete only, in which case I is the *equivalent moment of inertia* (or *second moment of area*) of the cross section. Hence

$$M = \varepsilon_1 E_c I \tag{3.2}$$

and comparing this with equation 3.1

$$I = \Sigma \delta A_{c1} d_{c1}^2 + \Sigma \alpha_e \delta A_{s1} d_1^2 \tag{3.3}$$

The area of steel δA_{s1} can be regarded as equivalent to an area of concrete $\alpha_e \cdot \delta A_{s1}$. In other words $\alpha_e \cdot \delta A_{s1}$ is the *equivalent area* of the area of reinforcement δA_{s1}. This means that to obtain I we just multiply each steel area by α_e and then obtain the moment of inertia of the section as though it were all of concrete. It is often convenient when considering compression steel to consider the gross section of concrete and, as the area of the compression steel has not been subtracted, to multiply each of the steel areas by $(\alpha_e - 1)$ instead of α_e. These give the ideas, in excess of the gross area, due to steel.

Example 3.1. The section shown in Figure 3.2(a) resists a bending moment of 56 kN m. Determine the maximum stress in the concrete and the stress in the steel if $\alpha_e = 15$.

Equivalent area of steel $= 15 \times 2 \times 0.7854 \times 25^2 = 14\,730$ mm². Figure 3.2(b) shows equivalent area of section, and centroid of this is the neutral axis XX. Equating moments of equivalent areas about axis XX

$$(150x)(x/2) = 14\,730(450 - x)$$

$$\therefore x^2 + 196.4x - 88\,380 = 0$$

$$\therefore x = 214.9 \text{ mm}$$

Taking moments of (equivalent) area about XX

$$I = (150x^3/3) + 14\,730 \times (450 - x)^2$$

$$= 1310 \times 10^6 \text{ mm}^4.$$

From equation 3.2

$$\varepsilon_1 = M/(IE_c) = 56/(1310E_c) = 0.04275/E_c$$

Figure 3.2(c) gives distribution of strain and Figure 3.2(d) gives corresponding distribution of stress. Therefore

$$f_c = E_c(\varepsilon_1 x) = 0.04275 \times 214.9 = 9.187 \text{ N/mm}^2$$

and

$$f_s = E_s \varepsilon_1 (450-x) = 0.04275 \alpha_e (450-x) = 150.8 \text{ N/mm}^2$$

These last two equations are sometimes expressed as

$$f_c = Mx/I \text{ and } f_s = \alpha_e M(450-x)/I \qquad (3.4)$$

Example 3.2. If the beam of Example 3.1 were simply supported over an (effective) span (l) of 9.75 m and all the loading was uniformly distributed (q), determine the central deflection. Assume that the bending moment of

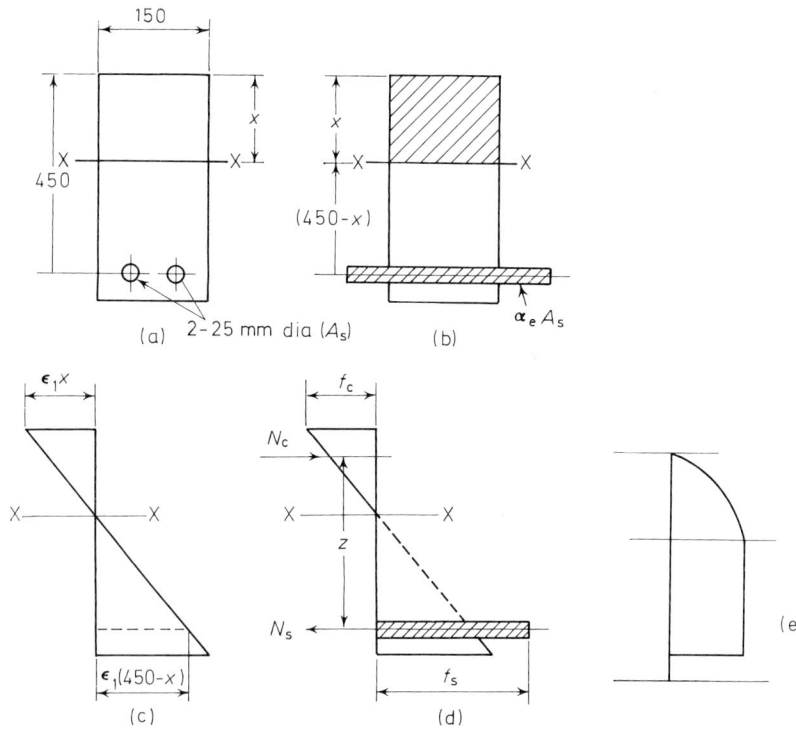

Figure 3.2

Elastic analysis for bending moments

56 kN m was at mid span. Take $E_s = 200$ kN/mm^2; then $E_c = 200/\alpha_e = 13.33$ kN/mm^2.

Central deflection at mid span $= a_1 = (5/384)(ql^4/EI)$
In this example $M = ql^2/8$

$\therefore q = (56 \times 8)/9.75^2 = 4.713$ kN/m (or N/mm)

$$\therefore a_1 = \frac{5 \times 4.713 \times 9750^4}{384 \times 13330 \times 1310 \times 10^6} = 31.76 \text{ mm}$$

Example 3.3. Determine the moment of resistance of the section shown in Figure 3.2(a) if the permissible stresses (i.e. the stresses allowed at working loads) are 10.5 N/mm^2 and 210 N/mm^2 for the concrete and steel, respectively, and the modular ratio is 15.

From Example 3.1, $x = 214.9$ mm and $I = 1310 \times 10^6$ mm^4.
If concrete is the criterion, from equation 3.4
Moment of resistance $= f_c(I/x) = 10.5 \times 1310 \times 10^6/214.9$ N mm
$= 64.01$ kN m

If steel is the criterion, from equation 3.4

$$\text{Moment of resistance} = \frac{f_s}{\alpha_e} \cdot \frac{I}{(450-x)} = \frac{210 \times 1310 \times 10^6}{15(450-214.9)} \text{ N mm}$$

$$= 78.01 \text{ kN m}$$

Therefore according to the assumptions of this design the moment of resistance of the beam is limited by the compressive strength of the concrete to 64.01 kN m.

3.2.3 *Method for tabulating calculations for x and I*

Table 3.1 illustrates the method. A is the equivalent area of a portion, y is the distance of the centroid of A from any chosen axis, say YY for Figure 3.3(a), I_n is the second moment of area for the portion about its neutral axis. Then taking moments of area about YY

$$\Sigma Ay = x\Sigma A \tag{3.5}$$

$$\therefore x = \Sigma Ay/\Sigma A \tag{3.6}$$

Second moment of area of whole section about YY

$$= I_y = \Sigma Ay^2 + \Sigma I_n \tag{3.7}$$

Table 3.1

Portion	Area	A	y	Ay	Ay^2	I_n
Concrete	$(45-16) \times 15 = 435$	435	7.5	3236	24470	$29 \times 15^3/12 = 8156$
Concrete	$16x$	$16x$	$0.5x$	$8x^2$	$4x^3$	$16x^3/12 = 1.333x^3$
Compression steel	$4 \times 3.142 = 12.57$	$(\times 14 =) 175.9$	6	1055	6332	
Tensile steel	$6 \times 4.909 = 29.45$	$(\times 15 =) 441.8$	83.7	36979	3095000	
Totals		$16x +$ 1053		$8x^2 +$ 38030		$5.333x^2 +$ 3101000

Elastic analysis for bending moments 93

If I is second moment of area of the whole section about its neutral axis XX, and x is depth of neutral axis below YY, then

$$I_y = x^2 \Sigma A + I \qquad (3.8)$$

From equations 3.7 and 3.8

$$I = \Sigma A y^2 + \Sigma I_n - x^2 \Sigma A \qquad (3.9)$$

Supposing we wish to obtain the lever arm z. Then considering the tensile steel, area A_s, effective depth d_1, the moment of resistance $= f_s A_s z = f_s I/(d_1 - x)$

$$\therefore z = I/[A_s(d_1 - x)] \qquad (3.10)$$

Example 3.4. The section shown in Figure 3.3 is through an external counterfort to a tank. The reinforcement bars comprise six of 25 mm diameter in tension and four of 20 mm diameter in compression and have 50 mm cover of concrete, $\alpha_e = 15$, the permissible stresses (for BS 5337:1976 strength calculations for Grade 25 concrete and Class A exposure for plain bars of steel) are: concrete in compression 9.15 N/mm², and steel in tension 85 N/mm². Determine the moment of resistance of the section at working loads, and the stress in the compression steel.

Figure 3.3

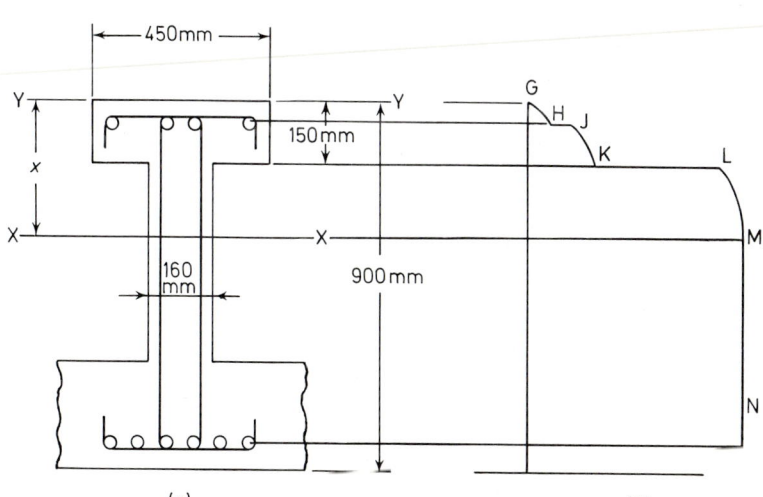

Table 3.1 shows the calculation (dimensions are in centimetres for convenience). Then from equation 3.5

$$8x^2 + 38030 = 16x^2 + 1053x$$

$$8x^2 + 1053x - 38030 = 0$$

$$\therefore x = 29.50 \text{ cm} = 295.0 \text{ mm}$$

From equation 3.9

$$I = 5.333x^2 + 3101000 - x^2(16x + 1053)$$

$$= 3101000 - 29.5^2(16 \times 29.5 + 1048)$$

$$= 1.778 \times 10^6 \text{ cm}^4 = 17780 \times 10^6 \text{ mm}^4$$

Referring to equations 3.4

$$\text{Moment of resistance (concrete)} = 9.15 \times \frac{17780 \times 10^6}{295} \text{ N mm}$$

$$= 551.5 \text{ kN m}$$

$$\text{Moment of resistance (steel)} = \frac{85}{15} \times \frac{17780 \times 10^6}{(837 - 295)} \text{ N mm}$$

$$= 185.9 \text{ kN m}$$

The latter is therefore the criterion, and the stress in the compression steel will be

$$= 15 \times \frac{185.9}{17780} \times (295 - 60) = 36.86 \text{ N/mm}^2$$

The latter is well within the permissible stress of 125 N/mm^2 given by BS 5337, and 551.5 is much greater than 185.9, hence a designer might reduce the diameter of the 20 mm bars – unless their robustness is required to support the reinforcement cage. Their number cannot be reduced because of the stirruping system.

NB. BS 5337:1976 has been superseded by BS 8007:1987 which does not now use elastic design for strength but does for limiting crack widths at limit state of serviceability.

3.2.4 Popular formulae for slabs and rectangular beams (elastic theory)

For a rectangular section such as shown in Figure 3.2, if b is its breadth and d the effective depth of the tension steel, then moments of areas about XX give

$$bx^2/2 = \alpha_e A_s(d-x)$$

Dividing throughout by bd^2 and substituting $\rho = A_s/bd$ and $x_1 = x/d$

$$x_1^2/2 = \alpha_e \rho(1-x_1) \tag{3.11}$$

$$x_1^2 + 2\alpha_e \rho x_1 - 2\alpha_e \rho = 0$$

$$x_1 = -\alpha_e \rho + \sqrt{[(\alpha_e \rho)^2 + 2(\alpha_e \rho)]} \tag{3.12}$$

As strain is linear, from Figure 3.2

$$\varepsilon_1 = f_c/(E_c x) = f_s/[E_s(d-x)]$$

$$\therefore f_s/f_c = \alpha_e(d-x)/x = \alpha_e(1-x_1)/x_1 \tag{3.13}$$

Let $\alpha_r = f_s/f_c$, then

$$x_1 = \alpha_e/(\alpha_e + \alpha_r) \tag{3.14}$$

From equations 3.11 and 3.13

$$\rho = x_1/2\alpha_r \tag{3.15}$$

In Figure 3.2(d) N_c is the total force ($= 0.5 f_c bx$) of the compressive stress in the concrete, and N_s is the force ($= A_s f_s$) in the tension steel. The distance between these two forces z is called the *lever arm* or *moment arm*, and $z_1 = z/d$. Thus

$$z = d - x/3 \text{ or } z_1 = 1 - x_1/3 \tag{3.16}$$

Resolving longitudinally $N_c = N_s$. If M is the bending moment resisted by the section then $M = N_c z = N_s z$, thus

$$M = N_c z = 0.5 f_c bxz = (0.5 f_c x_1 z_1) bd^2 = Kdb^2 \tag{3.17}$$

where $K = 0.5 f_c x_1 z_1 = M/bd^2$. Also

$$M = N_s z = A_s f_s z_1 d \tag{3.18}$$

Designers make use of the full permissible stresses of concrete and steel (unless other factors (for example deflection) dictate otherwise), and then the previous equations give useful design formulae. For example, for water containers from BS 5337:1976, the permissible stresses in concrete of Grade 25 and steel (plain bars, exposure Class A) are 9.15 N/mm² and 85 N/mm², respectively, and $\alpha_e = 15$. Substituting these figures in the previous equations gives $\alpha_f = 9.29$, thus $x_1 = 0.6175$, $z_1 = 0.7942$ and $\rho = 0.03325$. Then in equation 3.17 the coefficient $0.5 f_c x_1 z_1 = 2.2437$ N/mm².

The last paragraph did not make use of equation 3.12. This equation is most useful for obtaining x_1 when the section is fully defined.

Example 3.5. A cantilever wall of a shallow rectangular tank contains a 5 m head of water. Design the cross section at the bottom of the wall in accordance with the elastic method of BS 5337:1976.

Distribution of water pressure on wall is triangular, its maximum being $5 \times 10 = 50$ kN/m². For 1 m run of wall, bending moment at base of wall $= (50/2) \times 5 \times (5/3) = 208.3$ kN m/m. Let h = wall thickness.

Using a Grade 25 concrete the permissible tensile concrete stress for designing against cracking is 1.84 N/mm², thus $(1 \times h^2/6) = 208.3/1840$ $\therefore h = 0.824$ m, say 0.8 m as we have ignored the reinforcement. Using 50 mm cover and 20 mm diameter bars, $d = 800 - 60 = 740$ mm.

Designing for strength, assume for speed that the permissible stresses of concrete and steel stated previously apply simultaneously, then using the previous formulae, for concrete

$$M = 2244 \times 1 \times 0.74^2 = 1229 \text{ kN m}$$

This is more than required. For steel

$$A_s = 208.3/(85\,000 \times 0.7942 \times 0.74) \text{ m}^2 = 4170 \text{ mm}^2$$

From Table 3.2 use 20 mm diameter bars at 75 mm centres. We need to check that the increased I for an uncracked section, due to the steel, makes h satisfactory. Had we taken $h = 0.824$ or more this last check would be unnecessary.

$$I \simeq 1 \times 0.8^3/12 + 0.004\,19 \times 14 \times (0.74 - 0.4)^2 = 0.049\,45 \text{ m}^4$$

$$\therefore M \simeq 1840 \times 0.049\,45/0.4 = 227.5 \text{ kN m which is } > 208.3$$

A more economical method of obtaining the above steel, because of 1229 being much greater than 208.3, is

$$K = 208.3/(1 \times 0.74^2) \text{ kN/m}^2 = 0.380 \text{ N/mm}^2$$

Elastic analysis for bending moments 97

From Table 3.3, take z_1 as 0.885

$$\therefore A_s = 208.3/(85\,000 \times 0.885 \times 0.74) \text{ m}^2 = 3742 \text{ mm}^2$$

From Table 3.2, spacing of bars $= (3140/3742) \times 100 = 84$ mm, say 80 mm.

$$A_s = 3140 \times 100/80 = 3925 \text{ mm}^2$$

Take this as our design and check precisely for h. Table 3.4 is as described in Section 3.2.3, using dm units for convenience. Therefore

$$x = 360.7/85.5 = 4.219 \text{ dm} = 0.4219 \text{ m}$$

$$I = 2008 - 85.5 \times 4.219^2 = 486.1 \text{ dm}^4 = 0.048\,61 \text{ m}^4$$

$$M = 1840 \times 0.048\,61/(0.8 - 0.4219) = 236.6 \text{ kN m}$$

Table 3.2

No. of bars	Cross-sectional areas of groups of bars, mm²							
1	28.3	50.3	78.5	113.1	201.1	314.2	490.9	804.2
2	56.5	100.5	157.1	226.2	402.1	628.3	981.7	1609
3	84.8	150.8	235.6	339.3	603.2	942.5	1473	2413
4	113.1	201.1	314.2	452.4	804.2	1257	1964	3217
5	141.4	251.3	392.7	565.5	1005	1571	2454	4021
6	169.6	301.6	471.2	678.6	1206	1885	2945	4826
7	197.8	351.9	549.8	791.7	1407	2199	3436	5630
8	226.2	402.1	628.3	904.8	1609	2513	3927	6434
9	254.5	452.4	706.9	1018	1810	2827	4418	7238
10	282.7	502.7	785.4	1131	2011	3142	4909	8043
d_b, mm	6	8	10	12	16	20	25	32
50	565.0	1005	1571	2262	4021	6284	9817	16085
75	377.0	670	1047	1508	2681	4189	6545	10723
100	283.0	503	785	1131	2011	3142	4909	8042
125	226.0	402	628	905	1608	2513	3927	6434
150	188.0	335	524	754	1340	2094	3272	5362
175	162.0	287	449	646	1149	1795	2805	4596
200	141.0	251	393	565	1005	1571	2454	4021
250	113.0	201	314	452	804	1257	1963	3217
300	94.3	168	262	377	670	1047	1636	2681
Pitch of bars, mm	Cross-sectional areas of bars per metre, mm²							

Table 3.3.

α_r	10	12	14	16	18	20	22	24	26	28	30	32	34
K	2.040	1.603	1.299	1.078	0.911	0.781	0.677	0.594	0.525	0.468	0.420	0.379	0.344
z_1	0.800	0.815	0.828	0.839	0.848	0.857	0.865	0.872	0.878	0.884	0.889	0.894	0.898

$K = M/bd^2 \text{ N/mm}^2$, $f_c = 85 \text{ N/mm}^2$, $\alpha_e = 15$

which is >208.3. Design could be recommended using a slightly thinner wall, but the reinforcement would be increased slightly so the alteration in cost would be fairly insignificant and may be more or less.

The deflection of the top of a container wall like this can be very important, particularly near corners of rectangular tanks, and the above I is suitable for use in such elastic analyses because there are more uncracked than cracked sections. Although the value of h was determined so that the wall would not crack, there will be some cracks because of shrinkage, temperature changes and small relative settlements – the design mainly ensures that cracks will be few and small.

The vertical stress due to the self-weight of the wall has been ignored because it is relatively small compared to the flexural tensile stress and it is compressive.

Example 3.6. A slab with $h = 0.8$ m, $d = 0.74$ m, and 20 mm diameter bars at 80 mm centres as tension reinforcement withstands a bending moment of 208.3 kN m/m. Taking $\alpha_e = 15$, determine the stresses in the steel and the extreme fibre of the concrete.

Consider 1 m width of slab. From Table 3.2, $A_s = 3140/0.80 = 3925$ mm², thus $\rho = 0.003\,925/(1 \times 0.74) = 0.005\,304$. From equation 3.12

$$x_1 = -0.079\,56 + \sqrt{(0.079\,56^2 + 2 \times 0.079\,56)} = 0.4863$$

From equation 3.16, $z_1 = 1 - 4863/3 = 0.8379$. From equations 3.17 and 3.18

$$f_c = 2 \times 208.3/(0.4863 \times 0.8379 \times 1 \times 0.74^2) \text{ kN/m}^2$$

$$= 1.867 \text{ N/mm}^2$$

$$f_s = 208.3/(0.003\,925 \times 0.8379 \times 0.74) \text{ kN m}^2 = 85.59 \text{ N/mm}^2$$

(This demonstrates that the final design of Example 3.5 is reasonable.)

Table 3.4

Portion	A	y	Ay	Ay^2	I_n
Concrete	$10 \times 8 = 80$	4	320	1280	$10 \times 8^3/12 = 427$
Steel	$0.3925 \times 14 = 5.495$	7.4	40.7	291	—
Totals	85.50		360.7		2008

3.3 Elastic theory for shear stresses

From the elastic theory for bending it is possible to compute the distribution of horizontal shear stresses. From classical elastic theory, the shear force is equal to the rate of change of the bending moment along a beam, and for this to occur the beam has to withstand horizontal shearing stresses. The section of a reinforced concrete beam shown in Figure 3.4(a) is symmetrical about a vertical axis. The distributions of bending stresses for two sections distance δx apart are shown in Figure 3.4(b), the bending moments causing the distributions between M and $(M + \delta M)$ respectively. The horizontal shear stress will now be determined at AB. The concrete stress on the small element of area $b\delta y$ is given by

$$f_{c1} = (M/I)y \tag{3.19}$$

at one section of Figure 3.4(b) and at the other section by

$$f_{c1} + \delta f_{c1} = [(M + \delta M)/I]y \tag{3.20}$$

Subtracting these quantities

$$\delta f_{c1} = (\delta M/I)y \tag{3.21}$$

Forces on strip at the two sections are

$$N_{c1} = f_{c1} b\delta y \tag{3.22}$$

$$N_{c1} + \delta N_{c1} = (f_{c1} + \delta f_{c1}) b\delta y \tag{3.23}$$

Subtracting these quantities

$$\delta N_{c1} = \delta f_{c1} b\delta y \tag{3.24}$$

Figure 3.4(c) shows the same two sections as Figure 3.4(b). It can be seen that the plane ABA'B' has to resist shear stresses due to all such quantities as $(N_{c1} + \delta N_{c1}) - N_{c1} = \delta N_{c1}$. Hence the total shear stress resisted by plane ABA'B' is given by

$$v = (\Sigma \delta N_{c1})/(b_1 \delta x) \tag{3.25}$$

Substituting from equations 3.24 and 3.21, equation 3.25 becomes

$$v = [\Sigma(\delta M/I) yb \cdot \delta y]/(b_1 \delta x) \tag{3.26}$$

Now from the well-known theory of bending, shear force

$$V = \delta M/\delta x \tag{3.27}$$

Elastic theory for shear stresses

Figure 3.4

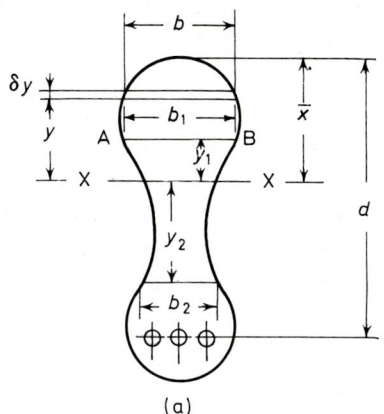

(a)

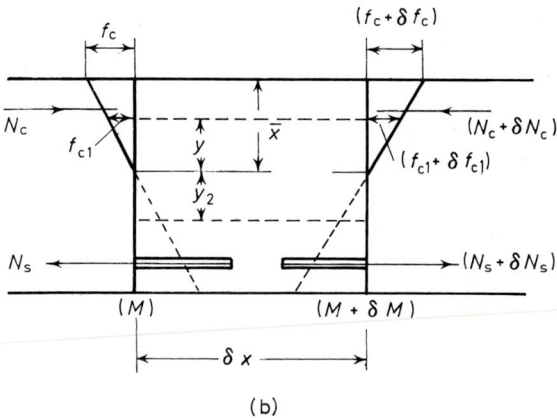

(b)

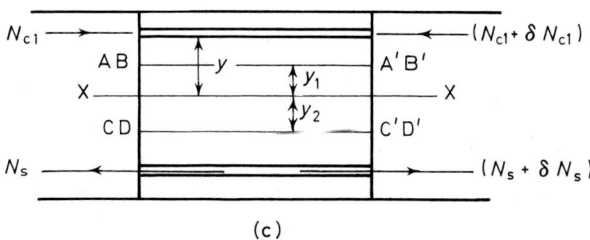

(c)

Therefore from equations 3.26 and 3.27

$$v = [\Sigma(V\delta x/I) yb \cdot \delta y]/(b_1 \delta x) = (V/Ib_1)\Sigma by \cdot \delta y \qquad (3.28)$$

Or more precisely

$$v = (V/Ib_1) \int_{y_1}^{\bar{x}} by \cdot dy \qquad (3.29)$$

This is the horizontal shearing stress at a point distance y_1 from the neutral axis XX. From classical theory of elasticity it is also the vertical shearing stress at this point. Equation 3.29 has been derived considering the rate of change of compressive stress in the concrete along the beam, and only concerns sections above the neutral axis. Considering a plane CDC'D' below the neutral axis, the horizontal shear stress resisted by this plane considering forces below it is given by

$$v = [(N_s + \delta N_s) - N_s]/(b_2 \delta x) = (1/b_2)(\delta N_s/\delta x) \qquad (3.30)$$

Now from Section 3.2.4, $M = N_s z$ and combining this with equation 3.27

$$V = \delta M/\delta x = z(\delta N_s/\delta x) \qquad (3.31)$$

From equations 3.30 and 3.31

$$v = V/zb_2 \qquad (3.32)$$

This equation is independent of y_2, hence the shear stress (vertical or horizontal) is constant below the neutral axis.

Equations 3.29 and 3.32 are expressions which apply to any section which is singly reinforced and symmetrical about its vertical axis. Applying these to a rectangular section as shown in Figure 3.2, $b_1 = b_2 = b$. Equation 3.29 therefore becomes

$$v = \frac{V}{I}\int_{y_1}^{\bar{x}} y \cdot dy = \frac{V}{2I}[(\bar{x})^2 - y_1^2] \qquad (3.32a)$$

This gives a parabolic distribution of stress above the neutral axis and the maximum value is at the neutral axis when $y_1 = 0$, thus

$$\max v = V(\bar{x})^2/(2I) \qquad (3.33)$$

Now from equations of Sections 3.2.3 and 3.2.4,

$$M = N_c z = (f_c/2)\bar{x}bz \qquad (3.34)$$

$$M = f_c(I/\bar{x}) \qquad (3.35)$$

Eliminating M between equations 3.34 and 3.35

$$(\bar{x})^2/(2I) = 1/(bz) \tag{3.36}$$

Substituting this in equation 3.33

$$\max v = V/(zb) \tag{3.37}$$

Below the neutral axis, applying equation 3.32,

$$v = V/(zb) \tag{3.38}$$

The distribution of shear stress is therefore as shown in Figure 3.2(e). As concrete is much stronger in compression and shear than it is in tension, the principal tensile stresses, often known as the *diagonal tensile stresses*, are the criterion as regards failure due to shearing forces. If the principal tensile stresses due to combining the stresses shown in Figures 3.2(d) and (e) are computed, below the neutral axis, there are no longitudinal concrete stresses in the diagram. As the horizontal shear stresses by classical theory have equal complementary vertical shear stresses, these combine to give principal diagonal tensile stresses at 45° to the horizontal and equal in magnitude to the horizontal shear stresses. Above the neutral axis the longitudinal compressive stresses reduce the diagonal stresses resulting from combining complementary horizontal and vertical shear stresses. Diagonal tensile stresses help shrinkage stresses in causing cracking. This diagonal cracking is sometimes simultaneous with shear failure for a beam with no web reinforcement.

For T-beams and beams with compression reinforcement, at and below the neutral axis the above applies, that is the maximum diagonal tensile stress is constant and equal to $V/(zb)$.

Example 3.7. From the previous discussion the distribution of horizontal (or vertical) shear stress in the concrete for the section in Figure 3.3(a) of Example 3.4 is as shown in Figure 3.3(b), being parabolic for GH, JK and LM. Determine the shear stresses represented by points H, J, K, L, M and N, if the shear force is 60 kN.

For points M and N, that is maximum at neutral axis XX, using equation 3.29 (and figures from Example 3.4) it is simpler to consider the section below the neutral axis for the moment of area term

$$v = [0.06/(0.01634 \times 0.16)] \times 0.04418 \times (0.847 - 0.3179) \text{ MN/m}^2$$

$$= 0.5365 \text{ N/mm}^2$$

As explained previously this is equal to $V/(zb)$. Thus alternatively from equation 3.10

$$z = 1.634 \times 10^6/[441.8(84.7 - 31.79)] \text{ cm}$$

$$= 699 \text{ mm}$$

then

$$v = 60\,000/(699 \times 160)$$

$$= 0.5365 \text{ N/mm}^2$$

For approximate preliminary design one would perhaps have guessed that the centre of compression was at about half the depth of the T-flange giving $z = 847 - 75 = 772$ mm, about 10% error on the dangerous side. The following shear stresses are not needed by the designer but are of academic interest.

For point H, using equation 3.29

$$v = \frac{0.06}{0.016\,34 \times 0.45} \times 0.45 \times 0.05(0.3179 - 0.025) \text{ MN/m}^2$$

$$= 0.053\,78 \text{ N/mm}^2$$

For point J

$$v = 0.053\,78 + \frac{0.06}{0.016\,34 \times 0.45} \times 0.017\,59 \times (0.3179 - 0.05)$$

$$= 0.053\,78 + 0.038\,45$$

$$= 0.092\,23 \text{ N/mm}^2$$

For point K

$$v = \frac{0.06}{0.016\,34 \times 0.45} \times 0.45 \times 0.15 \times (0.3179 - 0.075) + 0.038\,45$$

$$= 0.1722 \text{ N/mm}^2$$

For point L

$$v = 0.1722 \times 0.45/0.16 = 0.4843 \text{ N/mm}^2$$

3.4 Shear reinforcement

Generally speaking experimental research[1] shows that if design is based on ultimate strength in shear with suitable load factors, then diagonal crack widths at working loads are acceptable. The ultimate shear forces carried by beams with plain webs have been substituted, by researchers, in equation 3.38 to obtain ultimate values for v. The latter have varied with the many possible variables. Of these variables BS 8110 has selected the percentage of longitudinal reinforcement $100A_s/(bd)$ (where b_v = breadth of section of rectangular beam or of web of Tee beam etc. and d = effective depth) as the most important. BS 8110 Table 3.9 (shown in Table 3.4a) gives values of the design concrete shear stress, v_c, for various values of this percentage of longitudinal reinforcement and of d. When the design shear stress $v = V/b_v d$ does not exceed $(v_c + 0.4)$ N/mm² nominal links should be used to provide a design shear resistance[1,2] of 0.4 N/mm² so that

$$\frac{A_{sv}}{s_v} = \frac{0.4 b_v}{0.87 f_{yv}}$$

where A_{sv} is the cross-sectional area of the two legs of a link, f_{yv} is the characteristic strength of the links, and s_v is the spacing of the links $\not> 0.75d$. According to Ref. 4, f_{yv} should be the same for mild and high-

Table 3.4a *Values of v_c, design concrete shear stress*

$\frac{100A_s}{b_v d}$	Effective depth (in mm)							
	125	150	175	200	225	250	300	≥400
	N/mm²	N/mm²	N/mm²	N/mm²	N/mm²	N/mm²	N/mm²	N/mm²
≤0.15	0.45	0.43	0.41	0.40	0.39	0.38	0.36	0.34
0.25	0.53	0.51	0.49	0.47	0.46	0.45	0.43	0.40
0.50	0.67	0.64	0.62	0.60	0.58	0.56	0.54	0.50
0.75	0.77	0.73	0.71	0.68	0.66	0.65	0.62	0.57
1.00	0.84	0.81	0.78	0.75	0.73	0.71	0.68	0.63
1.50	0.97	0.92	0.89	0.86	0.83	0.81	0.78	0.72
2.00	1.06	1.02	0.98	0.95	0.92	0.89	0.86	0.80
≥3.00	1.22	1.16	1.12	1.08	1.05	1.02	0.98	0.91

Note: Allowance has been made in these figures for a γ_m of 1.25.
For characteristic concrete strengths greater than 25 N/mm², the values in Table 3.4a may be multiplied by $(f_{cu}/25)^{1/3}$. The value of f_{cu} should not be taken as greater than 40.

yield steel, as high yield stirrups are little more effective in resisting shear. When $v > (v_c + 0.4)$ the shear reinforcement is designed to resist $(v - v_c)$.

For say long continuous beams where temperature stresses assist shrinkage and diagonal tensile stresses, for want of research to the contrary, the writer[2,4] would suggest always using the above nominal links throughout the spans.

No matter how much shear reinforcement is provided, $V/(bd)$ must neither exceed $0.8\sqrt{f_{cu}}$ or 5 N/mm² according to BS 8110 because steel resists diagonal tension but not the diagonal compression.

Shear reinforcement can be links and/or inclined bars. BS 8110 favours a truss-analogy method for designing the latter. Research shows that beams do not act in this way (e.g. cracks prior to failure are inconsistent with it) but that the ultimate strength design is conservative with this method. BS 8110 favours at least 50% of the shear resistance provided by the reinforcement to be as links and not as bent up bars.

3.4.1 Design of shear reinforcement by BS 8110 truss analogy

The BS 8110 truss-analogy method has been judged conservative by research chiefly concerned with vertical stirrups, and stirrups[1] and bars inclined at 45° to the horizontal. Some work with reinforcement at 30° to the horizontal also supports the method. Outside this range one should seek experimental justification. In practice most stirrups are vertical and most bars inclined at 45°. However, BS 8110 limits the use of inclined bars to those inclined at 45 degrees and greater to the horizontal.

Bars belonging to the main tensile reinforcement are bent up at points such as C and E in Figure 3.5(a). Alternatively, independent shear bars (or stirrups) may be used as shown in Figure 3.5(b). A beam is considered to be a statically determinate truss as illustrated in Figure 3.5(a). The longitudinal tension reinforcement is analogous to tension members such as AC and CE in Figure 3.5(a); the concrete resisting longitudinal compression (due to bending) is analogous to compression members such as BD and DF; the bent-up bars are analogous to inclined tension members such as BC and DE, and the inclined compression members such as AB, CD and EF, required to complete the truss analogy, are provided by the concrete of the web. The forces in the analogous truss members AC, BC, DC and EC are as shown, namely N_{s2}, N_{sv}, N_c and N_{s1}, respectively. A vector diagram is drawn for these forces in Figure 3.5(c); as the bending moment increases for sections further away from the supports, N_{s1} will be greater than N_{s2} and their difference is represented by the vector KM;

Shear reinforcement

forces N_c and N_{sv} are represented by the vectors LK and LM respectively. If the area of tensile reinforcement which is analogous to member CE is A_s, and the area of the bars bent up is ψA_s, and if the bent-up bars are

Figure 3.5

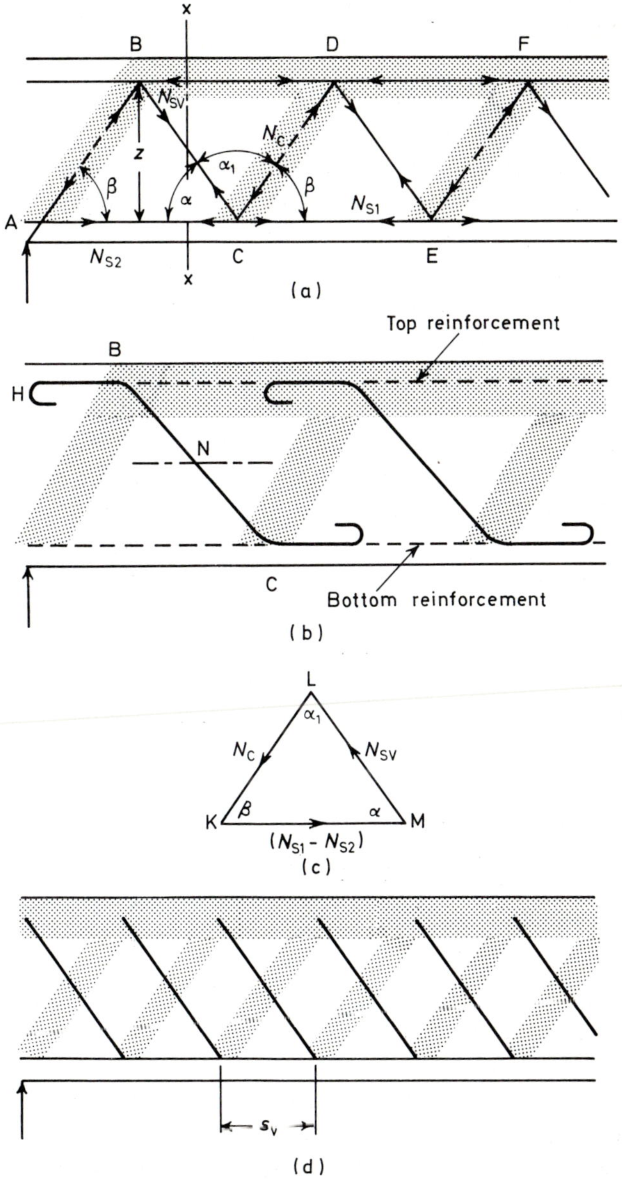

required to develop their full stress f_{sv}, then $N_{sv} = \psi A_s f_{sv}$. At the same time, if the stresses in the members CA and CE are not to exceed f_s, they are designed so that $N_{s1} = A_s f_s$ and $N_{s2} = (A_s - \psi A_s) f_s$. Hence, referring to Figure 3.5(c) the vector LM $= \psi A_s f_{sv}$ and the vector KM $= A_s f_s - (A_s - \psi A_s) f_s = \psi A_s f_s$. In the case of mild steel reinforcement $f_{sv} = f_s$ and therefore LM = KM; consequently in the vector diagram LKM,

$$\beta = \alpha_1 \tag{3.39}$$

For high-yield steel,[4] using $f_{sv} = 250$ N/mm² (BS 8110 says 460) and $f_s = 460$ N/mm², LM $= 250\psi A_s$ and KM $= 460\psi A_s$, and from the vector diagram

$$\sin\beta/\sin\alpha_1 = \text{LM}/\text{KM} = 250/460 \tag{3.40}$$

The inclined compression members are assumed to be sufficiently strong for all requirements. They are safeguarded by compliance with the penultimate paragraph of Section 3.4 previously. By Ritter's Method of Sections, assume the truss to be cut at the section xx shown in Figure 3.5(a). Then resolving vertically for, say, the left-hand side of this section

$$N_{sv} \sin\alpha = \text{Shear force at xx} = V \tag{3.41}$$

The principle of the superposition of trusses can be applied. For example, the system shown in Figure 3.5(d), where $s_v = \text{AC}/2$, is assumed to be twice as strong as the system of Figure 3.5(a); hence from equation 3.41

$$V = 2N_{sv} \sin\alpha \tag{3.42}$$

The inclined bars shown in Figures 3.5(a) and (d) are sometimes described as being in *single-shear* and *double-shear*, respectively. Extending this principle of superposition for any value of s_v in Figure 3.5(d), equation 3.42 becomes

$$V = (\text{AC}/s_v) N_{sv} \sin\alpha \tag{3.43}$$

From triangle ABC

$$\text{AC} = z(\cot\alpha + \cot\beta) \tag{3.44}$$

Now $z = d - d'$

$$\therefore \text{AC} = (\cot\alpha + \cot\beta)(d - d') \tag{3.44a}$$

Hence equation 3.43 becomes

$$V = (N_{sv}/s_v)\sin\alpha(\cot\alpha + \cot\beta)(d-d') \qquad (3.45)$$

This is the same as equation 4 of Section 3 of BS 8110

$$V_b = A_{sb}(0.87f_{yv})(\cos\alpha + \sin\alpha\cot\beta)\frac{d-d'}{s_b} \qquad (3.45a)$$

where $A_{sb} \times 0.87f_{yv} = N_{sv}$, A_{sb} = cross-sectional area of bent-up bars, and spacing of bent-up bars = s_b.

Applying equation 3.45 to mild steel reinforcement and hence using the equation 3.39, also from triangle KLM in Figure 3.5(c),

$$\alpha + \alpha_1 + \beta = 180° \qquad (3.46)$$

Therefore from equation 3.39

$$\beta = 90° - (\alpha/2) \qquad (3.47)$$

Substituting this in equation 3.45

$$V = (N_{sv}/s_v)\sin\alpha[\tan(\alpha/2) + \cot\alpha](d-d')$$
$$= (N_{sv}/s_v)[2\sin^2(\alpha/2) + \cos\alpha](d-d')$$
$$\therefore V = N_{sv}(d-d')/s_v \qquad (3.48)$$

Applying equation 3.45 to high tensile reinforcement, and hence using equation 3.40

$$\sin\beta = (250/460)\sin\alpha_1$$

Therefore from equation 3.46

$$\sin\beta = (250/460)\sin(180° - \beta - \alpha)$$
$$\therefore 1.84\sin\beta = \sin\beta\cos\alpha + \cos\beta\sin\alpha$$
$$\therefore \cot\beta = (1.84 - \cos\alpha)/\sin\alpha \qquad (3.49)$$

Substituting this in equation 3.45

$$V = (N_{sv}/s_v)(1.84 - \cos\alpha + \cos\alpha)(d-d')$$
$$= 1.84(N_{sv}/s_v)(d-d') \qquad (3.50)$$

For inclined bars BS 8110 recommends the truss analogy as described,

Table 3.5

f_{yv}, N/mm²	d_b, mm	\multicolumn{14}{c}{s_v, mm}													
		50	60	70	75	80	90	100	125	150	175	200	225	250	275	300
250	6	246	205	175.7	164	153	136	123	98	82	70	61	54	49	44	41
	8	437	364	312	291	273	242	218	174	145	124	109	97	87	79	72
	10	683	569	488	455	427	379	341	273	227	195	170	151	136	124	113
410	6	452	377	323	301	282	251	226	181	150	129	113	100	90	82	75
	8	804	670	574	536	502	446	402	321	268	229	201	178	161	146	134
	10	1257	1047	898	838	785	698	628	502	419	359	314	279	251	228	209

Values of V/d, two-arm stirrups, N/mm, $1/\gamma_m = 0.87$.

Shear reinforcement

but using α and $\beta \not< 45°$. For stirrups BS 8110 assumes that $z = d$ and $\beta = 45°$, so that equation 3.45 becomes

$$V = (N_{sv}/s_v)(\sin\alpha + \cos\alpha)d \tag{3.51}$$

From equation 3.51 and substituting $N_{sv} = A_{sv}f_{yv}/\gamma_m = A_{sv}0.87f_{yv}$

$$A_{sv}/s_v = V/[0.87f_{yv}(\sin\alpha + \cos\alpha)d] \tag{3.52}$$

For vertical stirrups $\alpha = 90°$, thus

$$A_{sv}/s_v = V/[0.87f_{yv}d] \tag{3.53}$$

Table 3.5 (upper half) is useful for designers, uses equation 3.53, and refers to mild steel stirrups with $f_{yv} = 250$ N/mm², from Table 2.10. According to Ref. 4 this also applies to all other stirrups. However, for those who wish to use BS 8110 for high-yield steel stirrups, $f_{yv} = 460$ N/mm², the lower half of Table 3.5 is provided.

For mild steel bars bent up at 45°, from equation 3.39, $\beta = \alpha_1 = 67.5°$. Table 3.6 gives shear resistances for single-shear systems for single bars, using equation 3.41 and $1/\gamma_m = 0.87$. Ref. 4 would use $f_{yv} = 250$ N/mm² for all other bars. BS 8110 allows $f_{yv} = 460$ N/mm² for high-yield steel bars and Table 3.6 gives shear resistances for $f_{yv} = 460$ N/mm².

Example 3.8. A beam of T-section has a rib of breadth 250 mm, $d = 600$ mm and $100A_s/(bd) = 1.2$. Design links (vertical stirrups) to resist an ultimate shear force of 200 kN if the characteristic strength of concrete = 25 N/mm².

$V/(bd) = 0.2/(0.25 \times 0.6)$ MN/m² = 1.333 N/mm², which is less than $0.8\sqrt{25}$ and 5 N/mm² and therefore satisfactory for shear reinforcement (see Section 3.4). From Table 3.9 of BS 8110, our Table 3.4a, shear resistance provided by concrete web alone

$$= [0.63 + (0.72 - 0.63)(1.2 - 1.0)/(1.5 - 1.0)]$$

$$\times 250 \times 600 \text{ N} = 99.9 \text{ kN}$$

Table 3.6

Single bars in single shear at 45°, $1/\gamma_m = 0.87$

f_{yv}, N/mm²	d_b, mm	10	12	16	20	25	32
250	V, kN	12.08	17.39	30.92	48.29	75.49	123.7
460	V, kN	22.23	32.01	56.89	88.90	138.9	227.6

112 Reinforced concrete beams

Hence shear reinforcement is required and it has to resist $200 - 99.9 = 100.1$ kN. Using stirrups the V/d required is $100.1/0.6$ kN/m $= 166.8$ N/mm. From Table 3.5 use 6 mm diameter mild steel two-arm stirrups at 70 mm centres ($175.7 > 166.8$).

Example 3.9. A beam of rectangular cross section has $b = 300$ mm, $d = 700$ mm, and $100A_s/(bd) = 1.87$. The ultimate shear force it has to resist is 642 kN. Design a suitable shear reinforcement system. Assume characteristic strength of concrete in compression $= 25$ N/mm².

$V/bd = 0.642/(0.3 \times 0.7)$ MN/m² $= 3.06$ N/mm², which is satisfactory as it is less than $0.8\sqrt{25}$ and 5 N/mm² and therefore satisfactory for shear reinforcement (see Section 3.4).

From Table 3.9 of BS 8110, our Table 3.4a, shear resistance provided by concrete web alone

$$= v_c = 0.72 + (0.8 - 0.72) \times (1.87 - 1.5)/(2.0 - 1.5)$$
$$= 0.7792, \text{ that is } 0.7792 \times 300 \times 700 = 163.6 \text{ kN}$$

Hence shear reinforcement is required and it has to resist $642 - 163.6 = 478.4$ kN. According to BS 8110 the shear force taken by bent-up bars must not exceed $0.5 \times 478.4 = 239.2$ kN. Using pairs of 20 mm diameter bent-up mild steel bars in double shear, from Table 3.6 this is worth $48.29 \times 4 = 193.2$ kN (< 239.2). Thus the amount to be resisted by stirrups is $478.4 - 193.2 = 285.2$ kN, giving $V/d = 285.2/700$ kN/mm $= 407$ N/mm. Using mild steel links with two arms, from Table 3.5, 10 mm diameter links at 80 mm centres give $V/d = 427.1$ N/mm (> 407).

3.5 'Bond' stresses due to shear (or flexural bond)

The theory expounded concerning shear stresses (Section 3.3) assumes perfect adhesion of the concrete to the tensile reinforcement, and therefore involves 'bond stresses' being developed between the steel and the concrete. Referring to Figure 3.4(b), the change of force in the tensile reinforcement between the sections shown is $(N_s + \delta N_s) - N_s = \delta N_s$. This can only be resisted by bond stresses which act on the contact area between the steel and the concrete of $\delta x \Sigma u_s$. Hence the bond stress at this locality is given by

$$f_{bs} = \delta N_s/(\delta x \Sigma u_s) = (1/\Sigma u_s)(dN_s/dx) \tag{3.54}$$

where $\Sigma u_s =$ sum of the perimeters of bars of tensile steel. Now

$$V = dM/dx = (d/dx)(N_s z) = z(dN_s/dx) \tag{3.55}$$

Hence from equations 3.54 and 3.55

$$f_{bs} = V/(z\Sigma u_s) = V/(z_1 d\Sigma u_s) \tag{3.56}$$

These bond stresses are known as *local bond stresses* and ultimate values of $V/(d\Sigma u_s) = z_1 f_{bs}$ are recommended for various types of concrete in Table 21 of CP 110, even though f_{bs} is derived from the elastic theory. Research on ultimate values has been related to $V/(z\Sigma u_s)$, however, and as the results are not very precise it is not unreasonable for CP 110 to take z_1 as constant. Designs need to ensure that ultimate local bond stresses are nowhere exceeded and this is the only requirement in this connection; such bond stresses are local effects and do not for instance require any anchorage.

Whereas all previous codes, the last one being CP 110, have recommended designers to check local bond stresses, BS 8110 does not do so. Sometimes the designer assumed that this was correct by his detailing procedure, for example knowing that if no less than about one quarter of the centre span reinforcement continued over the support then the maximum local bond stress namely at the support where the shear force was a maximum, would generally be satisfactory and the designer would possibly consider it a waste of time checking for local bond stress. BS 8110 perhaps believes its anchorage requirements of bars, for example requiring bars to extend at least 12 diameters beyond the centre line of a simply supported end of a member, to be sufficient to not require local bond stresses to be checked.

Example 3.10. The maximum tensile reinforcement in a beam consists of four 25 mm diameter plain bars, and $d = 600$ mm. The maximum ultimate shear force immediately adjacent to a support is 140 kN. If the ultimate local bond stress of Table 21 of CP 110 is 2 N/mm² ($= z_1 f_{bs}$), what is the least number of the reinforcement bars which must continue through to the support? Note that CP 110 calls our $z_1 f_{bs}$ just f_{bs}.

Applying equation 3.56, $\Sigma u_s = 140000/(2 \times 600) = 116.7$ mm. The circumference of one 25 mm diameter bar $= \pi \times 25 = 78.5$ mm. Number of bars required to continue through to support $= 116.7/78.5 = 2$, to nearest integer.

3.6   Torsion

Torques are usually calculated assuming a structure to be elastic and uncracked. This is true neither at working nor at ultimate loads, but there is no reliable alternative to this procedure. The monolithic nature of *in-situ* construction means that most sections inevitably experience torques, even if only very small, at some time or other. The experience of the designer usually enables him to provide for minor torques when detailing the reinforcement. For example, the external beams to a floor might be given nominal stirruping of say 10 mm diameter at 230 mm centres, as opposed to say 6 mm diameter at 300 mm centres for the internal beams (assuming the possibility of torques on the internal beams is negligible, that is a low ratio of live to dead load). This practice is obviously satisfactory in that torsional failures are extremely rare, yet the majority of structures are never overloaded and have been designed to more conservative past codes. Past practice has been that where the torsional resistance of members is ignored in the analysis of an indeterminate structure for bending moments and shearing forces, then nominal shear reinforcement should be used for torsion. If torsional resistance needs assessing, BS 8110 requires the torsional rigidity, $G \times C$, of a member to be such that $G = 0.42E_c$ and C, the torsional moment of inertia, equal to half the polar second moment of area based on the gross concrete section. This makes some allowance for the fact that plane cross sections warp under torsion, and the classical theory assumes plane sections remain plane. Torsion failures are very inconsistent and this leads to divergent views upon design by various researchers. In practice, torques often occur simultaneously with shear forces and bending moments, thus complicating the problem still further, especially as the design of members in shear is a difficult problem in itself. In this respect it is good practice to create structural systems so that torsion is always a subsidiary and negligible effect.

Design has been based on the classical work of St. Vernant[5] modified in the light of experimentation. The maximum shear stress due to torsion for a rectangular section is at the middle of the longer sides[5] according to St. Vernant, whereas BS 8110 assumes a plastic stress distribution, that is a uniform shear stress given by

$$v_t = 6T/[h_{min}^2(3h_{max} - h_{min})] \qquad (3.57)$$

where T is the torsional moment due to ultimate loads, h_{min} is the smaller dimension of the section, and h_{max} is the larger dimension of the section.

T-, L- or I-sections may be treated by dividing them into their component rectangles, so as to maximise the function $\Sigma(h_{min}^3 h_{max})$ which will generally be achieved if the widest rectangle is made as long as

Torsion

possible. The torsion shear stress carried by each component rectangle can be calculated by treating them as rectangular sections subjected to a torsional moment of

$$T[h_{min}^3 h_{max}/(\Sigma h_{min}^3 h_{max})]$$

Where the torsion shear stress, v_t, exceeds the value $v_{t,min}$ from BS 8110 Pt. 2 Cl. 2.4.5 that is our Table 3.6a, reinforcement should be provided. In no case should the sum of the shear stresses resulting from shear force and torsion $(v+v_t)$ exceed the value v_{tu} from Table 3.6a, nor, in the case of small sections ($y_1 < 550$ mm), should the torsion shear stress, v_t, exceed $v_{tu} y_1/550$, where y_1 is the larger dimension of a link in mm.

Torsion reinforcement should consist of rectangular closed links together with longitudinal reinforcement. BS 8110 requires this reinforcement to be additional to any requirements for shear or bending and to be such that:

$$0.87 f_{yv}(A_{sv}/s_v) \geq T/(0.8 x_1 y_1) \tag{3.58}$$

$$A_{sl} \geq (A_{sv}/s_v)(f_{yv}/f_{yl})(x_1+y_1)$$

$$= [T/(0.8 x_1 y_1)][(x_1+y_1)/0.87 f_{yl}] \tag{3.59}$$

where A_{sv} is the area of the legs of closed links at a section, A_{sl} is the area of longitudinal reinforcement, f_{yv} is the characteristic strength of the links, f_{yl} is the characteristic strength of the longitudinal reinforcement, s_v is the spacing of the links, x_1 is the smaller dimension of the links, and y_1 is the larger dimension of the links.

In the above formulae f_{yv} and f_{yl} are not to be taken as greater than 460 N/mm². (Ref. 4 would say 250 N/mm².)

Example. 3.11. Design links for the section shown in Figure 3.2, $h_{max} = 488$ mm, to resist an ultimate torsional moment of 3 kN m combined with

Table 3.6a *Values of $v_{t,min}$ and v_{tu}*

Concrete grade	$v_{t,min}$	v_{tu}
	N/mm²	N/mm²
25	0.33	4.00
30	0.37	4.38
40 or above	0.40	5.00

Note: Allowance is made for γ_m.

an ultimate vertical shear force of 60 kN. Concrete is of Grade 25, cover is 25 mm ($x_1 = 100$ mm, and $y_1 = 438$ mm), and $f_{yv} = 250$ N/mm². If $f_{yl} = 460$ N/mm², what extra longitudinal reinforcement is required?

From equation 3.57,

$$v_t = (6 \times 0.003)/[0.15^2(3 \times 0.488 - 0.15)] \text{ MN/m}^2$$
$$= 0.6088 \text{ N/mm}^2.$$

From Table 3.6a this is >0.33 so that torsional reinforcement is required.

$$V/(bd) = 0.06/(0.15 \times 0.45) \text{ MN/m}^2 = 0.8889 \text{ N/mm}^2$$

$$v_t + V/(bd) = 1.498$$

This is in order, as Table 3.6a limits this to 4.0.

As $y_1 < 550$ mm, v_t must not exceed $4.0 \times 438/550 = 3.185$ N/mm², which is all right as $v_t = 0.6088$ N/mm².

From equations 3.53 and 3.58

$$V/d = 0.87 f_{yv}(A_{sv}/s_v) = 3/(0.8 \times 0.1 \times 0.438) \text{ kN/m}$$
$$= 85.62 \text{ N/mm}$$

Using Table 3.2, $100 A_s/(bd) = (100 \times 982)/(150 \times 450) = 1.455$. From Table 3.4a

$$v_c = 0.63 + (0.72 - 0.63)(1.455 - 1.0)/(1.5 - 1.0) = 0.712 \text{ N/mm}^2$$

Hence shear reinforcement (two-arm links) is required to resist a value of $V/(bd) = 0.8889 - 0.712 = 0.1769$ N/mm².

This is less than $0.8\sqrt{25}$ and 5 N/mm and therefore satisfactory according to BS 8110.

$$V/d = 0.1769 \times b = 0.1769 \times 150 = 26.54 \text{ N/mm}$$

$$\therefore \text{Total } V/d = 85.62 + 26.54 = 112.2$$

From Table 3.6, use 8 mm diameter two-arm links at 200 mm centres. From equation 3.59,

$$A_{sl} = 85.62(100 + 438)/(0.87 + 460) = 115.1 \text{ mm}^2$$

Referring to BS 8110, the longitudinal reinforcement should be distributed evenly around the inside perimeter of the links (stirrups). The clear distance between the longitudinal bars used should be less than 300 mm and at least four bars, one in each corner, should be used. In our

Torsion

example $y_1 > 300$ mm so use two bars in the top corners of the stirrups, two at the half-depth (of y_1) of stirrups (wired to the inside of the stirrups) and two in the bottom corners of the stirrups. The latter can be catered for by just increasing the sizes of the tension steel but not in this example because if these bars are increased in size their cover will have to be increased and x_1 altered, etc. requiring revision of the example. Neither can a bar be placed between these tension bars because of the spacing required between bars (assuming $h_{agg} = 19$ mm). The bottom two bars for torsion will therefore be placed above the tension bars, a clear distance of $19 \times 2/3$ (according to BS 8110) = 13 mm above them. Thus, using Table 3.2, six 6 mm diameter bars will be used as the longitudinal torsion bars.

Example 3.12. An L-shaped beam has: depth and overall breadth of top flange 120 mm and 300 mm, respectively, thickness of web and overall depth of beam 100 mm and 600 mm, respectively. The ultimate vertical and horizontal shear forces are 20 kN and 10 kN, respectively and the ultimate torque is 2 kN m. Determine the reinforcement required for resisting shear and torsion. Concrete is of Grade 30. Cover to longitudinal steel is 20 mm.

Taking the gross web as one rectangle,

$$\Sigma h_{min}^3 h_{max} = 1^3 \times 6 + 1.2^3 \times (3-1) = 9.46 \text{ dm}^4$$

Taking the gross flange as one rectangle

$$\Sigma h_{min}^3 h_{max} = 1.2^3 \times 3 + 1^3 \times (6-1.2) = 5.184 + 4.8 = 9.984 \text{ dm}^4$$

Hence the latter is the way to consider the section as two rectangles.
For the gross flange, torque = $2 \times 5.184/9.984 = 1.038$ kN m.
For the web, torque = $2 - 1.038 = 0.962$ kN m.
From equation 3.57:

$$\text{for gross flange } v_t = \frac{6 \times 0.001\,038}{0.12^2(3 \times 0.3 - 0.12)} \text{ MN/m}^2$$

$$= 0.5545 \text{ N/mm}^2$$

$$\text{for web } v_t = \frac{6 \times 0.000\,962}{0.1^2(3 \times 0.48 - 0.1)} \text{ MN/m}^2 = 0.4307 \text{ N/mm}^2$$

From Table 3.6a, these are > 0.37 so that torsional reinforcement is required for both gross flange and web.

118   Reinforced concrete beams

For gross flange

$$V/(bd) = 0.01/(0.12 \times 0.27) = 0.3086 \text{ N/mm}^2$$

(assuming $d = 300 - 30 = 270$ mm).

For web

$$V/(bd) = 0.02/(0.1 \times 0.57) = 0.351 \text{ N/mm}^2$$

(assuming $d = 600 - 30 = 570$ mm).

For gross flange

$$v_t + V/(bd) = 0.5545 + 0.3086 = 0.8631 \text{ N/mm}^2$$

For web

$$v_t + V/(bd) = 0.4307 + 0.351 = 0.782 \text{ N/mm}^2$$

These are in order as Table 3.6a limits this value to 4.38.

For gross flange

$$y_1 = 300 - 40 = 260 \text{ mm}.$$

For web

$$y_1 = 600 - 40 = 560 \text{ mm}.$$

As $y_1 < 550$ for the gross flange, v_t for it must not exceed $4.38 \times 260/550$ = 2.071 N/mm², which is all right as $v_t = 0.5545$ N/mm².

The design is continued, treating the gross flange and web, respectively, as in Example 3.11, as though each were an independent member.

3.7 Plastic analysis

A material is in a plastic condition when stresses cause permanent deformations, that is when stress is no longer directly proportional to strain (as in Hooke's law). A section of a beam experiences such conditions when realising its ultimate moment of resistance. The plastic method of design predicts the ultimate moment of resistance, and this is required to equal the ultimate bending moment derived from the working loads multiplied by suitable load factors.

3.7.1   Assumptions of plastic design methods

Plastic design concerns two ideas. Firstly, with regard to the assessment of the bending moments in a redundant frame, plasticity is the

Plastic analysis

ability of highly stressed sections to what might be termed yield, and allow a redistribution[6] of the bending moments towards failure. Secondly, plastic design can be employed in the design of individual sections of structural members. In the latter instance the following assumptions are employed.

It is assumed that plane sections subjected to bending remain plane after bending, which means that the distribution of strain is linear. Some relationship is then assumed between this strain, and stress. This is where the methods differ. Concrete is assumed to have no resistance in flexural tension, perfect bond is assumed between the steel and the concrete, the depth of the steel reinforcement is assumed to be small compared with its effective depth, and normally temperature and shrinkage stresses are ignored in the stress analysis of sections.

3.7.2 Plastic design in bending

The term *balanced design* refers to the situation when the beam is designed to fail simultaneously in flexural compression and tension. *Under-reinforced* sections will fail in flexural tension and *over-reinforced* sections will fail in flexural compression. An under-reinforced section fails owing to yielding (or straining excessively in the case of high-yield steel) of the tensile reinforcement; this causes the cracks to open so that the depth of the beam available to resist flexural compression is reduced, and final collapse occurs by the crushing of the compression zone. This is not, however, a *flexural compression failure*, since the failure has actually been precipitated by the inadequacy of the tensile reinforcement and the final failure in apparent *flexural compression* is a secondary effect; it could be described as part of the disintegration of the beam after failure.

Figure 2.10 shows a typical relationship between stress and strain for

Figure 3.6

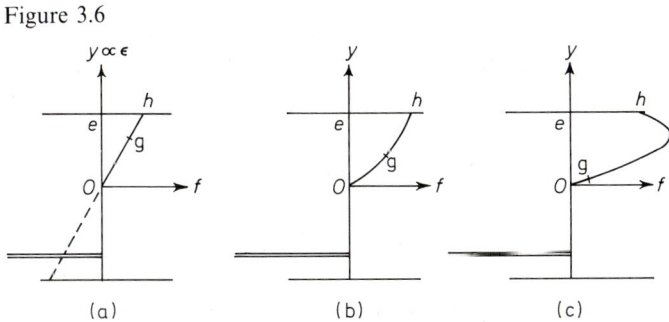

concrete in compression. As described in Section 2.3.17, this will vary in shape according to the speed of loading, the strength of the concrete, etc. Considerable plasticity is experienced towards failure, i.e. stress is not linearly proportional to strain near failure. It is assumed that the distribution of strain due to bending is linear. The strain is therefore proportional to the distance from the neutral axis. Curves such as those illustrated in Figure 2.10 can therefore be plotted on the axes Of and Oy as shown in Figure 3.6. For example Figure 3.6(a) illustrates the elastic stress distribution at working loads at a section where there is a crack. For higher loads the stress distribution becomes as shown in Figure 3.6(b), and just before failure the stress distribution will be as shown in Figure 3.6(c). The point denoted by g is at the same position on all of Figures 2.10, 3.6(a), (b) and (c). Different scales are used for the strains plotted on the axes Oy. The diagrams $ehgO$ in Figure 3.6 are termed *stress blocks*.

For estimating the ultimate moments of resistance of beams, the shape of the stress block just before failure must be known. This is assessed empirically, and shapes suggested for the stress block just before failure have included parabolas, cubic parabolas, trapeziums, ellipses, and many unusual shapes; some theories have even assumed that part of the concrete just below the neutral axis resists tensile stresses. This idea is not justified by experiments, because the cracks penetrate too far, so as to reduce the compression zones at the critical sections. C. S. Whitney, in 1937, suggested considering the stress block as equivalent to a rectangular shape. This leads to a simple theory which has often been found to be more accurate than other methods, for example see Ref. 7.

3.7.3 Plastic design of 'under-reinforced' rectangular sections

The distribution of stress at failure is shown in Figure 3.7. A general shape is considered for the stress block, the average compressive stress of which is equal to f_{cm}, and the centroid is at a depth of $k_2 x$. Equating longitudinal forces, $N_c = N_s$

$$f_{cm} xb = A_s f_s$$

$$\therefore x = A_s f_s / (f_{cm} b) \tag{3.60}$$

Taking moments about the line of action of N_c the ultimate resistance moment

$$M_u = N_s z = N_s (d - k_2 x) \tag{3.61}$$

Substituting for x from equation 3.60 this becomes

$$M_u = N_s[d - k_2 A_s f_s/(f_{cm} b)] \tag{3.62}$$

$$\therefore M_u = A_s f_s d[1 - k_1 \rho f_s/f_{cm}] = f_s \rho b d^2 [1 - k_2 \rho f_s/f_{cm}] \tag{3.63}$$

where $\rho = A_s/(bd)$. Whitney and the simplified method of BS 8110 use a rectangular stress block such that $f_{cm} = 0.85 f'_c$ (where $f'_c =$ U.S.A. cylinder strength $\simeq 0.84 f_{cu}$) and $0.45 f_{cu}$, respectively, $k_2 = 0.5$ for both, and f_s is f_y for Whitney and f_y/γ_m for BS 8110 where $\gamma_m = 1.15$. With the equivalent (unlike actual) stress block of Whitney the depth of the stress block x_1 is less than the depth of the neutral axis x. The above equations would use x_1 instead of x in this instance, viz.

$$x_1 = A_s f_s/(f_{cm} b) \tag{3.60a}$$

$$M_u = N_s(d - k_2 x_1) = A_s f_s(d - k_2 x_1) \tag{3.61a}$$

and equation 3.63 remains the same for Whitney and BS 8110. Whitney gives a good prediction of how a beam will actually fail[7] and has been the basis of British and A.C.I. codes of practice. The coefficients quoted for Whitney's theory in this chapter assume $f_{cu} \leq 33.33$ N/mm². For higher values of f_{cu} refer to Section 8.4.5. The simplified rectangular stress block of BS 8110 is chosen to have $x_1 = 0.9x$. BS 8110 gives a reliably conservative prediction of failure, distorted to ensure that flexural tension rather than compression failures will occur. The former failure gives plenty of warning – large deflections and cracks before failure – whereas the latter failure is very sudden.

The method claimed by BS 8110 to be more precise than its simplified

Figure 3.7

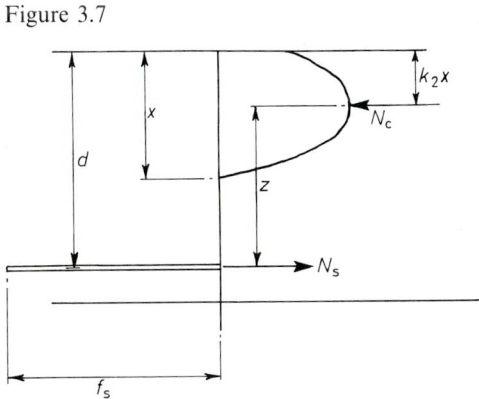

method uses a stress block as shown in Figure 3.8(b) and the distribution of strain, for precise and simplified methods, is shown in Figure 3.8(a). BS 8110 specifies $\varepsilon_0 = \{\sqrt{f_{cu}}\}/5000$, $f_1 = 0.45 f_{cu}$ and curve AB to be a parabola. Thus, considering the shape ABD, its area is AD × BD/3, C is its centroid and CE = BD/4. The compression force N_c is

$$f_{cm} xb = (\text{area ABGF}) b$$

$$\therefore f_{cm} x = \text{area ADGF} - \text{area ADB} = f_1 x - f_1 x_0/3$$

$$\therefore f_{cm} = f_1[1 - x_0/(3x)] = f_1[1 - \varepsilon_0/(3\varepsilon_1)]$$

$$\therefore f_{cm} = 0.45 f_{cu}[1 - \{\sqrt{(f_{cu})}\}/52.5] \tag{3.64}$$

Taking moments for compression force about F

$$N_c k_2 x = b[(\text{area ADGF}) 0.5x - (\text{area ADB})(x - CE)]$$

$$f_{cm} xk_2 x = 0.5 f_1 x^2 - (f_1 x_0/3)(x - x_0/4)$$

$$\therefore k_2 = (f_1/f_{cm})[0.5 - \{x_0/(3x)\}\{1 - x_0/(4x)\}]$$

$$= \{0.45 f_{cu}/(2 f_{cm})\}[1 - \{2\varepsilon_0/(3\varepsilon_1)\}\{1 - \varepsilon_0/(4\varepsilon_1)\}]$$

$$\therefore k_2 = (0.225 f_{cu}/f_{cm})[1 - \{(\sqrt{f_{cu}})/26.25\}[1 - (\sqrt{f_{cu}})/70]] \tag{3.65}$$

Equations 3.64 and 3.65 are the basis of the design charts of BS 8110.

Figure 3.8

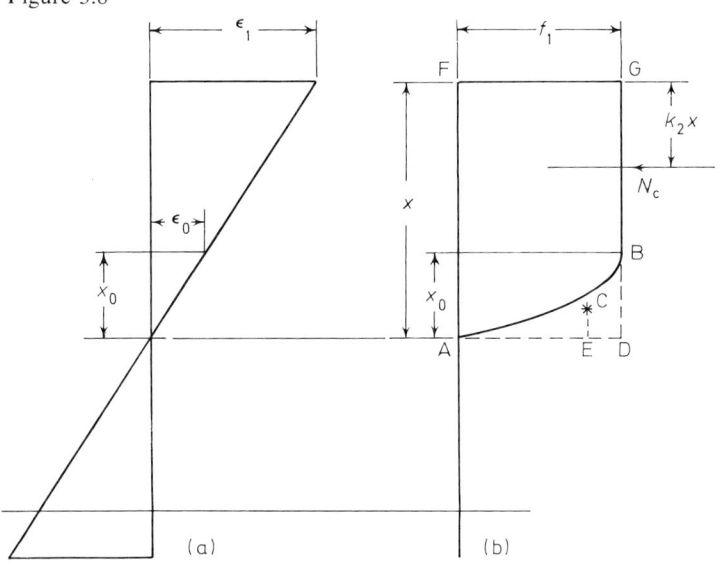

Plastic analysis

3.7.4 *'Balanced' plastic design of rectangular sections*

The equations of Section 3.7.3 apply. With these equations, as A_s increases x increases and M_u increases, but experimentally we find that x cannot increase beyond a certain amount and increasing the reinforcement further gives no increase in M_u, the section being known as *over-reinforced*. When x has its maximum value, and A_s corresponds to this, the section is in its 'balanced design' condition, the maximum flexural compression being balanced by the minimum A_s to give a maximum M_u for the section.

For balanced design Whitney gives $x_1 = 0.537d$, and BS 8110, for design purposes, gives $x = 0.5d$. Using BS 8110, see Section 3.7.3 previously, from equation 3.60a

$$x_1 = 0.9x = 0.9 \times 0.5d = 0.45d = A_s f_s/(f_{cm} b)$$

$$\therefore \rho = \frac{A_s}{bd} = 0.45 \times \frac{f_{cm}}{f_s} = 0.45 \times 0.45 \times \frac{f_{cu}}{f_s} = 0.2025 \frac{f_{cu}}{f_s} \quad (3.66)$$

Strictly speaking BS 8110 gives $f_{cm} = (\tfrac{2}{3})f_{cu}/\gamma_m$ which equals $(\tfrac{2}{3})f_{cu}/1.5 = (\tfrac{4}{9})f_{cu}$. Using this instead of the BS 8110 figure, approximated to two decimal places, equation 3.66 conforms more accurately to the basic wishes of the code

$$\rho = 0.2 \frac{f_{cu}}{f_s} \quad (3.66a)$$

Equation 3.61a becomes

$$M_u = A_s f_s(d - 0.5 x_1) = A_s f_s(d - 0.5 \times 0.9x)$$

$$\therefore M_u = A_s f_s(d - 0.45 \times 0.5d) = 0.775 A_s f_s d \quad (3.67)$$

Table 3.7

f_y, N/mm²	f_s, N/mm²	f_{cu}, N/mm²				
		20	25	30	40	50
250	217	1.843	2.304	2.765	3.687	4.608
460	400	1.000	1.250	1.500	2.000	2.500 ρ%
K_1,		N/mm² 3.12	3.9	4.68	6.24	7.8

124 *Reinforced concrete beams*

and substituting in this ρf_s from equation (3.66)

$$M_u = 0.775\rho bd^2 f_s = 0.775 \times 0.2025 f_{cu} bd^2$$

$$\therefore M_u = 0.1569 f_{cu} bd^2$$

The figure of 0.2025 in equation 3.66 is carried forward as 0.201 by BS 8110 for this latter calculation (0.201 is obtained by BS 8110 using 0.67, that is $\frac{2}{3}$ approximated to two places of decimal) to give 0.1558 (which BS 8110 calls 0.156) in lieu of 0.1569, so that last equation becomes

$$M_u = 0.156 f_{cu} bd^2 = K_1 bd^2 \tag{3.68}$$

with the BS 8110 proviso that in the case of a continuous beam the redistribution of bending moment does not exceed 10%.

Equations 3.66 and 3.68 are used for design Table 3.7. Without tables, equation 3.68 is usually used to decide the size of the member as limited by the strength of the concrete. Then A_s is often obtained from equation 3.67 thus:

$$A_s = \frac{M_u}{0.775 df_s} = \frac{M_u \gamma_m}{0.775 df_y} = \frac{1.15 M_u}{0.775 df_y} = \frac{1.484 M_u}{df_y} \tag{3.69}$$

Example 3.13. A slab 160 mm thick is reinforced in tension with 16 mm diameter bars having 30 mm cover. Determine the spacing of the reinforcement if the slab is designed in accordance with BS 8110 for an ultimate resistance moment of 27.6 kNm, and if $f_{cu} = 25$ N/mm² = 25000 kN/m², $f_y = 250$ N/mm² = 250000 kN/m² and γ_m for the steel = 1.15.

Using the simplified BS 8110 method, from equation 3.68 considering 1 m width of slab, for balanced design

$$M_u = 0.15 \times 25000 \times 1 \times (0.160 - 0.038)^2 = 58.05 \text{ kN m}$$

This is greater than 27.6 kN m, hence section is under-reinforced. From equation 3.62 (or 3.63), using $f_s = 250000/1.15 = 217400$ kN/m²

$$27.6 = 217400 A_s [0.122 - 0.5 \times 217400 A_s / (0.45 \times 25000 \times 1)]$$

$$\therefore A_s = 0.001144 \text{ m}^2 = 1144 \text{ mm}^2$$

From Table 3.2, use 16 mm diameter bars at 175 mm centres.

Plastic analysis 125

Example 3.14. Repeat Example 3.13 using the method preferred by BS 8110.

Equations 3.64 and 3.65 give

$$f_{cm} = 0.45 \times 25[1 - \{(\sqrt{25})/52.5\}] = 10.18 \text{ N/mm}^2$$

$$k_2 = (0.225 \times 25/10.18)[\![1 - [\{(\sqrt{25})/26.25\} \{1 - (\sqrt{25})/70\}]]\!] = 0.4548$$

From equations 3.60 and 3.62 for 1 m width of slab for balanced design

$$0.5 \times 122 = A_s 217.4/(10.18 \times 1000)$$

$$M_u = A_s \times 217.4[122 - 0.4548 A_s 217.4/(10.18 \times 1000)]$$

$$\therefore M_u = 61 \times 10\,180(122 - 0.4548 \times 61) \text{ N mm} = 58.53 \text{ kN m}$$

This is > 27.6, hence section is under-reinforced. Hence from equation 3.62

$$27.6 = 217\,400 A_s[0.122 - 0.4548 \times 217\,400 A_s/(10\,180 \times 1)]$$

$$\therefore A_s = 0.001\,145 \text{ m}^2 = 1145 \text{ mm}^2$$

From Table 3.2, use 16 mm diameter bars at 175 mm centres.

Example 3.15. The slab of Example 3.13 is reinforced in flexural tension with 16 mm diameter bars at 175 mm centres (that is, $A_s = 1149$ mm² per metre) and is to be tested to destruction. Predict its ultimate resistance moment using Whitney's theory.[7]

To determine whether it is under- or over-reinforced, apply equation 3.60 (referring also to Sections 3.7.3 and 3.7.4).

$$x_1 = [1149 \times 250/\{0.85(0.84 \times 25) \times 1000\}] = 16.09 \text{ mm}$$

For balanced design $x_1 = 0.537 \times 122 = 65.5$ mm, hence section is under-reinforced. Hence applying equation 3.62

$$M_u = 1149 \times 250[122 - 0.5 \times 1149 \times 250/\{0.85(0.84 \times 25) \times 1000\}] \text{ N mm}$$
$$= 32.73 \text{ kN m}$$

This is considerably greater than the 27.6 kN m used in Example 3.14, indicating the conservativeness built into the BS 8110 design method.

Example 3.16. Design a section of a beam, using the simplified BS 8110 method, to have an ultimate resistance moment of 200 kN m, using $f_{cu} = 20$ N/mm², $f_y = 250$ N/mm² and γ_m for steel $= 1.15$.

From equation 3.68,

$$200 \times 10^6 = 0.156 \times 20bd^2$$

$$\therefore bd^2 = 64.10 \times 10^6$$

If $b \simeq 0.5d$ say, then $d^3 = 128.2 \times 10^6$ and $d = 504$. So $b = 253$, say, choose $b = 250$ mm. Then $d = \sqrt{(64.10 \times 10^6/250)} = 507$ mm. From equation 3.69 (or 3.66 or 3.67)

$$A_s = 1.484 \times 200 \times 10^6/(507 \times 250) = 2342 \text{ mm}^2$$

From Table 3.2 use three 32 mm diameter bars.

Example 3.17. Repeat Example 3.16 using Table 3.7.

From Table 3.7, $K_1 = 3.12$ N/mm² and $\rho = 1.843\%$. Using equation 3.68, $bd^2 = 200 \times 10^6/3.12 = 64.1 \times 10^6$. As in Example 3.15, choose $b = 250$ mm, then $d = 507$ mm. Then $A_s = 0.01843 \times 250 \times 507 = 2336$ mm². From Table 3.2 use three 32 mm diameter bars.

3.7.5 Plastic design of any shape of 'under-reinforced' section

For the section of Figure 3.4(a), using a rectangular concrete stress block of average stress f_{cm} (see Section 3.7.3), equating longitudinal forces

$$f_{cm} A_c = A_s f_s \tag{3.70}$$

where A_c = area of concrete in compression. Taking moments about the line of action of N_c

$$M_u = A_s f_s z \tag{3.71}$$

where z = lever arm = distance between lines of action of N_c and N_s. N_c acts at centroid of A_c.

Whitney specifies $f_{cm} = 0.85 f'_c \simeq 0.85 \times 0.84 f_{cu} = 0.714 f_{cu}$ as before. BS 8110 specifies $f_{cm} = 0.45 f_{cu}$ for simplified design method.

3.7.6 'Balanced' plastic design of any shape of section

For balanced design (see Section 3.7.4) the depth of the stress block x_1 obtained from equation 3.70 is $0.537d$ for Whitney's theory and $0.45d$ for BS 8110.

Example 3.18. A T-beam has a flange of breadth 750 mm and depth 130 mm. The width of its rib or web is 300 mm and the tensile

Plastic analysis

reinforcement comprises one layer of five 25 mm diameter bars having an effective depth of 500 mm. Determine its ultimate resistance moment from the simplified design method of BS 8110, assuming $f_{cu} = 20$ N/mm², $f_y = 460$ N/mm² and $\gamma_m = 1.15$ for the reinforcement.

From equation 3.70 and Table 3.2,

$$0.45 \times 20[300x_1 + (750 - 300) \times 130] = 2454 \times (460/1.15)$$

$$\therefore x_1 = 168.6 \text{ mm}$$

As $0.45 \times 500 = 225$ the section is under-reinforced. Also $x_1 >$ depth of flange, hence beam is designed as a T-beam and not a rectangular beam.

If depth of centroid of A_c is $k_3 x_1$ then taking area moments about the top of the beam for A_c

$$A_c k_3 x_1 = 300 x_1^2/2 + (750 - 300) \times (130^2/2)$$

$$\therefore k_3 x_1 = [8\,826\,000/\{300x + (750 - 300) \times 130\}] = 77.8 \text{ mm}$$

From equation 3.71

$$M_u = 2454 \times (460/1.15) \times (500 - 77.8) \text{ N mm} = 414.4 \text{ kN m}$$

3.7.7 Plastic design of any shape of 'under-reinforced' section containing compression steel

It might be said that compression reinforcement is only required in a beam when the balanced design condition applies. Whilst this is often true, there are cases where compression steel is available even though not required to assist flexural compression, for example sometimes at the supports of continuous beams. In such cases the compression steel can increase the ultimate bending moment of the section and sometimes economises in tensile steel.

For a section like Figure 3.4(a) but including compression steel in the top, using a rectangular concrete stress block (see Section 3.7.3):

Compression force for gross area of concrete in compression = $A_c f_{cm}$

Compression force for compression steel over and above that included at this position above = $A'_s(f_{sc} - f_{cm})$

Therefore equating longitudinal forces

$$A_c f_{cm} + A'_s(f_{sc} - f_{cm}) = A_s f_s \quad (3.72)$$

where A_s = gross area of concrete in compression, A'_s = area of compression steel and f_{sc} = stress in compression steel (usually characteristic strength because the strain is high in the concrete and thus the steel as flexural concrete failure occurs). Taking moments about the line of action of N_c

$$M_u = A_s f_s z_1 = A_s f_s (d - k_2 x_1) \tag{3.73}$$

Whitney specifies $f_{cm} \simeq 0.714 f_{cu}$ as before. BS 8110 specifies $f_{cm} = 0.45 f_{cu}$ for its simplified design method. Whitney gives f_{sc} as yield stress of compression steel and BS 8110 gives f_{sc} as f_y/γ_m, where $\gamma_m = 1.15$, which it suggests can be simplified to $0.85 f_y$ for ease of calculation. These comments on f_{sc} depend upon the strain in the compression steel being at least that corresponding to its yield stress. The strain at the level of the compression steel needs to be assessed and related to the stress–strain relationship for the steel (for example BS 8110, Fig. 2.2) – see Section 3.7.10. Note $k_2 x_1$ is the depth to the total compression force resulting from concrete and compression steel forces.

3.7.8 'Balanced' plastic design for any shape of section containing compression steel

For balanced design (see Sections 3.7.4 and 3.7.6) the depth of the stress block x_1 is $0.537d$ for Whitney's theory and 0.45 for BS 8110.

Example 3.19. Determine the ultimate resistance moment from the simplified design method of BS 8110 of the beam section shown in Figure 3.3 where the reinforcement bars are 10 mm diameter in compression and 32 mm diameter in tension and have 40 mm cover of concrete. Assume $f_s = 250/1.15 = 217.4 \text{ N/mm}^2$, $f_{sc} = 217.4 \text{ N/mm}^2$, and $f_{cu} = 0.45 \times 25 = 11.25 \text{ N/mm}^2$.

From equation 3.72 and Table 3.2

$$[160 x_1 + (450 - 160) \times 150] \times 11.25$$
$$+ 314(217.4 - 11.25) = 4826 \times 217.4$$

$$\therefore x_1 = 275.0 \text{ mm}$$

This is > 150, hence beam is designed as a T- and not a rectangular beam. For balanced design, whether T- or rectangular section, $x_1 = 0.45 \times (900 - 56) = 379.8$ mm. Hence section is under-reinforced. Taking moments about top of beam for compression forces

Plastic analysis

$$k_2 x[\{160 \times 275 + (450 - 160) \times 150\} \times 11.25$$
$$+ 314(217.4 - 11.25)]$$
$$= [160 \times (275^2/2) + (450 - 160) \times (150^2/2)] \times 11.25$$
$$+ 314 \times (217.4 - 11.25) \times 45$$
$$\therefore k_2 x_1 = 102.6 \text{ mm}$$

From equation 3.73,

$$M_u = 4826 \times 217.4 \times [(900 - 56) - 102.6] \text{ N mm} = 777.9 \text{ kN m}$$

This assumes that the compression steel is not near the bottom of the stress block. Effective depth of compression steel $= d' = 45$ mm, whereas $x_1 = 275.0$. From BS 8110 (see Section 3.7.10) this matters when $d' > 0.43x = 0.478x_1$. In this example $d'/x_1 = 45/275 = 0.164$. If the compression steel is near the neutral axis (rather an unusual case) refer to Section 3.7.10.

3.7.9 Design of compression steel for a rectangular section

In practice the commonest place where compression steel is required is at the supports of continuous *in-situ* T-beams. The bending moments at mid span and supports are of similar magnitude; the T-section at mid span enables the rib (or stem) there to be small compared with the size of a rectangular beam; then at the support the bending moment is reversed and the beam is designed as a rectangular beam, with the small rib as its compression zone. In these circumstances the section here may require compression steel. Thus a rectangular section has to be designed to take a bending moment in excess of its balanced design moment by the addition of compression steel.

Example 3.20. Design a rectangular section 300 mm wide by 600 mm deep to have an ultimate resistance moment of 400 kN m in accordance with BS 8110. Assume $f_{cu} = 25$ N/mm², $f_y = 250$ N/mm² and γ_m for steel $= 1.15$.

For balanced design (with no compression steel) see Section 3.7.4, and applying equation 3.68, estimating $d = 560$ mm, $M_u = 3.9 \times 300 \times 560^2$ N mm $= 366.9$ kN m.

Hence section needs compression steel. An estimate of $d' = 35$ mm. Then z for compression steel $= d - d' = 525$ mm and z for concrete in compression $= 0.775 \times 560 = 434$ mm, because depth of stress block is 0.45×560. Then

$$A'_s = [(400 - 366.9) \times 10^6 / (0.87 \times 250 \times 525)] = 289.9 \text{ mm}^2$$

From Table 3.2 use say two 16 mm diameter bars. Resolving forces longitudinally (that is using equation 3.72),

$$(250/1.15) \times A_s = 300 \times 0.5 \times 560 \times 0.45 \times 25$$
$$+ 289.9 \times (0.87 \times 250 - 0.45 \times 25)$$
$$\therefore A_s = 4622 \text{ mm}^2$$

From Table 3.2 use say ten 25 mm diameter bars. These will need to be in two layers, say five in the bottom and five in the layer above. Using 19 mm down coarse aggregate the vertical distance between the layers of bars $= 13$ mm, say 15 mm. This will mean that, using 25 mm cover to the tension steel, $d = 600 - 25 - 25 - 7 = 546$ (not 560). Using 25 mm cover for the compression reinforcement, $d' = 25 + 8 = 33$ (not 35) mm. This design can be repeated with these more accurate values of d and d', but it should not alter the results as the reinforcement is on the generous side because of the limitation of bar sizes, and the initial estimates of d and d' were not too inaccurate. $d' \not> 0.43x$ ($= 0.215d$), hence (see Section 3.7.10) the stress we have taken in the compression steel does not need reducing.

3.7.10 Compression steel near to neutral axis

In practice this can hardly ever arise, as when compression steel is required it is placed as far from the neutral axis as possible for economic reasons. At failure in flexure the maximum strain in the concrete is about 0.0035 and the distribution of strain is approximately linear. Hence the strain at the level of the compression steel is $0.0035 (x - d')/x$. According to Fig. 2.2 of BS 8110, if this strain is less than $f_y/(\gamma_m E_c) = 0.001\,087$ if $f_y = 250 \text{ N/mm}^2$ and 0.002 if $f_y = 460 \text{ N/mm}^2$ then the stress–strain curve of Fig. 2.2 should be used to determine the design stress in the compression reinforcement. Hence for BS 8110 we do not have to concern ourselves in Sections 3.7.7, 3.7.8 and 3.7.9 with reducing the stress in the compression steel if $(x - d')/x \not< 10.87/35 = 0.3106$ for $f_y = 250 \text{ N/mm}^2$ and $20/35 = 0.5714$ for $f_y = 460 \text{ N/mm}^2$, that is $x \not< 1.45d'$ for $f_y = 250 \text{ N/mm}^2$ and $2.333d'$ for $f_y = 460 \text{ N/mm}^2$. In the case of balanced design $x = 0.5d$, and this becomes $0.5d \not< 1.45d'$ for $f_y = 250 \text{ N/mm}^2$ and $0.5d \not< 2.33d'$ for $f_y = 460 \text{ N/mm}^2$, (BS 8110 quotes $d'/x \not> 0.43$ and

Limit state of deflection

$d'/x = d/(0.5d)$, so $0.5d \not< (1/0.43)d' = 2.34d'$) that is $d' \not> 0.34448d$ for $f_y = 250$ N/mm² and $0.2143d$ for $f_y = 460$ N/mm².

3.7.11 Further points about compression steel

Compression steel, even if available in a section, should not be relied upon in design if not prevented by adequate anchoring from buckling out of the faces of the member; each bar should be anchored at right-angles to the outer surface of the concrete according to CP 114, but BS 8110 has reduced this requirement in its Clause 3.12.7. Both codes specify diameter and spacing of suitable stirrups. For example, framing bars in a beam are not always suitably anchored for compression steel when evaluating ultimate resistance moment.[8]

Compression steel, even if available in a section, should not be relied upon in design without adequate compression laps. For example, steel in the bottom of a continuous T-beam over a support with nominal lapping can only be used to the strength of the lapping in compression.

3.8 Limit state of deflection

Deflections can be calculated as in Example 3.2. This assumes the gross concrete section to be homogeneous and the deflection is obtained with elastic theory. The value assumed for E_c or α_e (as $E_c = E_s/\alpha_e$) can vary considerably (see Section 2.3.17). For accurate work it is best to obtain E_c from laboratory tests on specimens of the concrete. In loading tests on *in-situ* buildings with concrete perhaps about 2–3 months old, the writer has experienced α_e of about 10, that is due to the live load applied. In reinforced concrete design it is useful to divorce the live and dead loadings and take $\alpha_e = 8$ for strong concretes to 10 for weak concretes for calculating deflections due to live loads (that is of short duration; not developing much creep), and take $\alpha_e = 15$ for deflections due to dead loading (this will be realised over several years of creep).

Ignoring the reinforcement and including concrete in tension, which at the positions of cracks will not exist, is usual practice. In the writer's experience troubles with deflection arising from design are usually due to no calculations of deflections, on at least these lines, being made. In the laboratory, obtaining E_c and E_s from tests of specimens of the concrete and steel respectively, and allowing concrete to take tensile stresses and allowing for reinforcement to obtain I, deflections of beams can be predicted very accurately[1] before cracks about 0.01 mm wide occur. For greater loads the deflection often approaches the deflection calculated in the same way but excluding concrete in tension. Just before failure the deflection often becomes greater than this calculated amount.

For a rectangular beam (span l) carrying uniformly distributed loading q and if the breadth is a constant proportion of its depth and if E_c is constant, then maximum deflection $\sim ql^4/bd^3 \sim ql^4/d^4$. Thus the l/d ratio can be a guide to deflection, but only in conjunction with q. The tables restricting l/d in CP 114 for beams and slabs were inadequate in that q was ignored. Table 3.10 of BS 8110 is similar but requires modification by factors given in Tables 3.11 and 3.12. But for spans greater than 10 m if limitation of deflection is required because of not wishing to damage partitions, the figures from Table 3.10 should be multiplied by 10 m divided by the span in metres, apart from cantilevers where deflections should be calculated. From Table 3.11 the greater the bending moment the less the factor, which reduces the allowable l/d ratio, so some allowance is made for q. Table 3.12 allows for the fact that when compression steel is present it restrains the tendency of the shrinkage in this location to increase deflections.

Deflections must be limited so as not to cause trouble to internal partitions and finishes. Beams obviously sagging are aesthetically undesirable – the deflection can be calculated and the beam given an upward camber of at least this amount. Slightly hogging beams are aesthetically acceptable. Consideration should be given to each particular case, and BS 8110 gives general guidance on limitation of deflection.

3.9 Limit state of cracking

Research has indicated that rainwater cannot penetrate to the reinforcement to cause corrosion if cracks are not greater than 0.25 mm wide. This figure can vary with the concrete grade, cover, etc., and BS 8110 uses a figure of 0.3 mm, specifying other figures for various exposures. BS 8110 considers that its reinforcement detailing recommendations take care of undesirable cracking. For example, smaller diameter bars at closer centres resist cracks much better than the converse. When this problem is of particular importance because, say, of severe exposure, or where groups of bars are used, an empirical formula is given in BS 8110 for assessing crack widths.

References

1. Wilby, C. B., 'Strength of Reinforced Concrete Beams in Shear', Ph.D. Thesis, University of Leeds (1949).

References

2. Wilby, C. B., Permissible Shear Stresses of the 1957 British Code of Practice. *Journal of the American Concrete Institute*, June (1958).
3. Lessons from failures of concrete structures. *A.C.I. Monograph No. 1* (1965).
4. Wilby, C. B., Overstrain in high-tensile reinforcing bars at bends and in stirrups. *Indian Concrete Journal*, Jan. (1962).
5. Evans, R. H. and Wilby, C. B., *Concrete – Plain, Reinforced, Prestressed and Shell*, Edward Arnold (1963).
6. Wilby, C. B. and Pandit, C. P., Inelastic behaviour of reinforced concrete single bay portal frames. *Civil Engineering and Public Works Review*, Mar. (1967).
7. Evans, R. H., The plastic theories for the ultimate strength of reinforced concrete beams. *Journal of the Institution of Civil Engineers*, Dec. (1943).
8. Wilby, C. B., Buckling of compression reinforcement in reinforced concrete beams. T.N. 502, *Proc. Inst. Civil Engrs.*, London, Dec. (1988).

4

Reinforced concrete slabs

4.1 Slabs spanning 'one way'

These are designed per unit width as rectangular beams (see examples in Chapter 3).

One-way spanning slabs have always been designed as beams of considerable width. This involves secondary distribution reinforcement being provided which has been specified as various amounts by different codes of practice over the years. The specifications have been based on practical experience. This 'distribution reinforcement' is provided to distribute temperature and shrinkage effects, to assist in fixing and spacing the main steel, and to act as distribution steel for concentrated loads.

4.2 Slabs spanning 'two ways'

These are, for example, *in-situ* rectangular slabs supported on four, three or two adjacent sides. Originally they were designed by ascertaining bending moments and shear forces by elastic theory and then designing sections for these by elastic theory. Subsequently it was possible (CP 114) alternatively to design the resistance to bending moments by plastic theory. This seems to have been satisfactory, but is very illogical, as towards failure the distribution of the bending moments will be different to that given by the elastic theory.

Bending moments from elastic theory, sometimes adjusted slightly for plastic action, can be calculated from simple formulae in BS 8110 for rectangular slabs carrying uniformly distributed loads. These together with formulae for slabs with triangularly distributed loadings (for walls of tanks) and with concentrated loads are given in Reynolds' Handbook.

A later step to the elastic theory has been to design slabs by assessing the bending moments at collapse by Johansen's yield-line[1] or Hillerborg's strip method.[2]

Generally shear stresses are low and usually found to be satisfactory when checked. Slab thicknesses, particularly for roof slabs, are often dictated by deflection considerations and sometimes slabs have to have a minimum practical thickness of preferably 125 mm.

With regard to the serviceability limit states of deflection and cracking, BS 8110 allows these to be dealt with as follows:

1. Deflections are controlled by limiting span to depth ratios. But if one wishes to make more accurate calculations it recommends a method based upon the elastic theory.

2. Crack widths are controlled by attention to detailing of the reinforcement. For example small bars closely spaced result in a greater number of cracks of smaller widths than larger bars spaced wider apart, which produce fewer wider cracks, and it is the maximum crack width which matters as regards appearance and allowing the entry of corrosive elements such as acidic rainwater. If the detailing recommendations of BS 8110 are altered in any way, and there is not normally any reason to do so, or if there is special cause for concern, then BS 8110 requires calculations to be made in a way it recommends of maximum crack widths.

4.2.1 General discussion of design of two-way spanning slabs[2a]

British Standard CP 114, 1957 gave bending moment coefficients for two-way spanning rectangular slabs with simply supported edges in its Table 16. These coefficients were determined from Grashof–Rankine formulae (developed independently by Grashof in Germany and Rankine in the U.K.), which were derived by equating the central deflections of two strips of slab, each of unit width, at right-angles to each other, and each bisecting the slab. Ref. 3 claimed that this method gives greater bending moments than exist. The problem of corners tending to lift was considered complex. The neglect of torsion at the corners was justification for overestimating bending moments. Ref. 3 also considered that test results justified the omission of corner reinforcement.

Table 17 of CP 114, 1957 gave bending moment coefficients for slabs restrained along all edges; with hinged (discontinuous) and fixed (continuous) edge conditions. These coefficients were obtained from U.S.A. regulations based on a mathematical analysis by Westergaard[4,5] and supported by test data. Some plastic redistribution of bending moments was assumed to occur to reduce the number of coefficients to a minimum to help designers. At corners where at least one of the two sides meeting was discontinuous, reinforcement was specified by CP 114.

CP 114, 1957 allowed an alternative method to the above to be used, namely a 'purely theoretical analysis' based on the elastic theory with Poisson's ratio = 0, provided the sections were designed elastically using a modular ratio = 15. The basis of the exact elastic theory of plates spanning in two directions was established by Lagrange and Navier in the nineteenth century, but most of the problems having practical importance have been solved in the past sixty-five years or so, when the names of Neuber, Bubnov, Timoshenko, Galerkin, Vlassov, Kalmanok and Girkmann have been inseparably associated with the fundamentals of the classical theory of plates. These analyses, prior to the availability to designers of computers and finite element and other methods suitable for the computer, were considered to be out of the question for designers of reinforced concrete slabs. Thus Ref. 3 recommends the use of Marcus's[6] method (proposed and used in Germany) which is similar to the Grashof–Rankine method but includes a simple correction to allow for restraint at corners and for assistance given by torsion. The results of Marcus's method were considered[3] to deviate by only 1–2% from a rigorous elastic analysis based on the elastic theory of plates. Marcus's method was also used by the German reinforced concrete regulations. Bending moments in continuous panels were determined by a method provided by Loser[7] based on Marcus's method. Ref. 3 gives design tables using Loser's method for various ratios of dead to live load. Prior to CP 114, 1957 Reynold's *Reinforced Concrete Designer's Handbook* of 1948 recommended the same as the above except Pigeaud's method instead of Westergaard's method. CP 114, 1957 allowed a further alternative method of design to the above two methods, namely Johansen's[8,9] yield-line method. A load factor of 1.8 was recommended but a restriction was placed on using the full concrete cube strength for extra safety. However Ref. 3 expressed worries about the 'upper bound' (see Section 4.4.3) nature of yield-line designs and the method's inability to give 'stress conditions away from the yield lines and hence information on how to distribute reinforcement'. Ref. 3, however, anticipated that the yield-line method might be popular for 'complex slab systems for which computations according to the elastic theory are impracticable'. Alongside the British practice just described, Westergaard[4,5] was a pioneer in elastic analysis of two-way reinforced concrete slabs in the U.S.A. In Ref. 5 Westergaard recommended moment coefficients that gave considerable weight to the non-elastic (plastic) readjustments in slab moments which take place before failure. In recognition of these favourable adjustments, his recommended coefficients were established at 28% below strictly elastic values. The A.C.I. Standard Building Code Requirements for Reinforced Concrete (ACI 318-63) 1963 recommended three alternative designs. Two fundamentally stemmed from Westergaard although work by Van Buren, Di Stassio[10] and Bertin[11] was also recognised, whilst the other was based on the work of Marcus.

It will be noticed that the yield-line method was permitted by the British code in 1957 but not by the U.S.A. code of 1963. It was not permitted by the A.C.I. Building Code of 1971 but permitted by the Code of 1977. In 1962 Ref. 12 was published in the English language. This gave formulae for calculating the ultimate bending moment at collapse for many differently-shaped slabs.

In 1964 Ref. 13 was published in the Czech language. A German and English edition was published in 1969 and an enlarged German/English edition was published in 1971. This work gave about 600 pages of tables of coefficients for bending moments, shear forces, and deflections at many points of square, rectangular and skew slabs with many combinations of restraint and free edge conditions, and of reactions at various points along the supports.

About 1960 many bridges were beginning to be designed for a large programme of motorways in the U.K. There was considerable demand for methods of designing two-way deck slabs of rectangular and sometimes skew shapes. Computers were not easily available, nor easy to use by most designers of reinforced concrete slabs. The Grashof–Rankine method was extended by some so that a slab analysis was considered as a grillage analogy. Then the grillage could be analysed by the method of Hendry and Jaeger,[14] or later the method of Bareš and Massonnet,[15] prior to computers and then eventually by computers and computer packages. Also a great interest in both research and design developed concerning Johansen's yield-line method. Subsequently, and to date, finite difference and finite element methods for elastic design have been developed considerably for use with computers. BS 8110:1985 allows 'elastic analysis' for bending moments (and shear forces) as in CP 114. Likewise it again allows the use of Johansen's yield-line method. It also allows Hillerborg's strip method to be used. However, it now restricts these methods with the proviso that 'the ratios between support and span moments are similar to those obtained by the use of elastic theory.' The code previous to BS 8110, namely CP 110, suggested simply choosing a value of this ratio between 1.0 and 1.5. Otherwise this recommendation requires some sort of elastic analysis to be made as well as the ultimate strength analysis. This mitigates against the advantage of ultimate collapse mechanism analysis previously quoted; that is its advantage when elastic analyses are complex and a computer program is not available.

CP 110 recommends the use of the coefficients in its Table 12 for two-way spanning slabs which are simply supported along their edges, and have inadequate torsional resistance at their corners to prevent them lifting. This table is derived from the Grashof–Rankine formulae previously mentioned. It recommends the use of the coefficients in its Table 13 for slabs which are rectangular and cast monolithically with their supports. These coefficients have been derived from yield-line analysis and calculated from values given by Taylor et al.[16]

Hillerborg's method for designing for ultimate strength was published in Sweden (in Swedish) in 1956 and 1959. It received much more attention after a critical analysis of the method and a comparison of it with tests were published by Wood and Armer[17,18,19] in 1968. They found (mathematically) that the 'strip' method did not suffer from the disadvantage of being 'upper bound' as did Johansen's method. The strip method gave the designer wide freedom of choice in his design approach. It is easier to curtail reinforcement than is the case with Johansen's method. Wood and Armer pointed out that a design using moments approaching those from elastic analysis was an efficient design and to be preferred. The suitability of Hillerborg's method for slabs with openings is a strong point in its favour.

The most difficult slabs for Hillerborg's strip method are those supported on columns. For such cases Hillerborg developed what Crawford[20] calls the advanced strip method, using a rectangular element (in lieu of a strip) carrying load in two directions to a support at one corner of the element. Wood and Armer report that they could not prove a mathematical basis for this type of element even though they devoted a considerable time to this investigation. For irregular shapes the advanced Hillerborg method also uses elements of triangular shape. An alternative to Hillerborg's advanced method is Kemp's[21] method, which is also much easier to understand, but very laborious for practical cases which of course always involve uniformly distributed loads due to self-weight.

The methods just mentioned, namely Hillerborg (strip and advanced) and Kemp can, particularly in the hands of an inexperienced designer, produce designs which are very unsatisfactory for limit state of deflection and cracking. The less the design departs from elastic theory the more efficient the design in these respects as mentioned by Wood and Armer (see previously). As regards the design of an individual strip with one or both ends fixed, the distribution of bending moments obtained from an elastic analysis can be altered to say increase or decrease the mid-span moment in accordance with the plastic theory but not making this alteration can still be considered as one possible plastic analysis. That is this one particular plastic analysis choice does not conflict with elastic analysis and thus helps to control stresses (thus cracks) and deflections at working loads. Thus the method of Fernando and Kemp[22] was developed to control the freedom of choice of Kemp's[21] method so as not to depart too greatly from the elastic method of matching up the deflections of an element in the two directions at right-angles; this is similar to the method of Grashof–Rankine for elastic theory, but more rigorous, complicated and difficult (requiring computer assistance) than Grashof–Rankine in that deflections of all elements are dealt with, whereas Grashof–Rankine dealt only with the central point. In some ways the Fernando and Kemp method is similar to using a beam-analogy method for a slab and solving as a grillage with a computer program but ignoring torsional resistance of the beams. In this case normal flexibility coefficients would be used. These are simpler to derive than the special flexibility coefficients needed for the Fernando and Kemp method and which deal with short loads instead of point loads.

Wilby[23] wrote computer programs (which are essential) for using the strip-deflection method,[22] for any size of rectangular slab with any type of support conditions, loading and any number of strips taken in each direction. These programs were used to produce many design tables.[24] As the equations given in Ref. 22 are mainly incorrect they are fully developed in Ref. 24.

4.2.2   Design tables for two-way slabs[26]

Various design tables which have been in use over approximately the past decade are listed below. Many of these tables are still in use. They are all based on the limit state of ultimate strength, except for those based on CP 114 and even these are modified because of ultimate strength considerations.

1. Taylor, S. R., Hayes, B. and Mohamedbhai, G. T. G.[16] The coefficients presented are derived from the yield-line theory and apply to the full width of the slab. It is recommended[16] that the loading used should be the design load of 1.4 times the dead load plus 1.6 times the live load. It is also suggested[16] that although in theory the full width of the slab should be used, in practice only a middle strip (three-quarters of the width) of the slab might be reinforced in accordance with the moments produced from these coefficients, and similarly for the length of the slab. In the derivation of the coefficients, yield lines have been extended to the corners of the slab and corner levers (see later) have been ignored.

2. BS 8110 and CP 110. Coefficients in Table 13 of CP 110 and Table 3.15 of BS 8110 are based on work done by Taylor, Hayes and Mohamedbhai[16] but have been modified to some extent. They give coefficients for the full width of the slab with a suggestion of reinforcing only a middle strip (see 1. previously). Similarly, BS 8110 and CP 110 define a middle strip of three-quarters of the full width of the slab and state that the steel area, obtained from the moments calculated from the moment coefficients, is used only to reinforce this middle strip. Edge strips are then reinforced by using the minimum area of steel given in clause 3.11.4.1 of CP 110 and clause 3.12.5 of BS 8110.

3. Thakkar, M. C. and Rao, J. K. S.[26] In this method the average moment distribution per unit width of the slab is derived for uniform orthotropic reinforcement throughout the whole width of the slab. That is, the slab is analysed by Hillerborg's strip method and then the average of moments for all strips along each edge is taken and this value is the moment per metre width quoted by the tables.

4. CP 114. These tables have been obtained from a theoretical elastic analysis and adjusted in the light of experimental data. This code separates each direction of the slab into a middle strip, of width three-quarters the width of the slab, and edge strips one-eighth of the width of the slab. Where slabs have aspect ratios greater than four the middle strip in the short direction could be taken to have a width of $l_y - l_x$ and each edge strip a width of $l_x/2$, where l_x and l_y are the short and long spans of the slab, respectively. The coefficients given in the table are used for the middle strip of the slab only.

5. Wilby. C. B. Wilby has produced tables[24,27] with his computer program for the strip-deflection[22] method for eight strips in each of two mutually perpendicular directions, namely length and width. To obtain

coefficients for comparison with the previously mentioned tables, the mean values for the full widths of the slabs of bending moment per unit length have been taken. As the methods compare reasonably well these tables may be used in lieu of those in CP 110 and BS 8110 and they consider cases not considered by CP 110 and BS 8110. Also they give deflections.

4.3 Flat slabs

These are slabs without beams supported only by columns. Flared column heads usefully reduce the high shear stresses in the slabs around the column heads. Flat slabs generally give a heavier construction than beam and slab systems; they require more concrete and steel but the shuttering is much less expensive. For longer spans of flat slabs *dropped panels* are sometimes used to make the construction lighter in weight. This usually means dropping the soffits of rectangular portions of slab around the column heads. Flat slabs are described further in Section 7.3 which also describes 'waffle' slabs.

Design has been based on empirical formulae which are limited to systems with rectangular panels, length-to-width not exceeding 4/3, with at least three continuous spans in both directions. Such formulae are given in CP 110 and BS 8110 and are simple to use. The alternative method allowed by CP 110 and BS 8110 is more arduous but is useful when the empirical formulae do not apply. It consists of dividing the structure longitudinally and transversely into frames consisting of columns and the connecting strips of the slabs, and then elastically analysing these frames for bending moments and shear forces. This is well enunciated in CP 110 and BS 8110. More recently they might be designed using Johansen's yield-line[1] or Hillerborg's strip method.[2]

4.4 Yield-line theory of slab analysis

Having read the previous Sections 4.2 to 4.3 the reader might still ask, 'Why use the yield-line instead of the elastic theory?' Even though the basic equations of the theory of elasticity are simple enough, it can be extremely difficult to solve these equations for complex structural formations. This difficulty can also apply to finite element methods. Also the yield-line theory gives a more realistic representation of the behaviour of slabs at ultimate limit states than the elastic theory. There are more and more finite element programs available for the elastic analysis of slabs.

142 Reinforced concrete slabs

Many of these however require a work input to apply to complex shapes with various supports.

Generally slabs in practice are under-reinforced because of considerations of deflection and minimum thickness. The yield-line theory applies to slabs which are under-reinforced. There are two different methods of yield-line analysis, perhaps most simply introduced by the following two examples.

Example 4.1. A square *isotropically* reinforced slab (this means the slab is reinforced identically in orthogonal directions, which means that its ultimate resisting moment is the same in these two directions and along any line in any other direction – see proof in Section 4.4.1) is simply supported along all of its sides. Determine by the *equilibrium method of analysis* the ultimate resisting moment m per unit length of yield line balancing an ultimate uniformly distributed load q kN/m² (this includes the self-weight of the slab).

It is easy to imagine that the slab will essentially fail by the diagonals of Figure 4.1(a) becoming yield lines. That is, cracks will occur along these lines in the soffit of the slab and they will open as the tensile steel yields. Steel can maintain its yield stress as the steel yields, so the section rotates for no increase in moment, but eventually the rotation becomes excessive (extreme fibre strain $\simeq 0.0035$) and the concrete compression zone disintegrates. As the rotation at the centre of each yield line becomes considerable, but not excessive, due to yielding of the steel there, the rotation near the corners of each yield line eventually becomes sufficient for the steel to have yielded there also. Failure is precipitated therefore when each unit of length of each yield line has reached its ultimate bending

Figure 4.1

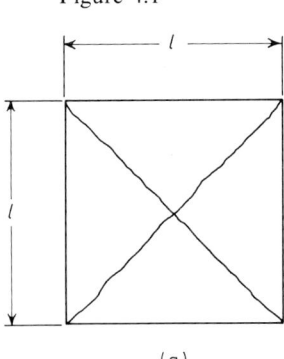

(a)

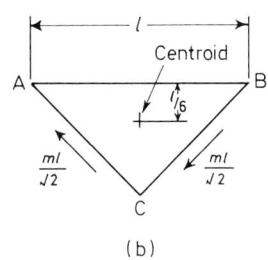

(b)

Yield-line theory of slab analysis

strength. Generally the rotation at the centre of a yield line will not have been sufficient to cause failure there before the ultimate bending moments near the ends of the yield line have been realised. Thus m is constant along each yield line and is the same for each because of symmetry. Considering the equilibrium of the moments of any one of the identical slab segments about the support (see Figure 4.1(b)) the total bending moment along CA is $ml/\sqrt{2}$, and is shown by a vector such that the moment acts in a clockwise direction when viewed along the vector arrow. This vector has a component parallel to BA of $(1/\sqrt{2})(ml/\sqrt{2})$. Similarly for BC. Hence taking moments about AB

$$(ql^2/4)(l/6) = 2(l/\sqrt{2})(ml/\sqrt{2})$$

$$\therefore m = ql^2/24 \text{ kN m/m}$$

Example 4.2. The rectangular isotropically reinforced slab shown in Figure 4.2 is simply supported along all of its sides. Determine by *virtual-work analysis* the ultimate resisting moment m per unit length of yield line balancing a total ultimate uniformly distributed load of q kN/m².

It is easy to imagine that the slab will essentially fail by the yield lines shown in Figure 4.2(a). The distance l_1 is unknown but will be such as to maximise the ultimate resistance moment required to balance the ultimate loading. A simple procedure, which lends itself to solution by computer, is by trial and error. In Figure 4.2(a) angle ACD is 90°, AC = $\sqrt{(1.5^2+l_1^2)}$. Triangles CBA and DBC are similar, thus CD:CA = CB:AB, and triangles ACE and ACB are also similar, thus EC:AB = AC:CB. That is

$$l_2 = (1.5/l_1)\sqrt{(2.25+l_1^2)} \text{ and } l_3 = (l_1/1.5)\sqrt{(2.25+l_1^2)}$$

Figure 4.2

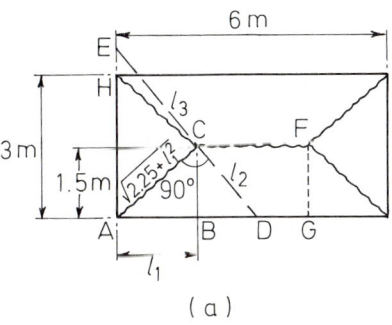

(a)

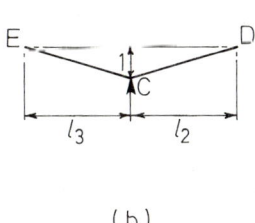

(b)

Considering AC, from Figure 4.2(b), the total angle of rotation at this yield line for a small unit increase of deflection at C in radians is

$$\frac{1}{l_3}+\frac{1}{l_2}=\frac{1.5}{l_1\sqrt{(2.25+l_1^2)}}+\frac{l_1}{1.5\sqrt{(2.25+l_1^2)}}=\frac{1}{\sqrt{(2.25+l_1^2)}}\left(\frac{1.5}{l_1}+\frac{l_1}{1.5}\right)$$

Similarly for this unit deflection at C the total rotation of the yield line CF is $1/1.5+1/1.5 = 1.333$. For our first trial let $l_1 = 2.1$ m. Then AC $= \sqrt{(2.25+4.41)} = 2.58$ m. The rotation at AC $= (1/2.58)(1.5/2.1+2.1/1.5) = 0.8195$.

The internal work done (bending moment × angular rotation) as the unit incremental deflection occurs at yield is

$$m \times 2.58 \times 0.8195 \times 4 + m(6-2\times 2.1) \times 1.333 = 10.86m$$

The external work done (making use of symmetry) whilst this incremental deflection occurs is

$$2(\text{Load on AHC}) \times \tfrac{1}{3} + 2(\text{Load on CFGB}) \times \tfrac{1}{2}$$
$$+ 4(\text{Load on ABC}) \times \tfrac{1}{3}$$

$$= 0.667 \times (1.5 \times 2.1q) + (6-2\times 2.1) \times 1.5q$$
$$+ 1.333(0.75 \times 2.1q)$$

$$= 6.9q$$

Equating internal and external works done $m = (6.9/10.86)q = 0.6354q$.

Trying other values for l_1 and summarising for values of l_1 of 1.8, 1.95, 2.1 and 2.25 the corresponding values of m/q are 0.635, 0.637, 0.635 and 0.632, respectively. For a given q the maximum $m = 0.637q$ kN m/m corresponding to the yield pattern when $l_1 = 1.95$ m.

4.4.1 Reinforcement

If a slab is not isotropically reinforced (see Example 4.1), its ultimate strengths are different in two perpendicular directions and it is *orthogonally anisotropically* or simply *orthotropically* reinforced. When isotropically reinforced (see Example 4.1), its ultimate resistance moment is the same in any direction. This will now be proved. As the lever arm is assumed constant, for the bending moment per unit length to be constant in any direction it is only necessary to prove that the force provided by the tensile reinforcement per unit length is constant in any direction. Referring to Figure 4.3 the reinforcement has the same spacing s in each rectilinear direction and the force in each bar is N_s. Considering the line CD the component of N_s at A perpendicular to this line is $N_s \cos \alpha$. Also AB = $s/\cos \alpha$. Thus the force per unit length of CD and perpendicular to CD due to the bars in the direction AE is $(N_s \cos^2 \alpha)/s$. The component of N_s at D perpendicular to CD is $N_s \sin \alpha$. Also CD = $s/\sin \alpha$. Thus the force per unit length of CD and perpendicular to CD due to the bars in the direction DF is $(N_s \sin^2 \alpha)/s$. Thus the total force per unit length of CD and perpendicular to CD is $(\cos^2 \alpha + \sin^2 \alpha) N_s/s = N_s/s$ which is the same as the force per unit length in either of the two rectilinear directions of the reinforcement, Q.E.D.

Slabs that are orthotropically reinforced can be dealt with by altering the dimensions for design purposes.[28]

In the above example the sections are assumed to be under-reinforced – this is normally the case for slabs, because of deflection and minimum thickness requirements. The analyses are dependent upon all yield lines being able to develop fully before say the initial portion loses its moment-carrying capacity due to excessive rotation – the extreme fibre strain reaching about 0.0035. The reinforcement per unit length can be different in each of two rectilinear directions, but must be constant along any line, otherwise in the above examples m would not be constant along each line. The analysis is most convenient for slabs of difficult shapes and slabs with holes or openings, where an elastic analysis is difficult. R. H. Wood[2] says that a slab designed elastically, stopping off all bars whenever one could, would generally be more economic than if it were designed by yield line (or strip method), and was to be preferred as a design. The reinforcement

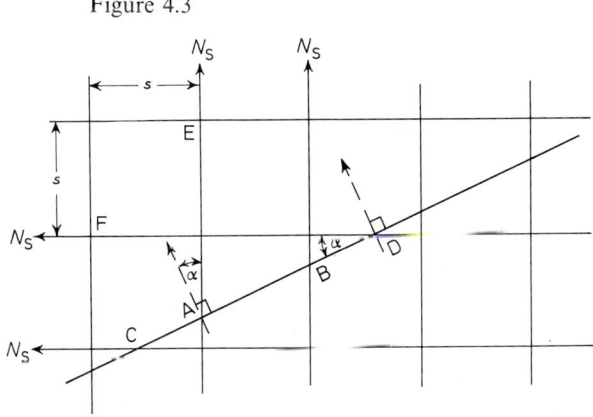

Figure 4.3

146 *Reinforced concrete slabs*

would not however be so simple a system. The yield-line analysis offers no information on the best distribution of the steel, but can be used to analyse a slab where the steel has been distributed according to some other method (for example Hillerborg's). Curtailing bars means that yield lines have to be considered at sections where the bars are discontinued.

If cracks need to open considerably the bars across the crack tend to kink to endeavour to be at right-angles to the cracks; this gives a slightly stronger resistance moment (up to about 14%). Hence the designer can ignore kinking and his design will be slightly safer.

Membrane action can help the strength of a slab when the deflections are large towards failure. It is reasonable for the designer to ignore it and have slightly extra safety.

4.4.2 *Further points on yield-line analyses*

Both of the previous methods of analysis give what is termed 'upper-bound' solutions (see Section 4.4.3) in that it might always be possible to think of some other pattern of yield lines which might require a greater m to balance a given loading.

Essentially there is not much difficulty in choosing various possible sensible yield-line patterns and thus in practice there is no great need to worry about this upper-bound problem. One chooses from experimental experiences of failure or from imagining how failures might occur. For example, the tank wall (Refs. 30 to 36) shown in Figure 4.4 may fail by either of the yield-line mechanisms shown, and to design the wall both have to be investigated.

Yield lines are generally straight, lie along lines of encastré supports, pass over columns, and pass through the intersection of rotating adjacent slab elements. Strictly speaking when a yield line meets an unsupported

Figure 4.4

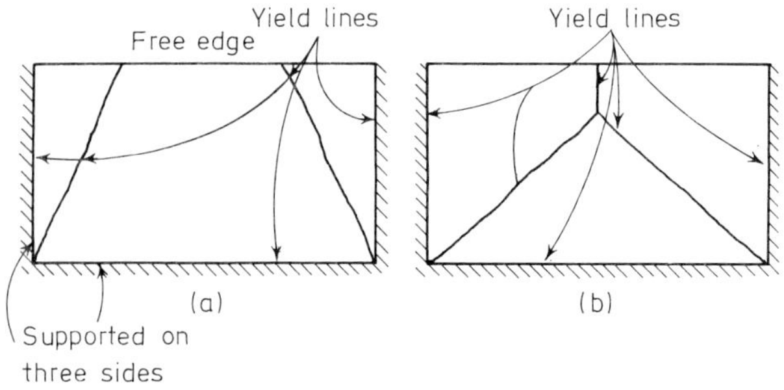

edge it must do so perpendicularly, as yield-line moments are maximum moments. However, if the yield line away from this edge is skew and straight, it is usually continued in a straight line to the edge.[28] This it is said usually makes negligible error in the calculations.

In the previous examples an alternative possibility is for the yield-line pattern to be as shown in Figure 4.5. If the corners are not held down each corner element such as ABCD will rotate about AC lifting B from the support. If the corners are held down, at each corner, lines such as AC will become yield lines. In this case reinforcement is required perpendicularly to lines such as AC in the top of such a slab. The slab spans between AD and CD, and AB and BC, and sometimes supplementary reinforcement is desired to take care of this, the bars being parallel to the direction AC and in the bottom of the slab. The yield lines at corners are called 'corner levers'. Although their effect is adverse they are often neglected in yield-line analyses for simplicity. For right-angled corners this causes an error of about 9% with bottom steel only and much less when there is top as well as bottom steel. The error is particularly high[37] for acute angles with free edges, about 26% with bottom steel only and 14% when there is top and bottom steel.

For non-rectangular shapes Ref. 12 is useful.

4.4.3 'Upper-bound' and 'lower-bound' solutions

Most engineering analyses or designs make approximations which cause them to give conservative solutions. In yield-line terminology these are called 'lower-bound' solutions. The yield-line theory as developed to date gives only 'upper-bound' solutions which are dangerous.

It was of course a worry, which delayed the acceptance of yield-line

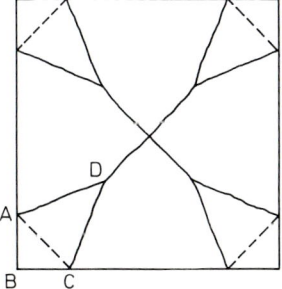

Figure 4.5

148   Reinforced concrete slabs

analysis for design, that one might consider a reasonably sensible yield-line pattern for failure and then the solution should turn out to be upper bound and one might sometimes wonder with a complicated slab how many crack patterns one might need to consider to obtain a solution at least felt to be insignificantly upper bound.

The critical pattern was easy to obtain in Example 4.2 by varying one parameter, namely l_1, to obtain a maximum value of m/q. Had one of the short sides been fixed (or encastré) then there would have been a second similar variable locating the distance of F from one of the shorter sides. Had two adjacent sides been fixed then using a similar crack pattern there would have been four variables, namely the co-ordinates locating C and F. If one programmed the analysis on say a microcomputer then one could make many trials of various combinations of guessed values of the four co-ordinates and this would be a speedy and easy way of obtaining a solution, which could be difficult without the computer.

In Example 4.2 the maximised or critical yield-line pattern with $l_1 =$ 1.95 m still gives an upper-bound solution because other patterns, often for example with a greater number of yield lines, will give solutions which are less upper bound. (NB The trials with l_1 not equal to 1.95 m do not satisfy equilibrium and are therefore invalid.)

4.4.4   Further consideration of the 'equilibrium method'

This has been introduced by giving Example 4.1 for a simple symmetrical case. For a less symmetrical case, suppose a region of a slab enclosed by yield lines is as shown in Figure 4.6. Each arrow along each edge indicates vectorially the bending moment for the yield line. Each arrow perpendicular to each edge and in the plane of this region of the slab

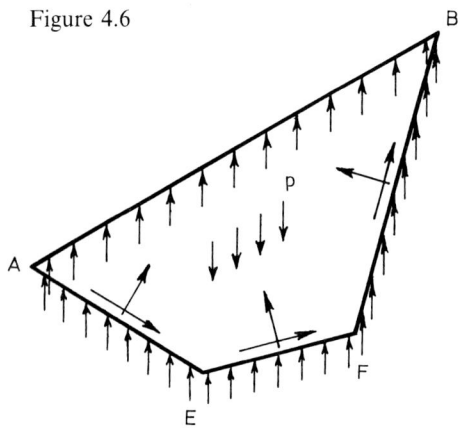

Figure 4.6

indicates vectorially the twisting moment for this yield line. Along each edge there will be a resultant shear force normal to the plane of the slab.

Johansen considered that for each edge the total twisting moment and the total shear force could be replaced by two forces (normal to the plane of the slab), one at each end of this edge. These forces are referred to as 'nodal forces'. For equilibrium it can be shown[5] that:

1. If several yield lines converge to a point called a 'node' then all the nodal forces at this node must vectorially add up to zero.

2. When three yield lines meet at a point (that is a node) and the reinforcement is isotropic (that is identical in orthogonal directions) the nodal force for each yield line at this point is zero.

3. When a yield line intersects a free edge (see Figure 4.7) and the reinforcement is orthotropic and the m-moment key lines are as shown, then the nodal (or knot or edge) forces are as shown in Figure 4.7 and are given by

$$K_{12} = +\mu m \cot \psi \tag{4.1}$$

$$\text{and } K_{31} = -\mu m \cot \psi \tag{4.2}$$

where a 'moment key line' is a line giving vectorially the ultimate moment of resistance of the slab (for the relevant reinforcement) and where $\psi < 90°$.

4. When a yield line intersects a free edge and the reinforcement is isotropic (that is $\mu = 1$ in case 3, previously) then,

$$K_{12} = +m \cot \psi \tag{4.3}$$

$$\text{and } K_{31} = -m \cot \psi \tag{4.4}$$

where $\psi < 90°$.

Figure 4.7

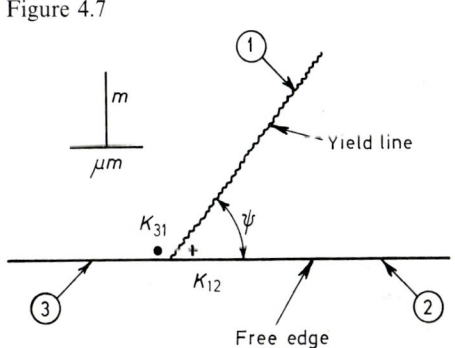

5. In cases 3 and 4 above when $\psi = 90°$ the nodal forces K_{12} and K_{31} will be zero.

The next example shows how to use the equilibrium method for a more complicated analysis than that of Example 4.1 in that nodal forces have to be considered.

Example 4.3. Determine the ultimate moment of resistance m per unit length of yield line balancing a total uniformly distributed loading of 5 kN/m² for the isotropically reinforced slab shown in Figure 4.8. The shading indicates that side AB is not supported, sides AD and BC are fixed and side DC is simply supported. The slab is under-reinforced and, assuming Figure 4.8 shows it in plan, the bottom reinforcement is such that the ultimate resisting moment in any direction (see Section 4.4.1) is m and the top reinforcement, at the supports, is such that the ultimate resisting moment is αm in any direction. These moments are shown vectorially in Figure 4.8. This example will consider a seemingly possible yield-line layout as shown in Figure 4.4(a).

The nodal force acting at E for the assumed rigid region AED is (from equation 4.3 and see Figure 4.9)

$$m \cot \psi = m \frac{x}{4} \tag{4.5}$$

The moment for the vector along ED is $m(DE)$ and resolving this into components in the directions EA and AD, respectively, the component in the direction AD is

$$m(DE) \sin \psi = m \cdot DE \cdot \frac{AD}{DE} = 4m \tag{4.6}$$

Figure 4.8

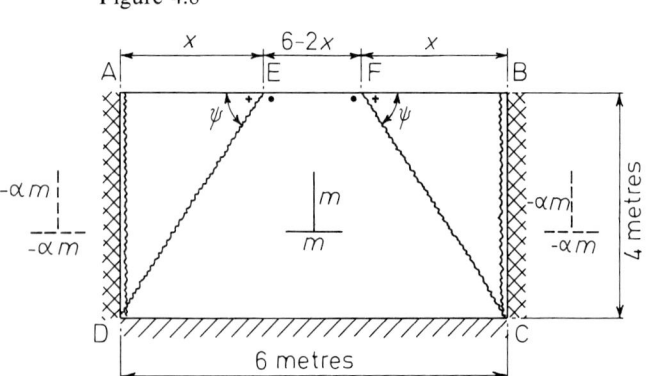

Yield-line theory of slab analysis

Taking moments about AD in Figure 4.9 (using equations 4.5 and 4.6)

$$4m + 4\alpha m - m\frac{x}{4}x - \frac{4x}{2}\cdot\frac{x}{3}\cdot 5 = 0$$

$$\therefore 4m\left(1+\alpha-\frac{x^2}{16}\right) = \frac{10x^2}{3}$$

$$\therefore m = \frac{40}{3}\cdot\frac{x^2}{(16+16\alpha-x^2)} \qquad (4.7)$$

Region FBC is similar to region ADE and will give the same equation.

The nodal forces at E and F for the region EFCD shown in Figure 4.10 are from equation 4.4; each (as in equation 4.5) equals

$$-m\frac{x}{4} \qquad (4.8)$$

The moment for the vector along DE is $m(\mathrm{DE})$ and resolving this into components in the directions DC and at right-angles to DC, respectively, the component in the direction DC is

$$m(\mathrm{DE})\cos\psi = m\cdot\mathrm{DE}\cdot\frac{x}{\mathrm{DE}} = mx \qquad (4.9)$$

For region EFCD, moments about CD (see Figure 4.10) give

$$2mx + 2\cdot\frac{mx}{4}\cdot 4 - [(6-2x)\times 4 \times 2 + 2\times\tfrac{1}{2}\times 4\times x\times x\times\tfrac{4}{3}]\times 5 = 0$$

$$\therefore m = \frac{20}{3}\cdot\left(\frac{9-2x}{x}\right) \qquad (4.10)$$

Figure 4.9

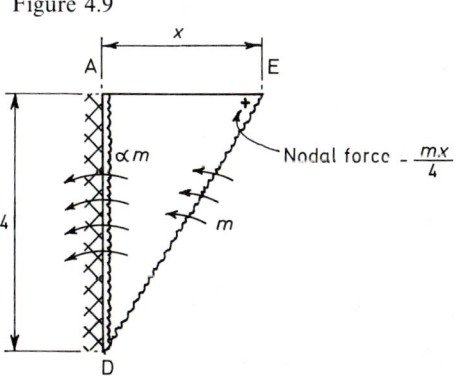

152   Reinforced concrete slabs

Equating the values of m in equations 4.7 and 4.10 gives

$$9x^2 + 32(1-\alpha)x - 144(1+\alpha) = 0 \tag{4.11}$$

Exactly the same equation can be obtained by using the virtual-work analysis (see Example 4.4), but not as simply and directly as a differentiation is involved to obtain the maximum value of m.

Equation 4.11 only applies to the yield-line pattern considered and which can only exist if $x \leqslant 3$, that is half the length of AB. From equation 4.11

$$x = \frac{-32(1+\alpha) + \sqrt{[32^2(1+\alpha)^2 + 36 \times 144(1+\alpha)]}}{18} \tag{4.12}$$

Now α is positive and it can be seen that in equation 4.12 the square root is of a larger amount than $32^2(1+\alpha)^2$ and so it will give x as positive which of course it is. It can also be seen that increasing the value of α increases the value of x. Therefore, the greatest value of α for this yield-line layout is when x is a maximum for it, namely 3. Putting $x = 3$ in equation 4.11, gives $\alpha = 0.687$. If $\alpha \leqslant 0.687$ then the yield-line pattern just considered can be used. If, however, $\alpha > 0.687$ then we are outside the range of this pattern and we shall need to consider a yield pattern the same as shown in Figure 4.9.

The above has been treated algebraically. For computer use it is better generally to analyse one yield pattern at a time and repeat the calculation by altering a relevant variable. In this case α would be chosen, a value of x guessed, and then m calculated from equations 4.7 and 4.10. The difference in these values is obtained. Then other values of x are chosen until the difference just mentioned is considered to be negligible.

Figure 4.10

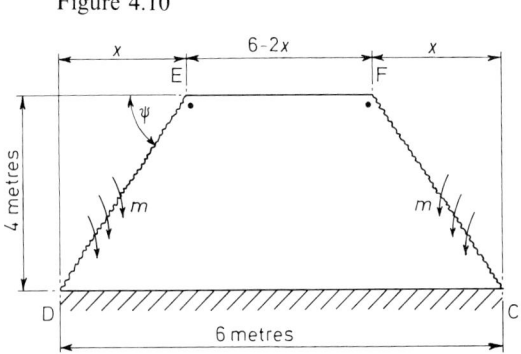

4.4.5 Further consideration of the virtual-work method

The problem of Example 4.3 will now be solved using the virtual-work method in the following example.

Example 4.4. Consider unit displacement, normal to the plane ABCD in Figure 4.8, at E and F. The expenditure of energy by applied loading is as follows:

1. Portion AED (same as BFC by symmetry)

$$(\tfrac{1}{2} \times 4 \times x) \times 5 \times \tfrac{1}{3} = \tfrac{20}{6}x$$

2. Portion EFCD

$$2 \times (\tfrac{1}{2} \times x \times 4) \times 5 \times \tfrac{1}{3} + \{(6-2x) \times 4\} \times 5 \times \tfrac{1}{2} = 60 - \tfrac{40}{3}x$$

The internal energy held in yield lines is as follows:

(a) Yield line AD (same as BC by symmetry)
angular rotation for unit displacement at $E = 1/x$ radians
moment $= 4\alpha m$
energy $= 4\alpha m/x$

(b) Yield line ED (same as FC by symmetry)
For convenience the moment vector along ED can be obtained by vectorially adding its components in directions parallel to DC and AD, respectively.
For component in former direction:
angular rotation for unit displacement at E relative to DC $= 1/4$
moment $= mx$
energy $= mx/4$
For component in latter direction:
angular rotation for unit displacement at E relative to AD $= 1/x$
moment $= 4m$
energy $= 4m/x$

The work equation is now obtained by equating (1) and (2) to (a) and (b) thus

$$2 \times \tfrac{20}{6}x + 60 - \tfrac{40}{3}x = 2 \times \left\{ \frac{4\alpha m}{x} + \frac{mx}{4} + \frac{4m}{x} \right\}$$

$$\therefore m = \frac{40}{3} \left\{ \frac{9x - x^2}{16(1+\alpha) + x^2} \right\} \tag{4.13}$$

154   Reinforced concrete slabs

For m to be a maximum $dm/dx = 0$, that is

$$(9-2x)\{16(1+\alpha)+x^2\} = (9x-x^2)2x$$

$$\therefore 9x^2 + 32(1+\alpha)x - 144(1+\alpha) = 0 \qquad (4.14)$$

This is the same as equation 4.11 enabling the reader to compare the analyses of Examples 4.3 and 4.4. This example would continue as Example 4.3, except that the last paragraph of Example 4.3 would not apply because the differentiation required to obtained equation 4.14 is effected from an algebraic equation.

4.4.6   Combination of equilibrium and virtual-work methods

These methods can be combined to speed design. Examples 4.3 and 4.4 solve the same problem by both methods. If to effect these solutions values of x are guessed each time an evaluation is made using, say, either a programmable hand calculator or desktop microcomputer, it will be found that if the value of x is a certain small amount different to its value corresponding to the critical value of m, then the value of m obtained using the virtual-work method will be very much nearer indeed to its critical value than that obtained using the equilibrium method. This illustrates that the latter is very much more sensitive to yield-line layout than the former method. Hence, a yield-line analysis can be effected relatively simply by combining the two methods as follows:

Step 1. Assume a suitable yield-line layout and use the equilibrium method to obtain moments in each of the rigid regions.

Step 2. If the moments obtained for the various rigid regions are such that the difference between the maximum and minimum is within about 50% of the minimum moment, then apply the work method to this layout and the moment thus obtained may be considered sufficiently accurate for design purposes, and no further calculations need be done.

Step 2A. If the difference between the maximum and minimum moments in Step 2 is more than about 50% of the minimum moment, then assume a second trial layout and apply the equilibrium method. Repeat this procedure if necessary until the difference is within 50% of the minimum moment and then proceed as in Step 2.

In most cases the above procedure gives sufficiently accurate results for design purposes with minimum effort. The examples that will be given now will use this procedure to obtain the design moments.

Example 4.5. A square isotropically reinforced slab shown in Figure 4.11 carries an ultimate uniformly distributed load of p/unit area. Determine

Yield-line theory of slab analysis

the corresponding ultimate moment of resistance per unit length, m, of yield line.

A possible yield line pattern is shown in Figure 4.11.

First trial

Let us assume that $x = 0.8l$. From geometry

$$AE = CG = \frac{l}{(2l-0.8l)}(0.8l) = \frac{2}{3}l$$

$$EB = BG = AB - AE = \frac{l}{3}$$

angle BEF = angle $BGF = \psi$ = angle AEJ

$$= \cot^{-1}\left(\frac{AE}{JA}\right) = \cot^{-1}\left(\frac{2}{3}\right)$$

$$\therefore \cot \psi = \tfrac{2}{3}$$

$\therefore$ The nodal force at E in the rigid section (c) (see Section 4.4.4)

$$= +m \cot \psi$$
$$= +\tfrac{2}{3}m$$

Nodal force at E in the rigid region (a) $= -\tfrac{2}{3}m$

Figure 4.11

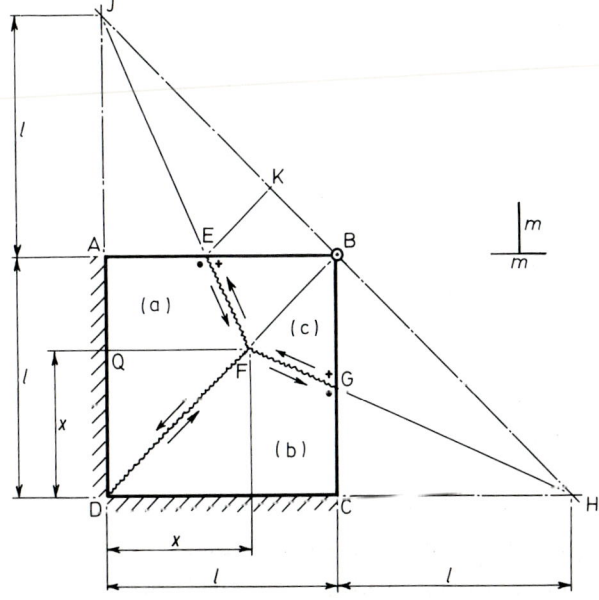

156 Reinforced concrete slabs

The nodal forces at F in each of the three regions are zero from symmetry (see Section 4.4.4). The equilibrium equations for the rigid regions are:

Region (a) or (b). Taking moments about AD

$$ml + \tfrac{2}{3}m \cdot AE - p \cdot [\tfrac{1}{2}(JD \cdot 0.8l) \cdot \tfrac{1}{3} \cdot 0.8l - \tfrac{1}{2} \cdot (JA \cdot AE) \cdot \tfrac{1}{3} \cdot AE] = 0$$

$$\therefore m = 0.0965 pl^2$$

Region (c). Referring to Figure 4.12, vector moments in the directions GF and FE added together give a vector moment which can be resolved into vector moments in the directions GB and BE added together. Taking moments about EB,

$$m\frac{l}{3} - \frac{2}{3}m \cdot BG - p \cdot \left[(\text{area RBNF}) \cdot \frac{l}{10} + (\text{area FNG}) \cdot \left(\frac{1}{5} + \frac{1}{3} \cdot \frac{2}{15}\right) \cdot l\right.$$

$$\left. + (\text{area ERF}) \cdot \frac{1}{3} \cdot \frac{l}{5}\right]$$

$$= 0$$

$$\therefore \frac{ml}{9} = p \cdot \left[\frac{l}{5} \cdot \frac{l}{5} \cdot \frac{l}{10} + \frac{1}{2} \cdot \frac{l}{5} \cdot \frac{2}{15} \cdot \frac{11}{45} l + \frac{1}{2} \cdot \frac{l}{5} \cdot \frac{2l}{15} \cdot \frac{l}{15}\right]$$

$$\therefore m = 0.0745 pl^2$$

Figure 4.12

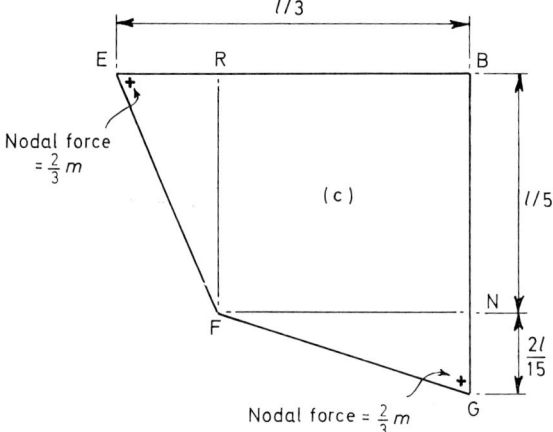

Yield-line theory of slab analysis

The difference in yield-line moments obtained from regions (a), or (b), and (c) is $(0.0965 - 0.0745)pl^2 = 0.022pl^2$ which is about 30% of the lesser value, $0.0745pl^2$. Therefore, referring to Step 2, as this is less than 50%, we can apply the work equation to this layout.

For Figure 4.11, taking the vertical deflection of F as unity

$$\text{Slope of (a) normal to AD} = \frac{1}{QF} = \frac{5}{4l}$$

$$\text{Slope of (c) normal to EG} = \frac{1}{BF} = \frac{1}{BD - FD} = \frac{5}{l\sqrt{2}}$$

The arrows show the directions of the yield-line moments vectorially.

Portion (a) rotates about an axis AD but not at all about an axis at right-angles to AD (for example in the direction AE). The moment vectors EF and FD give a resultant ED which can be resolved into EA and AD, in other words one can travel from E to D via F or A. The vector component AE does not rotate but the vector AD rotates by the slope of (a) just given. Therefore, the energy absorbed at the yield lines for portion (a) (or portion (b) from symmetry)

$$= (m \cdot AD) \times \frac{5}{4l} = \frac{5}{4}m$$

Portion (c) rotates about an axis EG (by the slope of (c) given previously) but not at all about an axis at right-angles to EG. The moment vectors GF and FE give a resultant GE. Therefore, the energy absorbed at the yield lines for portion (c)

$$= (m \cdot GE) \times \frac{5}{l\sqrt{2}} = m \cdot BE \cdot \sqrt{2} \times \frac{5}{l\sqrt{2}} = \frac{5}{3}m$$

Therefore, the total energy absorbed at yield lines

$$= 2 \times \tfrac{5}{4}m + \tfrac{5}{3}m = \tfrac{25}{6}m$$

The deflection at E

$$= \frac{AE}{QF} = \frac{\tfrac{2}{3}l}{0.8l} = \frac{5}{6}$$

The work done by the loading for region (a) (same as region (b) by symmetry)

$$= p \cdot \left[(\text{area JFD}) \times \frac{1}{3} - (\text{area JEA}) \times \frac{1}{3} \times \frac{5}{6}\right]$$

$$= p \cdot \left[\frac{1}{2} \cdot 2l \cdot \frac{0.8l}{3} - \frac{1}{2} \cdot l \cdot \frac{2}{3} l \times \frac{5}{18}\right] = \frac{47}{270} pl^2$$

Now $EK = EB/\sqrt{2} = l/(3\sqrt{2})$, $BF = 0.2l\sqrt{2}$ and $JB = l\sqrt{2}$
The work done by the loading for region (c)

$$= 2 \cdot p \cdot \left[(\text{area JBF}) \times \frac{1}{3} - (\text{area JBE}) \times \frac{1}{3} \times \frac{5}{6}\right]$$

$$= 2 \cdot p \cdot \left[\frac{1}{2} \cdot l\sqrt{2} \cdot \frac{0.2l\sqrt{2}}{3} - \frac{1}{2} \cdot l\sqrt{2} \cdot \frac{l}{3\sqrt{2}} \cdot \frac{5}{18}\right] = \frac{11}{270} pl^2$$

Equating total work done to total energy absorbed at yield lines

$$pl^2[2 \times \tfrac{47}{270} + \tfrac{11}{270}] = \tfrac{25}{6} m$$

$$\therefore m = 0.09333 pl^2$$

Example 4.6. The reinforced concrete slab shown in Figure 4.13 carries an ultimate distributed load of 6 kN/m². Determine the corresponding ultimate moment of resistance per unit length, m, of yield line.

A possible yield-line pattern is shown in Figure 4.13.

Figure 4.13

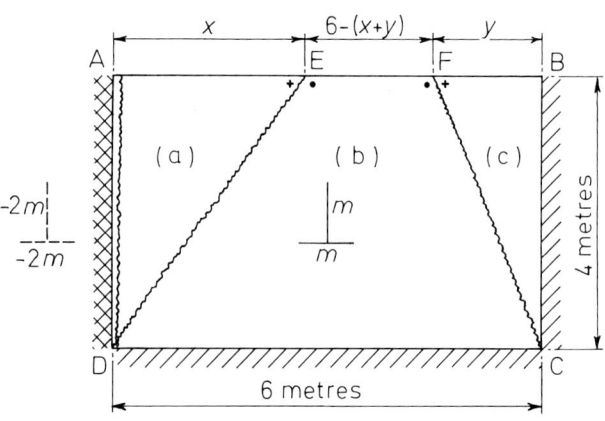

First trial

> Guess/estimate $x = 4$ and $y = 1$
>
> Nodal force at E (see Section 4.4.4) $= \dfrac{mx}{4} = m$
>
> Nodal force at F $= \dfrac{my}{4} = \dfrac{m}{4}$

For region (a), taking moments about AD:

> for yield line AD, $2m \times 4 = 8m$
> for yield line DE, vector ED can be resolved into vectors EA and AD, $m \cdot AD = 4m$
> for nodal force at E, $-m \cdot x = -4m$
> for loading, $-6 \times \tfrac{4}{2} \times 4 \times \tfrac{4}{3} = -64$
> $\therefore 8m + 4m - 4m - 64 = 0$,
> $\therefore m = 8$

For region (c), taking moments about BC:

> for yield line CF, vector CF can be resolved into vectors BF and CB, $m \cdot CB = 4m$
>
> for nodal force at F, $-\dfrac{m}{4} \cdot y = -\dfrac{m}{4}$
>
> for loading, $-6 \times \tfrac{1}{2} \times 4 \times \tfrac{1}{3} = -4$
>
> $\therefore 4m - \dfrac{m}{4} - 4 = 0$
>
> $\therefore m = \tfrac{16}{15}$

For region (b), taking moments about DC:

> for yield line ED, $mx = 4m$
> for yield line FC, $my = m$
>
> for nodal forces at E and F, $\left(m + \dfrac{m}{4}\right) \cdot 4 = 5m$
>
> for loading, $-6 \times [\tfrac{1}{2} \times 4 \times 4 \times \tfrac{4}{3} + 1 \times 4 \times \tfrac{4}{3} + \tfrac{1}{2} \times 4 \times 1 \times \tfrac{4}{3}] = 128$
> $\therefore 4m + m + 5m = 128$
> $\therefore m = 12.8$

These results indicate that regions (a) and (b) have been chosen too large and region (c) too small. A reduction in area (a) increases area (b) whilst an increase in area (c) reduces area (b). Because of the large difference in the value of m for regions (b) and (c) it is highly unlikely that the yield-line pattern shown in Figure 4.13 could give the same value of m for the three regions. Hence the alternative pattern shown in Figure 4.14 will now be considered. Let us guess/estimate $x = 4$ and $y = 1$. The nodal forces at points F and E will be zero (see rules 2 and 3 in Section 4.4.4). The equilibrium equations for the regions (a), (b) and (c) are:

For region (a), taking moments about AD:

for yield line AD, $(2m) \cdot AD = 8m$

for yield line EFD, vectors EF and FD add up to a resultant ED which can be resolved into vectors EA and AD, $m \cdot AD = 4m$

for loading, $-6 \cdot \left[x \cdot y \cdot \dfrac{x}{2} + \dfrac{x}{2} \cdot (4-y) \cdot \dfrac{x}{3} \right] = -96$

$\therefore 8m + 4m - 96 = 0$
$\therefore m = 8$

For region (b), taking moments about DC:

for yield line DFC, vectors DF and FC have a resultant vector DC, $m \cdot DC = 6m$

Figure 4.14

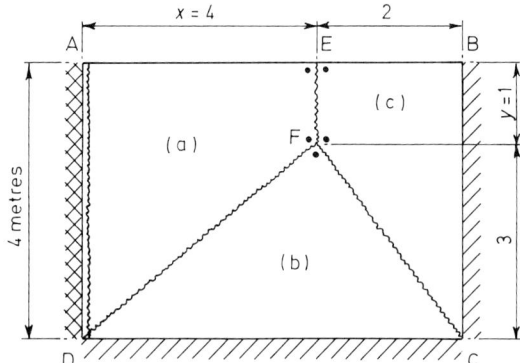

for loading, $-6 \cdot \left[\dfrac{6}{2} \cdot \dfrac{(4-y)^2}{3} \right] = -54$

$\therefore 6m - 54 = 0$
$\therefore m = 9$

For region (c), taking moments about BC:

for yield line EFC, vectors CF and FE add up to a resultant CE which can be resolved into vectors CB and BE, $m \cdot CB = 4m$

for loading, $-6 \cdot \left[y \cdot \dfrac{(6-x)^2}{2} + \dfrac{4-y}{2} \cdot \dfrac{(6-x)^2}{3} \right] = -24$

$\therefore 4m - 24 = 0$
$\therefore m = 6$

The maximum value of m, namely 9, does not exceed a 50% increase in the minimum value of m, namely 6, so from Step 2 previously, we can use the work equation for this yield-line pattern.

Taking the vertical deflection of line EF as unity, the slope of region (a) normal to AD is 1/4, the slope of (b) normal to DC is 1/3 and the slope of (c) normal to BC is 1/2.

Work done by loading for region (a)

$= 6 \cdot \left[x \cdot y \cdot \dfrac{1}{2} + \dfrac{x}{2} \cdot (4-y) \cdot \dfrac{1}{3} \right] = 24$

Work done by loading for region (b)

$= 6 \cdot \left[(4-y) \cdot \dfrac{6}{2} \cdot \dfrac{1}{3} \right] = 18$

Work done by loading for region (c)

$= 6 \cdot \left[(6-x) \cdot y \cdot \dfrac{1}{2} + \dfrac{(6-x)}{2} \cdot (4-y) \cdot \dfrac{1}{3} \right] = 12$

Energy absorbed at the yield lines:
For portion (c), it rotates about an axis BC only, an amount 1/2, see above. The resultant of the moment vectors CF and FE can be resolved into the moment vectors CB and BE. Hence energy absorbed at yield lines

$= (m \cdot CB) \cdot (1/2) = 2m$

For portion (b), it rotates about an axis DC only, an amount 1/3, see above. The resultant of the moment vectors DF and FC is moment vector DC. Hence energy absorbed at yield lines

$$= (m \cdot DC) \cdot (1/3) = 2m$$

For portion (a), it rotates about an axis AD only, an amount 1/4, see above. The energy absorbed at yield line AD

$$= (2m \cdot AD) \cdot (1/4) = 2m$$

The resultant of the moment vectors DF and FE is DE and this can be resolved into the moment vectors DA and AE. Hence energy absorbed at yield lines DF and FE

$$= (m \cdot DA) \cdot (1/4) = m$$

Equating energy absorbed to work done:

$$2m + 2m + 2m + m = 24 + 18 + 12$$

$$\therefore m = 7.714$$

With the increased use of computers a hand-held or desktop microcomputer might be programmed to solve this example by the method of equilibrium, or virtual work. Then various sets of x and y can be guessed until the solution has adequate accuracy. This was done for the yield-line pattern of Figure 4.14 and the results were $x = 3.804$, $y = 1.215$ and $m = 7.754$. So the above result of 7.714 has an error of only 0.52% supporting the effectiveness of the combined method advocated in this section.

4.4.7 Affine slab transformations

There are affinity theorems by Johansen for transforming certain slab problems into equivalent simpler ones to analyse (by the methods already described in this chapter). An affine slab and loading is devised to correspond to a given real slab and loading and the results for the former apply to the latter. Then the affine slab can be designed as though its reinforcement is the same in any direction and it is rectangular instead of skew. These theorems[1] are summarised as follows.

Affinity theorem for orthotropic reinforcement (the reinforcement in one direction gives an ultimate resisting moment, m, and in the other direction μm):

Yield-line theory of slab analysis 163

1. Multiply all relevant dimensions (defining slab shape or load position) in the direction of the μm reinforcement by $1/\sqrt{\mu}$.
2. Multiply each total load by $1/\sqrt{\mu}$.

Affinity theorem for skew reinforcement (the angle between the reinforcement in two directions being φ):

1. Define all relevant points by co-ordinates relative to axes parallel to the reinforcement.
2. Replot these points to orthogonal axes.
3. Multiply each total load by cosec φ.

In both the above cases, the support conditions are the same for the affine as the real slab.

Example 4.7. The slab shown in Figure 4.15(a) carries the point load W kN, a uniformly distributed load q kN/m² and a line load w kN/m. The reinforcement in the direction of dimension 'a' is obtained from the bending moment μm. Give details of the affine slab.

Figure 4.15(b) shows the affine slab, where, from the above

$$a' = a/\sqrt{\mu}$$

$$W' = W/\sqrt{\mu}$$

$$q'a'l = qal/\sqrt{\mu}$$

$$\therefore q' = q$$

$$y_3' = y_3/\sqrt{\mu}$$

$$y_1' = y_1/\sqrt{\mu}$$

$$y_2' = y_2/\sqrt{\mu}$$

length of line load $w = L = \sqrt{[(y_2-y_1)^2+(x_2-x_1)^2]}$

length of line load $w' = L' = \sqrt{[(y_2'-y_1')^2+(x_2-x_1)^2]}$

$$w'L' = wL/\sqrt{\mu}$$

$$\therefore w' = \frac{wL}{L'\sqrt{\mu}}$$

Now this affine slab, Figure 4.15(b), can be analysed as though the reinforcement in any direction were the same (see Section 4.4.1).

Example 4.8. The slab shown in Figure 4.16(a) has edges simply supported along AB and DC, fixed along AD and unsupported along BC. It carries a uniformly distributed loading of q kN/m² over the whole area and a line load of w kN/m along BC. Reinforcement is parallel to the edges and provides ultimate moments of resistance per metre as follows:

Top reinforcement:
bars parallel to shorter edges: $1.2m$
bars parallel to longer edges: $0.4m$
Bottom reinforcement:
bars parallel to shorter edges: m
bars parallel to longer edges: $m/3$

Figure 4.15

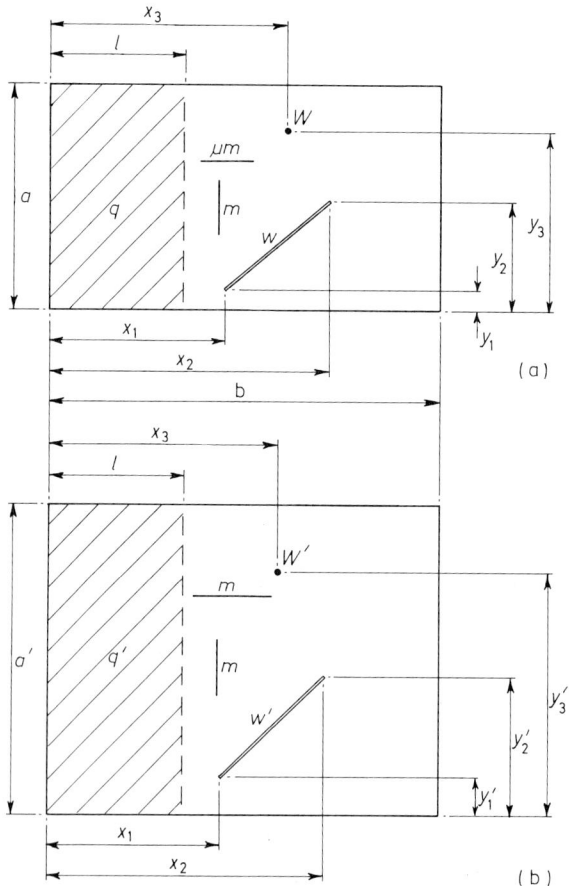

(a)

(b)

Give details of the affine slab.

Figure 4.16(a) is transformed to Figure 4.16(b) and this is transformed to give the final affine slab in Figure 4.16(c).

$$u = 1/3$$

$$b' = \frac{b}{\sqrt{1/3}} = b\sqrt{3}$$

total line load for slab (a) $= wb$

total line load for slab (b) $= wb \csc 60° = \frac{2wb}{\sqrt{3}}$

total line load for slab (c) $= \frac{2wb}{\sqrt{3}} \cdot \frac{1}{\sqrt{1/3}} = 2wb$

line load along unsupported edge of affine slab

$$= \frac{2wb}{b'} = \frac{2w}{\sqrt{3}} \text{ kN/m}$$

Total load/unit area for slab (a) $= qab \sin 60°$
Total load/unit area for slab (b) $= qab \sin 60° \csc 60°$
Total load/unit area for slab (c)

$$= qab \sin 60° \csc 60° \frac{1}{\sqrt{\frac{1}{3}}}$$

Figure 4.16

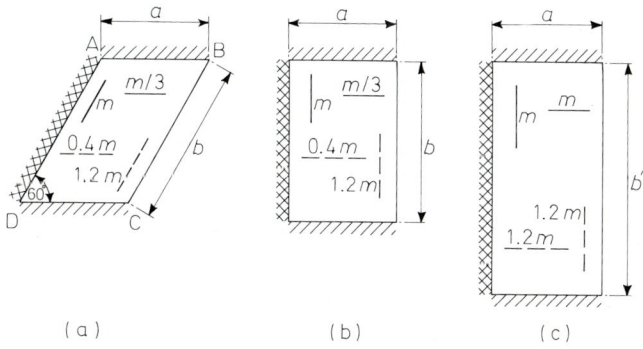

(a)   (b)   (c)

area of slab (c) $= ab' = ab\sqrt{3}$

load/unit area for slab (c) $= \dfrac{qab \sin 60° \operatorname{cosec} 60°}{\sqrt{\tfrac{1}{3}} ab \sqrt{3}} = q$

Now the affine slab Figure 4.16(c) can be analysed to obtain m using the previously explained methods.

4.5 Hillerborg's strip method of slab design

This is perhaps most simply explained by the following example.

Example 4.9. Design the slab shown in Figure 4.17 which has edges restrained against rotation and has to carry an ultimate uniformly distributed total load of 15 kN/m².

It seems sensible (based on our experience of elastic theory, or tests) to reduce the reinforcement parallel to supports towards supports from mid span. For simplicity in practice we shall design for six bands of reinforcement in each direction, so we have to divide the loading to give constant loading for the width of each band. In the x-direction, from symmetry we need to design only three bands, and the typical strips to be designed, as representative of each band, are LM, NP, and QR. The only load these strips are designed to carry is that on their shaded portions, there being none on LM in this instance. Similarly in the y-direction the strips ST, UV and WX are designed to carry the load on their shaded portions. Thus we have chosen that the two zones such as ABCDEFGHA, now called (A) and (B), are to be carried by strips in the x-direction and the remainder of the slab, zone (C), is to be carried by strips in the y-direction. This means that we have chosen that the load on the two zones (A) and (B) is carried by strips to the edges ac and bd, and that the load on the zone (C) is carried by strips to the edges ab and dc. This kind of loading on the edges is in line with much past practice for deciding loads on supporting peripheral beams, and is more recognisable if the internal *discontinuity lines*, such as ABCDEFGH, were ae, ec, bf and fd. The stepped discontinuity lines were chosen to approximate to ae, ec, bf and fd, because of the desire to have the reinforcement in bands. The discontinuity lines are chosen to be a sensible (with regard to one's experience of elastic theory, yield line, and/or experimentation) division of the areas of the slab likely to be carried by each support.

The distribution of bending moments and shear forces in each strip can be determined by either elastic or plastic theory. For example, for strip ST, taking a nominal breadth of 1 m, the total load along ST is $15 \times 4.5 = 67.5$ kN. By elastic theory, it is a fixed beam, so each support moment is $67.5 \times 4.5/12 = 25.31$ kN m and the mid-span moment is $25.31/2 = 12.66$ kN m. By plastic theory, suppose we choose to keep the maximum bending moment to a minimum, that is make the support and mid-span moments equal, then either of these $= 67.5 \times 4.5/16 = 18.98$ kN m.

4.5.1 Further points on Hillerborg's strip method

This method involves a tremendous amount of plastic action. If one's experience of elastic analysis, yield line or tests is severely gone against in deciding discontinuity lines and points of contraflexure of strips, then a very undesirable slab can be obtained, with regard to cracking. It is also possible that such a slab might not pass the British Standard loading test because of lack of recovery of deflection due to high plasticity, even though the ultimate strength might be satisfactory.

If the strips are designed elastically then the method is very illogical, in that deflections of strips are not matched up, as was done years ago by Grashof–Rankine.

In the past decade at least, there has been a tendency to use computers to design structures more accurately, so in one sense the strip method

Figure 4.17

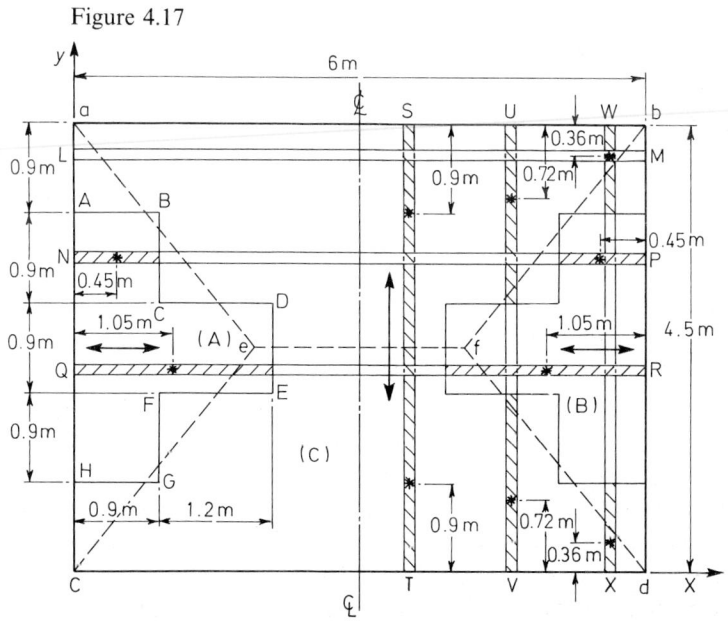

168   *Reinforced concrete slabs*

seems a retrograde step. It is the kind of method one has not been proud to use in design offices when precise elastic analyses[38] have been too difficult to attempt in the time available. But there are more computer packages available today.

The great advantage of the method is that it is easy to use and apply to any shape. It is probably best when there is not an elastic analysis available on a computer package.

For skew slabs the strips are taken as beams cranked in plan and the geometry involves different portions of a strip having different widths.[29]

It is an easy method to apply slabs containing holes.[29]

It is generally accepted[37] that Hillerborg's strip method is not upper bound which is a concern with yield-line analyses. If yield lines are chosen to be disposed where one would imagine from experience of tests or from common engineering sense then the analysis should not be significantly upper bound; however it will generally be the latter. For the strip method it is best to choose the discontinuity lines from thoughts (probably based on experience of elastic theories) of which areas would be sensibly carried by which supports.

Hillerborg's strip and advanced (see Section 4.7) methods are very economical in that generally each portion of loading is carried only once. This contrasts with methods used over the past half century in design offices where beams are formed within the slab thickness and a portion of loading is, for example, carried twice by slab and this then by a beam.

4.6   BS 8110 and CP 110 and yield-line and strip methods

BS 8110 and CP 110 recommend that yield-line and strip methods can be used provided that the ratios between support and span moments are similar to those obtained by elastic theory and CP 110 suggests choosing ratios between 1:1 and 1.5:1. This helps to safeguard against designing a slab which may crack badly at working loads.

Certain plastic methods used[2] choose the positions of the points of inflection or contraflexure at 0.2 of the span from each support for strips such as ST, 0.4 of the length of the loaded area from each support for strips such as UV and WX, 0.5 of the length of the loaded area from each support for strips such as NP and QR (strip LM having no loading). These points are marked with an asterisk in Figure 4.17. On this basis Figure 4.18(a) shows the loading on strip NP and its points of contraflexure. Figure 4.18(b) shows how the bending moments are to be calculated for the portion of NP between the points of contraflexure. Reaction $R_1 = R_2$

$= 0.45 \times 15 = 6.75$ kN. The bending moment diagram is shown in Figure 4.18(c) and its maximum bending moment is $R_1 \times 0.45 - 0.45 \times 15 \times 0.225 = 1.519$ kN m. Figure 4.18(d) shows how the bending moments are to be calculated for the portion of NP between the points of contraflexure and the supports. The bending moment for the portion shown is shown in Figure 4.18(e) and its maximum value is $R_1 \times 0.45 + 15 \times 0.45 \times 0.225 = 4.556$ kN m. Shear forces can be calculated correspondingly. Other strips can be treated similarly.

4.7 Hillerborg's advanced method[39]

This uses quadrilateral shapes, rectangles, triangles as well as strips. A basic rectangular element has a point support at one corner and two mutually perpendicular inplanar moment vectors – each one parallel to a side of the rectangle. The writer is considering moments as vector

Figure 4.18

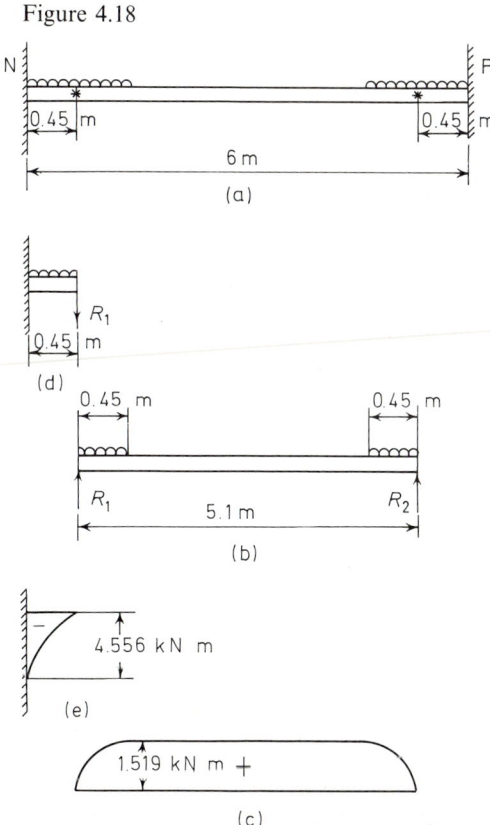

170 *Reinforced concrete slabs*

moments using the right-hand rule, as commonly used in classical mechanics.

Example 4.10. The slab shown in Figure 4.19(a) carries a uniformly distributed load q. It is supported by columns at A and B and by a simple line support along DC. Suggest a design solution using Hillerborg's methods.

Consider the rectangular element AGFE. It is in equilibrium as follows: a point support upwards at A, a loading $q \cdot AG \cdot AE$ downwards, a moment vector FG and a moment vector EF. From external considerations; the reactions at A and B are the same from symmetry; the

Figure 4.19

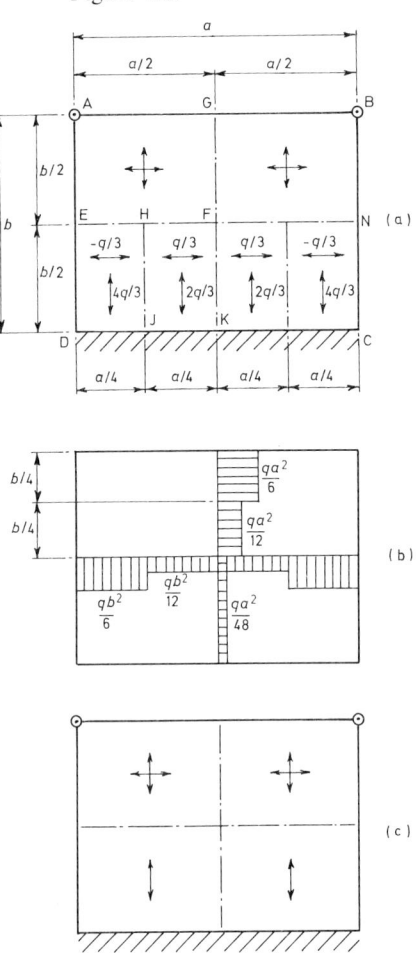

total of these reactions is the same as the total reaction due to the line-load support DC from symmetry considerations. Thus the shear forces on sections GK and EN are zero. Considering the whole slab

$$\text{Reaction at } A = R_A = qab/4$$

Therefore the element must be of the area shown so that resolving vertically for it: the downward load $= qab/4$ which equals the value of R_A above and which is therefore correct. For the element take moments about GF, then the vector moment GF

$$= R_A \cdot \left(\frac{a}{2}\right) - q \cdot \frac{a}{2} \cdot \frac{b}{2} \cdot \left(\frac{a}{4}\right) = \frac{qa^2b}{16}$$

So the vector moment GF per unit length

$$= \frac{qa^2b}{16} \div \left(\frac{b}{2}\right) = \frac{qa^2}{8}$$

Now it is sensible, bearing in mind serviceability conditions, to have a stronger strip of slab near to the free edge AB; hence the bending moment for design along GF is distributed as in Figure 4.19(b): the half nearer the edge is allocated a moment $qa^2/6$ per unit length and the other half a moment $qa^2/12$, so that the average moment per unit length is $qa^2/8$ as required above.

Again for the element take moments about EF, then the vector moment FE

$$= R_A \cdot \left(\frac{b}{2}\right) - q \cdot \frac{a}{2} \cdot \frac{b}{2} \cdot \left(\frac{b}{4}\right) = \frac{qab^2}{16}$$

So the vector moment FE per unit length

$$= \frac{qab^2}{16} \div \left(\frac{a}{2}\right) = \frac{qb^2}{8}$$

It is sensible, bearing in mind serviceability conditions, to have a stronger strip of slab near to the free edge AD so we halve EF, use a moment $qb^2/6$ per unit length for the outer half and a moment $qb^2/12$ per unit length for the other half. These then average $qb^2/8$ per unit length for FE.

As DC is a line-load support it should be split into several portions. In this case four portions are taken. The rectangular element EHJD is supported by a line load along DJ, there are vector moments EH and HJ and it carries a loading $q \cdot EH \cdot ED$. In Figure 4.17 each portion of loading was carried in either one direction of two mutually perpendicular directions. But with both Hillerborg's strip method and his advanced method any portion of loading can be carried by one proportion of it being carried in one direction and the remainder carried in the other direction. Furthermore one proportion can be greater than one so that the remainder is negative. In the case of a vertically downwards portion of loading if one proportion is greater than unity then the negative remainder would be positively upwards.

Suppose for the loading q on element EHJD its proportion carried in the DE direction is q_1. Then taking moments about DJ for the element (N.B. vector moment HE, which is part of vector moment FE, for the strip AGFE, has an equal and opposite vector moment EH for strip EHJD, also as this latter is a pure moment it is the same about any axis, i.e. the same about axis EH as about axis DJ):

$$\frac{qb^2}{6} \cdot \frac{a}{4} = q_1 \cdot \frac{a}{4} \cdot \frac{b}{2} \cdot \left(\frac{b}{4}\right)$$

$$\therefore q_1 = \frac{4q}{3}$$

Therefore the proportion of the loading q carried in the DJ direction is $q - q_1 = -q/3$ (that is an upwards loading). Now the element EHJD is not the proper Hillerborg rectangular element described at the beginning of this section, because there is not one corner load but a line load along DJ and therefore moments cannot be taken about HJ for this element. But ENCD can be treated as a strip spanning from ED to NC just as for the Hillerborg strip method of Section 4.5. This strip spans from end to end ignoring the support DC just as in Figure 4.17 strip WX spans from end to end ignoring the support bd. The main point with Hillerborg's methods is to carry towards collapse all loads and portions of loads somehow or other, no matter how badly cracked the slab is.

Suppose for the loading q on element HFKJ its proportion carried in the JH direction is q_2. Then taking moments about JK for the element:

$$\frac{qb^2}{12} \cdot \frac{a}{4} = q_2 \cdot \frac{a}{4} \cdot \frac{b}{2} \cdot \left(\frac{b}{4}\right)$$

$$\therefore q_2 = \frac{2q}{3}$$

Hillerborg's advanced method

Therefore the proportion of the loading q carried in the JK direction is $q - q_2 = q/3$.

Now for the Hillerborg strip ENCD taking moments about FK, the moment vector FK

$$= \frac{q}{3} \cdot \frac{a}{4} \cdot \frac{b}{2} \cdot \left(\frac{3}{8}a\right) - \frac{q}{3} \cdot \frac{a}{4} \cdot \frac{b}{2} \cdot \left(\frac{a}{8}\right)$$

$$= \frac{qa^2b}{96}$$

So the vector moment FK per unit length

$$= \frac{qa^2b}{96} \div \left(\frac{b}{2}\right) = \frac{qa^2}{48}$$

The bending moments per unit length and the directions with arrows of

Figure 4.20

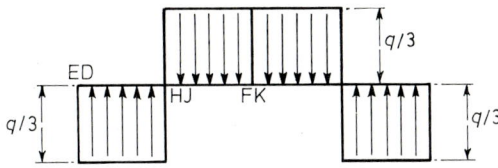

(a) Loading diagram

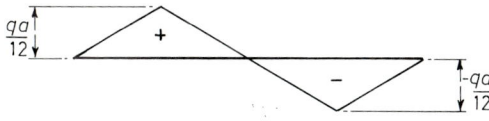

(b) Shear force diagram

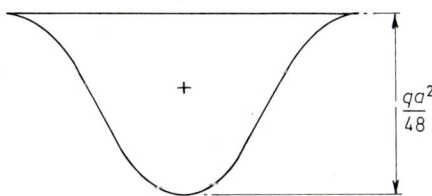

(c) Bending moment diagram

174 *Reinforced concrete slabs*

the way in which loading is carried are shown in Figure 4.19(b) and (a) respectively. Figure 4.20(a) shows the loading diagram for strip ENCD with corresponding shear force and bending moment diagrams in Figure 4.20(b) and (c), respectively.

Alternatively to the above, if one did not like distributing the bending moment vectors along GF and FE in an arbitrary fashion, then AGFE could have been split into four rectangular elements to obtain a similar result. Furthermore a greater number of rectangular elements could be used for this area and the rest of the slab. For this present example Hillerborg suggests using fewer elements for practical design as shown in Figure 4.19(c) and then distributing the moments along the edges of the rectangle in a reasonable way. He justifies this by saying 'that different theoretical solutions give somewhat different distributions of moments and that a reinforcing bar in practice is efficient in resisting moments occurring within a considerable relative distance from the bar itself'. The writer's comments on this would be that the greater the number of sensible (that is bearing in mind serviceability, viz. how it tends to act elastically at working loads) elements the better the solution, and the extra work involved is quite reasonable when Hillerborg's methods are solving problems outside existing design tables and which would be tremendously formidable by other analyses.

Hillerborg states 'For practical design the main condition is that the equilibrium is fulfilled for the elements as a whole and that the lateral distribution of moments chosen is reasonable'.

Example 4.11. The slab shown in Figure 4.21(a) carries a uniformly distributed load q. Suggest a design solution using Hillerborg's methods.

To obtain lines of zero shear force, if one considers serviceability/elastic considerations, towards the edge ABC the slab will tend to span like a continuous beam of two spans, so referring to Table 6.2 (page 218) it is reasonable to make FB say $0.6a$. Considering the direction AE the slab is simply supported at sides ED and AC so it is reasonable to make AH = $b/2$. The shear force is therefore assumed to be zero along lines FG, BK, and HJ.

Then for element AFGH: resolving vertically, the vertical reaction at A

$$= R_\mathrm{A} = q \cdot 0.4a \cdot \frac{b}{2} = 0.2qab$$

Figure 4.21

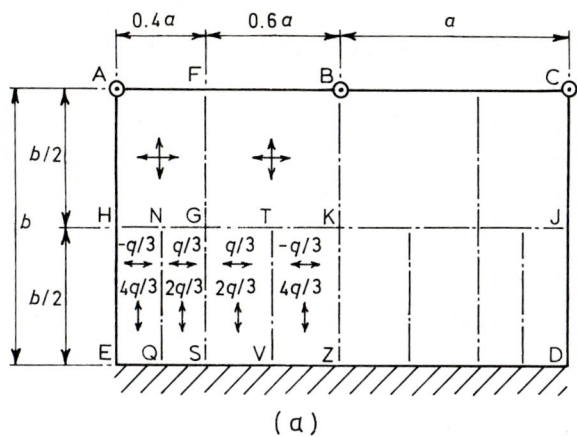

(a)

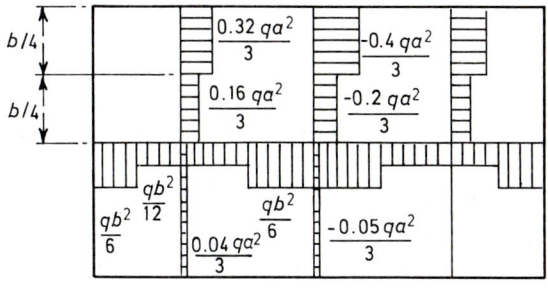

(b)

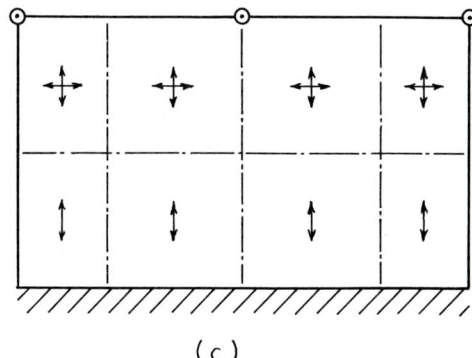

(c)

Taking moments about FG, the vector moment FG

$$= R_A \cdot 0.4a - q \cdot 0.4a \cdot \frac{b}{2} \cdot \left(\frac{0.4a}{2}\right) = 0.04qa^2b$$

So the vector moment for FG per unit length

$$= 0.04qa^2b \div \left(\frac{b}{2}\right) = 0.08qa^2 = \frac{0.24}{3}qa^2$$

Now it is sensible, bearing in mind serviceability conditions, to have a stronger strip of slab near to the free edge AB, hence the bending moment for design along FG is distributed as in Figure 4.21(b): the half nearer the edge is allocated a moment $0.32qa^2/3$ per unit length and the other half a moment $0.16qa^2/3$, so that the average moment per unit length is $0.24qa^2/3$ as required above.

Taking moments about HG the vector moment GH

$$= R_A \cdot \frac{b}{2} - q \cdot 0.4a \cdot \frac{b}{2} \cdot \left(\frac{b}{4}\right) = \frac{qab^2}{20}$$

So the vector moment GH per unit length

$$= \frac{qab^2}{20} \div (0.4a) = \frac{qb^2}{8}$$

It is sensible, bearing in mind serviceability conditions, to have a stronger strip of slab near to the free edge AE so HG is halved and a moment $qb^2/6$ per unit length used for the outer half and a moment $qb^2/12$ per unit length used for the other half. These then average $qb^2/8$ per unit length for GH.

Now the vector moment FG for element AFGH is equal to the vector moment GF for element FBKG. Hence for this latter element, taking moments about BK (N.B. taking moments about a reaction is useful in that it saves arithmetic and thought in that the reaction does not need to be assessed. This could have been done in the previous examples but it might have made a complication for the reader, by making short cuts, when the main objective was to explain the basis of the method) the vector moment KB

$$= q \cdot 0.6a \cdot \frac{b}{2} \cdot (0.3a) - 0.04qa^2b = 0.05qa^2b$$

So the vector moment for KB per unit length

$$= 0.05qa^2b \div \left(\frac{b}{2}\right) = \frac{qa^2}{10}$$

Similarly to before it is sensible to have a stronger strip of slab near to the free edge AB so BK is halved and a moment $0.4qa^2/3$ per unit length used for the outer half and a moment $0.2qa^2/3$ per unit length used for the other half. These then average $qa^2/10$ per unit length for KB. The ratio of moments per unit length between the halves of FG and BK is kept the same, namely 2 to 1 as chosen for FG. These are shown in Figure 4.21(b).

Again for element FBKG, taking moments about FB the vector moment KG

$$= q \cdot 0.6a \cdot \frac{b}{2} \cdot \left(\frac{b}{4}\right) = \frac{3}{40}qab^2$$

So the vector moment for KG per unit length

$$= \frac{3}{40}qab^2 \div (0.6a) = \frac{3}{24}qb^2$$

(N.B. The vector moment KG is a pure moment and has the same moment about any line parallel to KG, for example FB in this case.)

It is sensible, bearing in mind serviceability conditions, to have a stronger strip near to the support B so GK is halved and a moment $qb^2/6$ per unit length used for the half nearest to BK and a moment $qb^2/12$ per unit length used for the other half. These then average $3qb^2/24$ per unit length, as required above, for KG.

For element HNQE, if the portion of q to be carried in the HE direction is q_1 then taking moments about EQ

$$\frac{qb^2}{6} \cdot 0.2a - q_1 \cdot 0.2a \cdot \frac{b}{2} \cdot \left(\frac{b}{4}\right) = 0$$

(N.B. Vector moment GH for element AFGH is equal and opposite to vector moment HG for element HGSE.)

$$\therefore q_1 = 4q/3$$

Therefore the portion of q to be carried in the EQ direction is $q - q_1 = -q/3$.

178 *Reinforced concrete slabs*

For element NGSQ, if the portion of q to be carried in the NQ direction is q_2, then taking moments about QS

$$\frac{qb^2}{12} \cdot 0.2a - q_2 \cdot 0.2a \cdot \frac{b}{2} \cdot \left(\frac{b}{4}\right) = 0$$

$$\therefore q_2 = \frac{2q}{3}$$

Therefore the portion of q to be carried in the QS direction is $q - q_2 = q/3$.

For element GTVS, if the portion of q to be carried in the GS direction is q_3 then taking moments about SV

$$\frac{qb^2}{12} \cdot 0.3a - q_3 \cdot 0.3a \cdot \frac{b}{2} \cdot \left(\frac{b}{4}\right) = 0$$

$$\therefore q_3 = \frac{2q}{3}$$

Then the portion of q to be carried in the SV direction is $q - q_3 = q/3$.

For the element TKZV, if the portion of q to be carried in the TV direction is q_4 then taking moments about VZ

$$\frac{qb^2}{6} \cdot 0.3a - q_4 \cdot 0.3a \cdot \frac{b}{2} \cdot \left(\frac{b}{4}\right) = 0$$

$$\therefore q_4 = \frac{4q}{3}$$

Then the portion of q to be carried in the VZ direction is $q - q_4 = -q/3$.

Now for the Hillerborg strip spanning from HE to JD, taking moments about GS, the vector moment GS per unit length (considering a strip of unit width)

$$= \frac{q}{3} \cdot HN \cdot (0.3a) - \frac{q}{3} \cdot NG \cdot (0.1a) = \frac{0.04}{3} qa^2$$

Again for this strip, taking moments about ZK the vector moment ZK per unit length

Hillerborg's advanced method

$$= -\frac{q}{3} \cdot \text{HN} \cdot (0.9a) + \frac{q}{3} \cdot \text{NG} \cdot (0.7a) + \frac{q}{3} \cdot \text{GT} \cdot (0.45a)$$

$$-\frac{q}{3} \cdot \text{TK} \cdot (0.15a)$$

$$= \frac{0.05}{3} qa^2$$

The bending moments per unit length and the directions with arrows of the way in which loading is carried are shown in Figure 4.21(b).

The system of elements chosen decides the reactions at A and B, namely R_A already calculated and R_B which can be calculated by resolving vertically for element FBKG, viz.

$$\tfrac{1}{2} R_B = q \cdot 0.6a \cdot \frac{b}{2}$$

$$\therefore R_B = 0.6qab$$

What this means is that towards ultimate load all portions of loading can be carried by the reinforcement provided, from the above rectangular element and strip analysis, no matter how much cracking is involved to allow this to happen, and then the reactions will be as calculated above.

For Example 4.11 Hillerborg alternatively suggests using fewer elements for practical design as shown in Figure 4.21(c) and then distributing the moments along the sides of the rectangles in a reasonable way. However, the writer considers that more guidance is given by more elements as in Figure 4.21(a) and that the extra work is reasonable considering the immense work in solving this problem by other methods.

Example 4.12. The slab shown in Figure 4.22(a) carries a uniformly distributed load of 4 kN/m² and a point load $P = 60$ kN. Suggest a design solution using Hillerborg's methods.

From symmetry the reactions at A and B are the same and at D and C are equal.

Looking at Fig. 4.22(a) from the right-hand side there are two reactions at each end. Taking moments about AB, the two reactions at C and D are equal

$$(6 \times 5 \times 4 \times 5/2 + 60 \times 1)/5 = 72 \text{ kN}$$

180 Reinforced concrete slabs

The line of zero shear at distance y from DC is found from

$$72 = 6y4$$

$$\therefore y = 3\text{m}$$

For element AEJH, taking moments about AE, the vector moment JH

$$= 4 \times 3 \times \frac{(5-y)^2}{2} + \frac{60}{2} \times 1$$

Figure 4.22

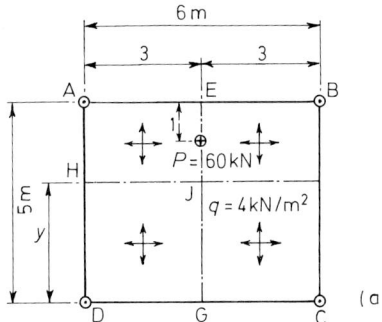

(a)

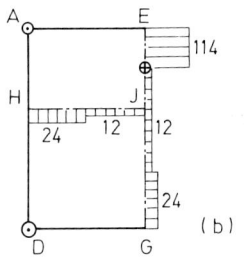

(b)

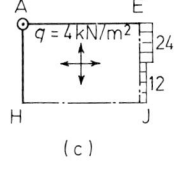

(c)

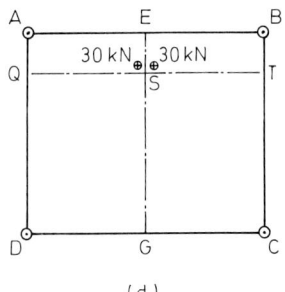

(d)

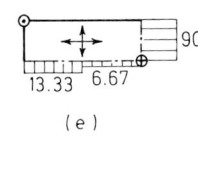

(e)

Hillerborg's advanced method

(N.B. The vector moment JH is a pure moment and has the same moment about any line parallel to JH, for example AE in this case.)

For element HJGD, taking moments about DG, the vector moment HJ

$$= 4 \times 3 \times \frac{y^2}{2} = 54 \text{ kN m and per unit length}$$

$$= \frac{54}{3} = 18 \text{ kN m/m}$$

It is desirable, see previous examples, to have a stronger strip near the outer edge AD, so we choose to distribute this moment as shown in Figure 4.22(b).

For element HJGD, taking moments about HD, the vector moment JG

$$= 4 \times \frac{3^2}{2} \times y = 54 \text{ kN m}$$

So the vector moment JG per unit length

$$= \frac{54}{y} = 18 \text{ kN m/m}$$

It is desirable, see previous examples, to have a stronger strip near the outer edge DC, so we choose to distribute this moment as shown in Figure 4.22(b).

Again it is desirable to have a stronger strip near the outer edge AB particularly because of the proximity of the point load to this edge. One way Hillerborg suggests dealing with this desirability is to consider separately the uniformly distributed loading and the point load. For the former, taking moments about AH for element AEJH, the vector moment EJ per unit length (ignoring the point load)

$$= 4 \times \frac{3^2}{2} = 18 \text{ kN m/m}$$

This is distributed sensibly as shown in Figure 4.22(c).

Figure 4.22(d) shows elements for carrying the point load. Looking at Figure 4.22(a) from the right-hand side there are two reactions at each end and now only the point load. Therefore the line QT through the point load is a line of zero shear force, as also is EG, from symmetry, and these lines bound Hillerborg rectangular elements. Elements AESQ and EBTS each

182 Reinforced concrete slabs

carry half of the point load at their corners at S as shown. For element AESQ, taking moments about AQ, the vector moment ES per unit length

$$= 30 \times 3 = 90 \text{ kN m/m}$$

Taking moments about AE, the vector moment SQ per unit length

$$= 30 \times 1/3 = 10 \text{ kN m/m}$$

These moments are shown in Figure 4.22(e), the 10 kN m/m being distributed sensibly. Figure 4.22(c) and (e) combined give the resultant distribution of moment along EJ and this is shown in Figure 4.22(b) which now gives the complete solution.

There is of course no need to consider separately the point load and the distributed load per unit area to obtain the bending moment per unit length along EJ and this latter, if the loadings are not separated, can still be distributed sensibly with more at the slab edge and underneath the point load and the writer prefers this simpler approach.

Example 4.13. The slab shown in Figure 4.23(a) carries a uniformly distributed load q. Suggest a design solution using Hillerborg's methods.

Hillerborg[39] recommends the use of triangular corner-supported elements with reinforcement parallel to the diagonals and analysing the slab as in Figure 4.23(b), as if it were supported at only two corners. (AD and DC just act to stabilise the slab from overturning about an axis AC. For example AD and DC could be built into a 110 mm outer skin of

Figure 4.23

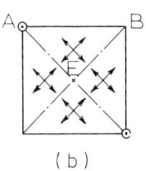
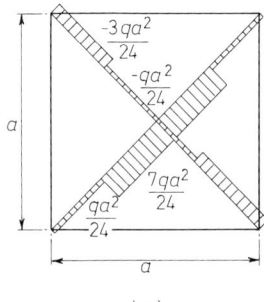

brickwork.) From symmetry lines AC and BC will be lines of zero force and therefore the boundaries of Hillerborg's triangular elements.

For triangular element ABE: Taking moments about AE, the vector moment AE

$$= \frac{qa^2}{4} \cdot \frac{EB}{3}$$

So the vector moment AE per unit length

$$= \frac{qa^2}{4} \cdot \frac{EB}{3} \div EA = \frac{qa^2}{12}$$

Taking moments about a line at right-angles to AE at A, to avoid involving the support reaction, the vector moment BE

$$= \left(\frac{qa^2}{4}\right) \cdot \frac{2}{3} \cdot AE$$

(N.B. The vector moment BE is a pure moment and has the same value about any line parallel to BE, for example the line through A in this case.)

So the vector moment BE per unit length

$$= \left(\frac{qa^2}{4}\right) \cdot \frac{2}{3} \cdot AE \div EB = \frac{qa^2}{6}$$

Hillerborg, bearing in mind which bands ought to be stronger, chooses to distribute the above moments as shown in Figure 4.23(c). As in the previous examples this choice is 'reasonable' and will vary to some extent with different designers. However the distribution is made the slab will crack accordingly towards failure so that all portions of loading can be carried according to the distribution of reinforcement provided.

Example 4.14. The author was once required to design a slab approximating to that shown in Figure 4.24(a) over a small petrol filling station. A lightweight kiosk was between two columns and a row of pumps was at right angles to this and in line with the individual column. The actual roof had curves instead of corners. The writer's solution involved shallow hidden beams and deflection checks so that a wavy periphery would not result. This was prior to the methods of Johansen's yield line, Kemp, Fernando and Kemp and Hillerborg being used in the U.K. The writer has consulted Professor Kemp and it was agreed that the

184 *Reinforced concrete slabs*

methods of Refs. 21 and 22 could not be used for this problem. The writer consulted Professor Hillerborg who kindly provided the following solution to what he regarded as a difficult problem. For simplicity unit uniformly distributed loading is considered.

The slab is split up into suitable Hillerborg elements and these are numbered 1 to 13 on Figure 4.24(a). Also on this figure suitable directions, generally parallel to the sides, are shown, with arrows, for the

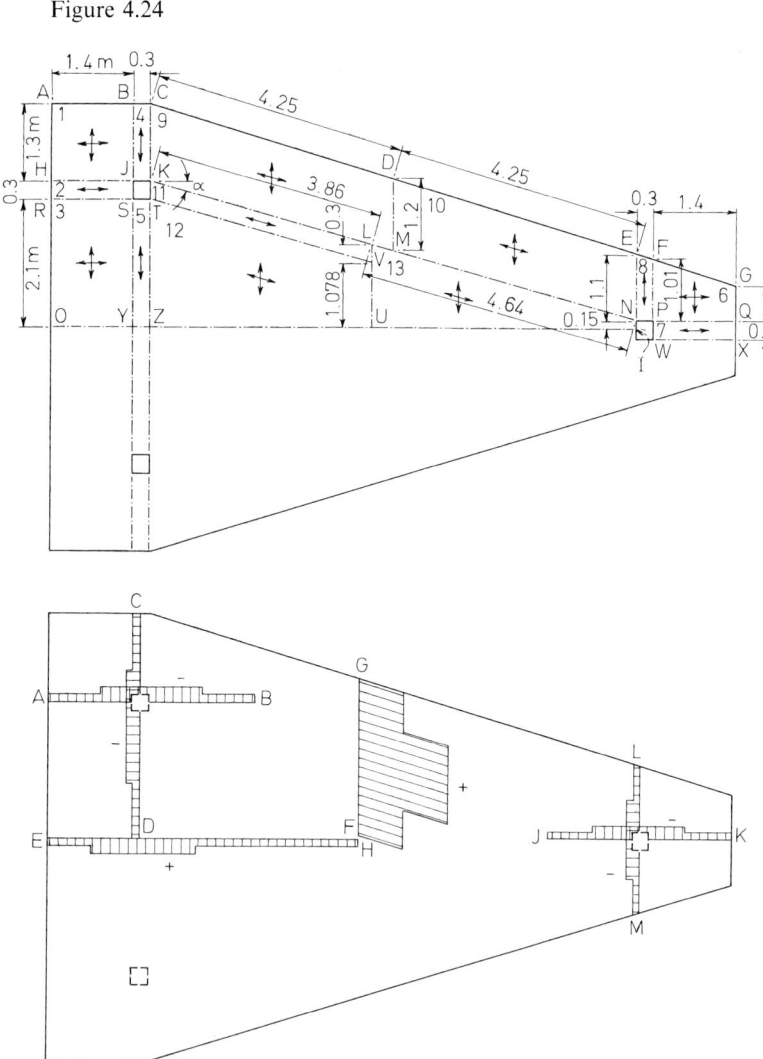

Figure 4.24

reinforcement. Negative bending moments refer to tension in the top of the slab, when recorded on Figure 4.24(b). The bending moments calculated are each for a complete side not per unit length. Angle $\alpha = 15.37°$.

Element No. 1: Taking moments about HJ, the vector moment HJ

$$= 1.4 \times \frac{1.3^2}{2} = 1.183$$

Taking moments about BJ, the vector moment JB

$$= 1.3 \times \frac{1.4^2}{2} = 1.274$$

Element No. 2: Taking moments about JS, the vector moment SJ

$$= 0.3 \times \frac{1.4^2}{2} = 0.294$$

Element No. 3: Taking moments about SY, the vector moment YS

$$= 2.1 \times \frac{1.4^2}{2} = 2.058$$

Taking moments about RS, the vector moment YO

$$= \frac{1.4 \times 2.1^2}{2} - 1.183 = 1.904$$

(N.B. The vector moment YO is a pure moment and has the same moment about any line parallel to YO, for example RS in this case.)

(N.B. The shear force at OY is zero from symmetry.)

Element No. 4: This element can usefully be designed to carry a moment required for equilibrium because the reinforcements at this junction in elements 1 and 9 are not in line, or the vector moments at right-angles to these reinforcements are not parallel. Figure 4.25(a) shows a triangle of forces for these vector moments. Thus the moment at BJ = 1.274, the monent at the end KC of element No. 9 at right-angles to the reinforcement direction CD = 1.322 and the moment to be carried by element No. 4 = 0.349. Taking moments about JK, the vector moment JK

$$= 0.3 \times \frac{1.3^2}{2} - 0.349 = -0.096$$

Thus the support moment is 'formerly' slightly positive, oppositely to the negative moments in the adjacent strips (that is along HJ and KM).

Element No. 5: Again the reinforcement at right-angles to SY in element No. 3 is not in line with the similar reinforcement in element No. 12. Element No. 5 can usefully be designed to carry a moment required to satisfy equilibrium similarly to element No. 4 and to carry the moment because of the reinforcements in elements Nos 2 and 11 being out of line – it may as well be carried here rather than in the column. The triangle of forces, similar to the one shown in Figure 4.25(a), is shown in Figure 4.25(b). Thus the moments at JS and SY total $0.294 + 0.2058 = 2.352$, the moment at the end TS of element No. 12 at right-angles to the reinforcement direction $TV = 2.440$ and the moment to be carried by element No. $5 = 0.644$. Taking moments about ST, the vector moment ZY

$$= 0.3 \times \frac{2.1^2}{2} + 0.644 + 0.096 \text{ (from element No. 4)}$$

$$= 1.402$$

Element No. 6: Taking moments about FP, the vector moment FP

$$= 0.59 \times \frac{1.4^2}{2} + (1.01 - 0.59) \times \frac{1.4}{2} \times \frac{1.4}{3} = 0.715$$

Figure 4.25

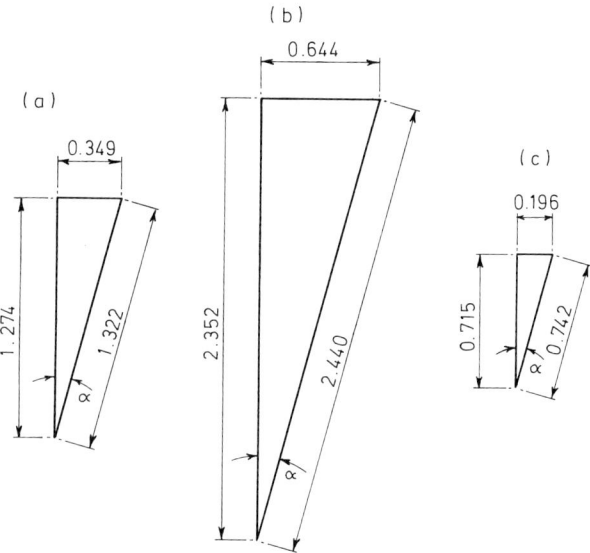

Taking moments about PQ, the vector moment PQ

$$= 1.4 \times \frac{0.59^2}{2} + (1.01 - 0.59) \times \frac{1.4}{2} \times \left(0.59 + \frac{0.42}{3}\right)$$

$$= 0.458$$

Element No. 7: Taking moments about PW, the vector moment PW

$$= 0.3 \times \frac{1.4^2}{2} = 0.294$$

Element No. 8: Again the reinforcement at right-angles to FP in element No. 6 is not in line with the similar reinforcement in element No. 10. Element No. 8 can usefully be designed to carry a moment required to satisfy equilibrium similarly to element No. 4. The triangle of forces, similar to the one shown in Figure 4.25(a), is shown in Figure 4.25(c). Taking moments about NP, the vector moment NP

$$= \frac{0.3}{2} \times \left(\frac{1.10 + 1.01}{2}\right)^2 + 0.196 = 0.363$$

Element No. 9: From element No. 4 the vector moment at the end KC of element No. 9 at right-angles to the reinforcement direction CD is 1.322. Taking moments about end KC the vector moment at right-angles to the reinforcement direction CD at end DM (assuming DM has been chosen so that the shear force at DM is zero)

$$= \left[1.2 \times \frac{4.25^2}{2} + (1.3 - 1.2) \times \frac{4.25}{2} \times \frac{4.25}{3}\right] \cdot \cos \alpha - 1.322$$

$$= 9.416$$

Taking moments about end KM, the vector moment at right-angles to the reinforcement in direction CK at end KM

$$= \left[4.25 \times \frac{1.2^2}{2} + \frac{4.25}{2} \times (1.3 - 1.2) \times \left(1.2 + \frac{0.1}{3}\right)\right] \cdot \cos \alpha$$

$$= 3.202$$

Element No. 10: From Figure 4.25(c) the vector moment at the end EN

188 *Reinforced concrete slabs*

of element No. 10 at right-angles to the reinforcement direction DE is 0.742. Taking moments about end EN the vector moment at right-angles to the reinforcement direction DE at end DM

$$= \left[1.2 \times \frac{4.25^2}{2} - \frac{(1.2-1.1)}{2} \times \frac{4.25^2}{3}\right] \cdot \cos\alpha - 0.742$$

$$= 9.415$$

(assuming DM has been chosen so that the shear force at DM is zero). This value needs to agree with the corresponding value for element No. 9, namely 9.416 earlier. This agreement is satisfactory. If the difference were more than 10% then the position of DM would have to be changed, that is an alteration would be required if the relative sizes of elements Nos 9 and 10. If the two values have a difference of less than 10% then they are averaged to give the design moment at this location.

Taking moments about end MN the vector moment at right-angles to the reinforcement direction DM at end MN

$$= \left[4.25 \times \frac{1.1^2}{2} + \frac{4.25}{2} \times (1.2-1.1) \times \left(1.1 + \frac{0.1}{3}\right)\right] \cdot \cos\alpha$$

$$= 2.711$$

Elements Nos 11 and 12: These can be usefully taken together when considering vector moments at right-angles to the reinforcement in the TV direction. From Figure 4.25(b) and earlier the vector moment for these two elements at end KZ at right-angles to the reinforcement in the direction TV is 2.440. Taking moments about end KZ, the vector moment at right-angles to the reinforcement direction TV at end LU

$$= \left[1.378 \times \frac{3.86^2}{2} + \frac{(2.4-1.378)}{2} \times \frac{3.86^2}{3}\right] \cdot \cos\alpha - 2.440$$

$$= 9.903$$

(assuming LU has been chosen so that the shear force at LU is zero). The vector moment at right-angles to the reinforcement in direction CK at end TV, from element No. 9, is 3.202. Taking moments, for element No. 12 about end TV, the vector moment at right-angles to the reinforcement in direction ZT at end ZU

$$= \left[3.86 \times \frac{1.078^2}{2} + \frac{3.86}{2} \times (2.1 - 1.078) \right.$$

$$\left. \times \left(1.078 + \frac{1.022}{3} \right) \right] \times \cos \alpha - 3.202$$

$$= 1.658$$

(N.B. The shear force at ZU is zero from symmetry.)
Element No. 13: From element No. 7 the vector moment at the end NI of element No. 13 at right-angles to the reinforcement direction LN

$$= \frac{0.294}{2 \cos \alpha} = 0.152$$

Taking moments about end NI the vector moment at right-angles to the reinforcement direction LN at end LU

$$= \left[0.15 \times \frac{4.64^2}{2} + (1.378 - 0.15) \times \frac{4.64^2}{3} \right] \cdot \cos \alpha - 0.152$$

$$= 9.921$$

(assuming LU has been chosen so that the shear force at LU is zero). This should agree with the corresponding value from elements Nos 11 and 12, namely 9.903. These are within a 10% difference so average these values giving 9.912. Had the difference been greater than 10% then the position of line LVU would have needed altering.

From element No. 10 the vector moment at the end LN of element No. 13 at right-angles to the reinforcement direction LU is 2.711. Taking moments about UI, the vector moment IU

$$= - \left[4.64 \times \frac{0.15^2}{2} + \frac{4.64}{2} \times (1.378 - 0.15) \right.$$

$$\left. \times \left(0.15 + \frac{1.228}{3} \right) \right] \cdot \cos \alpha + 2.711$$

$$= 1.125$$

All design moments are now known. To give a sensible, with regard to thoughts on serviceability, reinforcement distribution, the design moments

190 *Reinforced concrete slabs*

are distributed as shown in Figure 4.24(b). This is Hillerborg's suggestion. He also says that other distributions are of course possible. Naturally, different designers will propose slightly different distributions. Hillerborg says that there must always be enough reinforcement near the columns and all parts have to have some reinforcement. This latter is taken care of by using code minimum requirements. These latter are particularly important for those parts of Figure 4.24(b) to which no design moments are allocated. Hillerborg states that all bottom reinforcement shall continue to the column lines. In this example the dimensions for the actual structure were metres, but are not stated above. The loading was unit load per unit area but the dimensions are not stated above – it was naturally not unity for the actual structure.

On Figure 4.24(b) the bending moment for which reinforcement has to be calculated for DC $= -1.274 - 0.294 - 2.058 = -3.626$. The negative sign means that the reinforcement must be in the top of the slab. This bending moment for the section DC decides the total reinforcement which is then distributed as shown in Figure 4.24(b). The total design moments for the other sections shown on Figure 4.24(b) are as follows:

AB, $-1.183 + 0.096 - 3.202 = -4.289$

EF, $1.904 + 1.402 + 1.658 = 4.964$

GH, $9.416 + 9.912 = 19.328$

JK, $-2.711 - 0.363 - 0.458 = -3.532$

LM $-0.715 - 0.294 - 0.715 = -1.724$

Point F would be on the centre line. Points F and H are not necessarily the same point.

The reader should have thoroughly understood the previous examples in this section before studying this example.

4.8 Slab with hole using Hillerborg's strip method

The following example shows a way in which Hillerborg recommends using his strip method for a slab with a hole. This particular method is less economic than the previous Hillerborg examples in this book in that, for example, a portion of load on portion No. 2a (see Figure 4.26) is carried to portion No. 8 which then carries it to strips Nos 10 and 11 which then carry it to the supports. A normal economic advantage of Hillerborg's methods is that each portion of load is only carried once directly to the supports. However, for a slab with a hole the following is an economic method.

Example 4.15. The slab shown in Figure 4.26(a) carries a uniformly distributed load of 12 kN/m² except where the hole 2 m by 1.5 m occurs. Suggest a design solution using Hillerborg's methods.

The first step is to ignore the opening and determine moments as previously. There are generally two alternatives: one is to span each

Figure 4.26

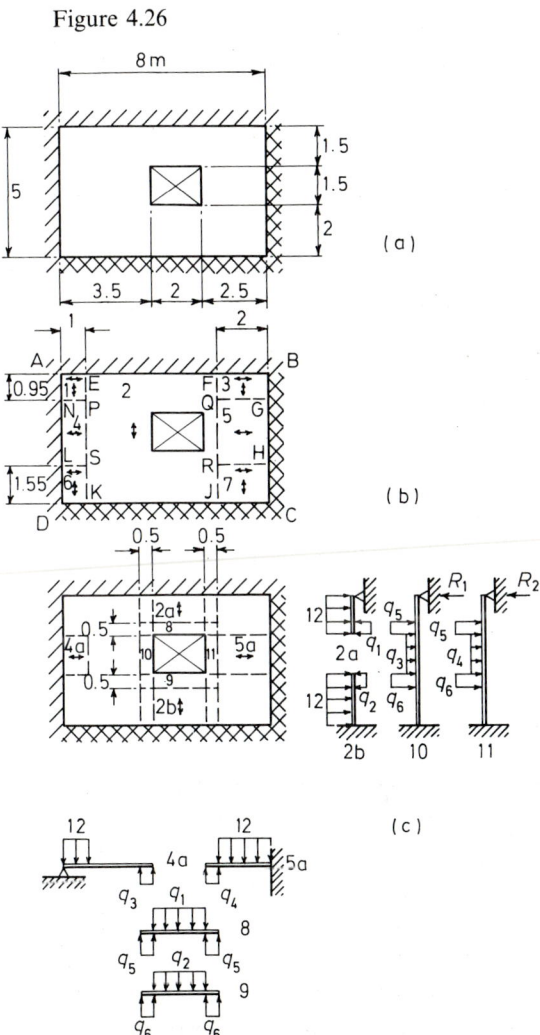

portion of load in only one direction (as in Figure 4.17), the other is to span each portion of load, part in one direction and the remainder in a direction perpendicular to it. Using this latter method Hillerborg suggests discontinuity lines as shown in Figure 4.26(b) and carrying the loads in the directions shown with arrows.

For strip No. 2, the strip spanning from EF to KJ, the writer would probably calculate the maximum moments from Table 7.2 (page 239): span $12 \times 5^2/14.2 = 21.13$ kN m/m and support $-12 \times 5^2/8 = -37.5$ kN m/m. Alternatively a position for the point of contraflexure can be chosen. For this example Hillerborg chooses this at 3.8 m from the free edge; this then gives a maximum span moment

$$= 12 \times 3.8^2/8 = 21.7 \text{ kN m/m}$$

and a maximum support moment

$$= -12 \times 1.9 \times (5-3.8) - \tfrac{12}{2} \times (5-3.8)^2 = -36 \text{ kN m/m}$$

Hillerborg has guessed, by his experience, a point of contraflexure which gives a similar solution to the elastic theory used above. With plasticity this departure from elastic theory is naturally permissible. If Hillerborg has a choice he normally prefers to increase the elastic support moment slightly whereas detailers normally prefer the support and span moments to be equal.

For strip No. 4–2–5, the strip spanning from NL to GH, Hillerborg chooses a point of contraflexure 1 m from GH. This makes the simple span between NL and the point of contraflexure symmetrically loaded and the reaction is $12 \times 1 = 12$ kN/m and maximum span moment

$$= 12 \times 1 - 12 \times 1^2/2 = 6 \text{ kN m/m}$$

Then the support moment

$$= -12 \times 1 - 12 \times 1^2/2 = -18 \text{ kN m/m}$$

In this case an elastic analysis gives a slightly lesser span moment.

For portions 1, 3, 6 and 7 Hillerborg, because they are corner portions, allows half the loading to be taken in each direction. Then for strips 1–2–3 and 6–2–7 the bending moments are half those of strip 4–2–5, and are therefore: maximum span moment $= 6/2 = 3$ kN m/m and support moment $= -18/2 = -9$ kN m/m.

Then for either strip 1–4–6 or 3–5–7 Hillerborg chooses the point of contraflexure 0.6 m from the fixed support. This makes the calculation simple in that the simply supported span between this point of contraflexure and the edge AB is symmetrically loaded and the end reaction

$$= 6 \times 0.95 = 5.7 \text{ kN/m}$$

The maximum span moment is the same all along the central portion and

$$= 5.7 \times 0.95 - 6 \times 0.95^2/2 = 2.7 \text{ kN m/m}$$

The support moment, at edge DC,

$$= -5.7 \times 0.6 - 6 \times 0.6^2/2$$
$$= -4.5 \text{ kN m/m}$$

Edge strips bounding the hole are shown in Figure 4.26(c) and the directions in which the loading will be carried in portions Nos 2a, 2b, 4a and 5a are shown with arrows. These portions are within the portions Nos 2, 4 and 5 shown in Figure 4.26(b).

Edge strip No. 8 supports portion No. 2a with a uniformly distributed reaction q_1 as shown in Figure 4.26(c). Taking moments about edge EF

$$q_1 \cdot 0.5 \times 1.25 = 12 \times 1.5 \times 0.75$$
$$\therefore q_1 = 21.6 \text{ kN/m}^2$$

Then the reaction at the support (EF)

$$= 12 \times 1.5 - 21.6 = 7.2 \text{ kN/m}$$

The distance from this support to the point of zero shear (that is maximum moment)

$$= 7.2/12 = 0.6 \text{ m}$$

Then the maximum span moment

$$= 7.2 \times 0.6 - 12 \times 0.6^2/2 = 2.16 \text{ kN m/m}$$

Edge strip No. 9 supports portion No. 2b with a uniformly distributed reaction q_2 as shown in Figure 4.26(c). Taking moments about edge KJ the support moment

$$= q_2 \cdot 0.5 \times 1.75 - 12 \times 2 \times 1 = 0.875 \cdot q_2 - 24$$

Now if this moment $= -24$ then q_2 is zero. The basic case (that is ignoring the hole) gave a moment at this support of -36 kN m/m. To maintain this value would give a negative value of q_2 meaning that portion No. 9 would not be supporting portion No. 2b but dragging down on it. Then portion No. 9 would be lifting portions Nos 10 and 11. The general idea was that portion No. 9 would support portion No. 2b and that portions Nos 10 and 11 would support portion No. 9. In this case Hillerborg decided not to allow q_2 to be negative and yet allow the support moment to be as near to -36 as possible. Thus this support moment is taken as -24 kN m/m and then q_2 is zero. The portions Nos 2b and 9 of the slab thus cantilever from the support (KJ).

Edge strip No. 10 supports portion No. 4a with a uniformly distributed reaction q_3 as shown in Figure 4.26(c). Taking moments about edge NL

$$q_3 = \frac{12 \times 1 \times 0.5}{0.5 \times 3.25} = 3.7 \text{ kN/m}^2$$

The maximum span moment will be much less than the basic of 6 kN m/m, see earlier. Hillerborg takes this latter figure.

Edge strip No. 11 supports portion No. 5a with a uniformly distributed reaction q_4 as shown in Figure 4.26(c). Taking moments about GH the support moment

$$= q_4 \cdot 0.5 \times 2.25 - 12 \times 2 \times 1 = 1.125 \cdot q_4 - 24$$

If this is made -18 to agree with the basic case, see earlier, then

$$q_4 = 6/1.125 = 5.3 \text{ kN/m}^2$$

Half of edge strip No. 8 is carried on edge strip No. 10 (and half on edge strip No. 11), see Figure 4.26(c), therefore

$$q_5 \cdot 0.5 \times 0.5 = q_1 \cdot 0.5 \times 1.0$$

$$\therefore q_5 = 2q_1 = 43.2 \text{ kN/m}^2$$

and the moment at mid span for edge strip No. 8 (taking moments about the mid span)

$$= 0.5 \cdot q_5 \cdot 1.25 - 1.0 \cdot q_1 \cdot 0.5 = 16.2 \text{ kN m/m}$$

Half of edge strip No. 9 is carried on edge strip No. 10 (and half on edge strip No. 11), see Figure 4.26(c), therefore

$$q_6 \cdot 0.5 \times 0.5 = q_2 \cdot 0.5 \cdot 1.0$$

$$\therefore q_6 = 2q_2 = \text{zero}$$

Slab with hole using Hillerborg's strip method

and the moment at mid span for edge strip No. 9 (taking moments about the mid span)

$$0.5 \cdot q_6 \cdot 1.25 - 1.0 \cdot q_2 \cdot 0.5 = 0$$

For strip No. 10, taking moments about edge KJ, the support moment

$$= 5R_1 - 43.2 \times 0.5 \times 3.75 - 3.7 \times 1.5 \times 2.75$$
$$= 5R_1 - 96.3$$

Hillerborg chooses this moment as -24 kN m/m. Then

$$R_1 = (96.3 - 24)/5 = 14.5 \text{ kN/m}$$

The distance from the free edge to the point of zero shear force (that is maximum moment)

$$= (14.5/43.2) + 1 = 1.336 \text{ m}$$

Then the maximum span moment

$$= 14.5 \times 1.336 - 43.2 \cdot (1.336 - 1)^2/2 = 16.9 \text{ kN m/m}$$

For strip No. 11, taking moments about edge KJ, the support moment per metre run

$$= 5R_2 - 43.2 \times 0.5 \times 3.75 - 5.3 \times 1.5 \times 2.75$$
$$= 5R_2 - 102.8$$

Hillerborg chooses this moment as -24 kN m/m because he chose this for portion No. 10, see earlier. Then

$$R_2 = (102.8 - 24)/5 = 15.8 \text{ kN/m}$$

Figure 4.27

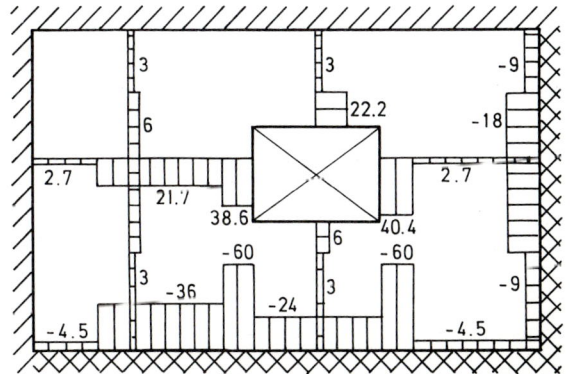

196   Reinforced concrete slabs

The distance from the free edge to the point of zero shear force (that is maximum moment)

$$= (15.8/43.2) + 1 = 1.336 \text{ m}$$

Then the maximum span moment

$$= 15.8 \times 1.366 - 43.2 \times (1.366 - 1)^2/2 = 18.7 \text{ kN m/m}$$

In Figure 4.27 the distribution of moments is shown, calculated by adding the moments of the basic case and the moments in the edge strip around the hole.

4.9   Hillerborg elements with shear forces along their edges

When shearing forces are introduced along the edges of corner supported elements, this gives a certain degree of freedom to choose the position of the boundaries of the elements. However it is advisable to choose the boundaries so that the shearing forces along them are as small as possible. In the following illustrative example it is natural to arrange the boundaries so that the average shearing force along a boundary across the slab is zero.

Example 4.17. The slab shown in Figure 4.28(a) carries a uniformly distributed load of 4 kN/m and a point load $P = 75$ kN. Suggest a design solution using Hillerborg's methods.

To commence; the lines across the slab of zero shearing force will be determined.

From Figure 4.28(b), taking moments about end AD

$$R \times 6 = 75 \times 2 + (5 \times 6) \times 4 \times 3, \text{ therefore } R = 85 \text{ kN}$$

Shear force at X−X for $x = 2$ to 6,

$$= 85 - (6 - x) \times 5 \times 4 = 20x - 35$$

This is linear, when $x = 2$, shear force $= 5$ kN

when $x = 6$, shear force $= 85$ kN

Shear force at X−X for $x = 0$ to 2,

$$= 20x - 35 - 75 = 20x - 110$$

This is linear, when $x = 0$, shear force $= -110$ kN

when $x = 2$, shear force $= -70$ kN

Hence shear force is zero at $x = 2$, because of sign change in shear force. From Figure 4.28(c), taking moments about end **AB**

$$S \times 5 = 75 \times 1 + (5 \times 6) \times 4 \times 5/2, \text{ therefore } S = 75 \text{ kN}$$

Shear force at $X-X$ for $x = 0$ to 4,

$$= 75 - (6x) \times 4 = 75 - 24x$$

This is linear, when $x = 0$, shear force = 75 kN

when $x = 4$, shear force = -21 kN

Shear force at $X-X$ for $x = 4$ to 5,

$$= 75 - 24x - 75 = 24x$$

Figure 4.28 (a)–(c)

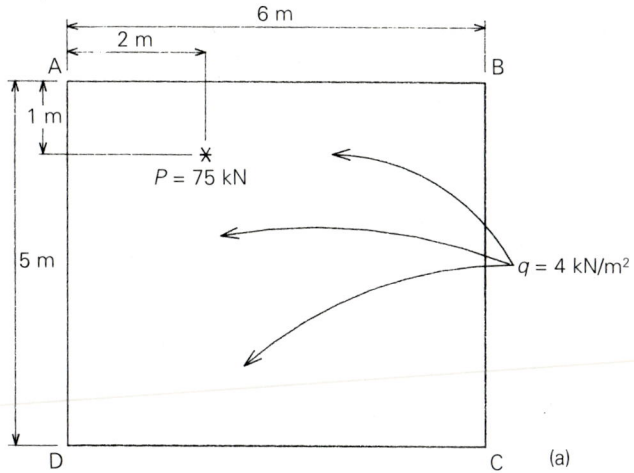

(a)

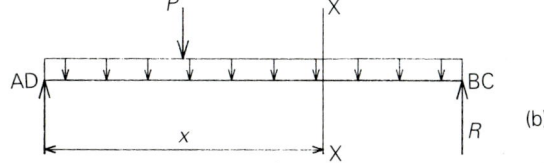

(b)

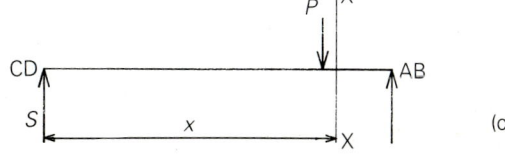

(c)

This is linear, when $x = 4$, shear force $= -96$ kN

when $x = 5$, shear force $= -120$ kN

Hence shear force is zero in the portion $x = 0$ to 4.

Put $75 - 24x = 0$, therefore $x = 3.125$ m for position of zero shear force. Therefore in Figure 4.28(d) the lines of average zero shearing force EH and FG can be used to divide the slab into four Hillerborg rectangular elements. The boundary which passes through the point load divides this load into two parts, P_1 and $75 - P_1$, supported by separate elements. The shearing force for boundary EJ namely V_1 is therefore zero. To explain this statement, for a simply supported beam, when there is a change in sign of the shear force at a point load, then the shear force at the point load is zero. At the other boundaries between the elements, the shear forces V_2,

Figure 4.28(d), (e)

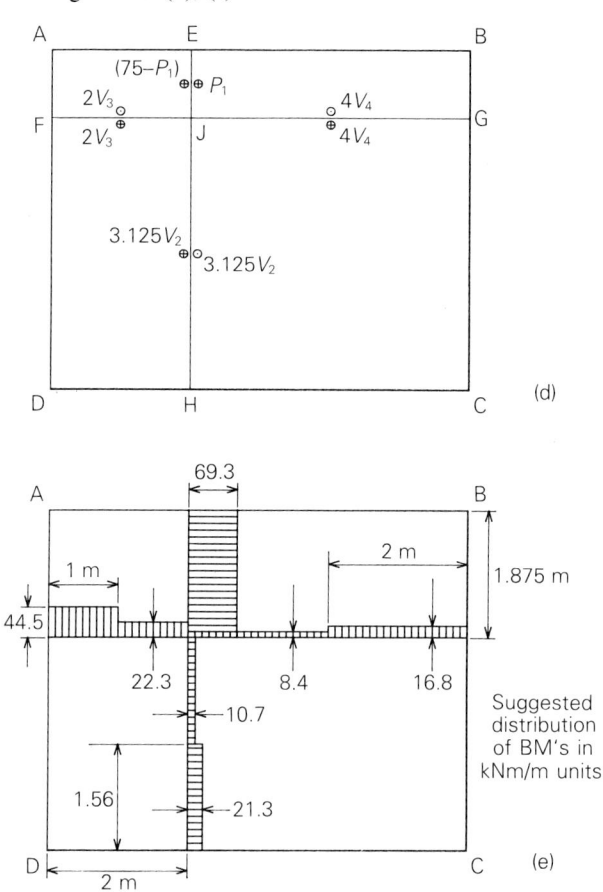

Elements with shear forces along their edges 199

V_3 and V_4, which are shear forces per unit length, are shown at their centroids in Figure 4.28(d).

For element AEJF, taking moments about AF, vector moment EJ

$$= -2V_3 \times 1 + (75 - P_1) \times 2 + (2 \times 1.875) \times 4 \times 1$$

$$= -2V_3 + 165 - 2P_1 \tag{4.15}$$

For element AEJF, taking moments about AE, vector moment JF

$$= (75 - P_1) + (2 \times 1.875) \times 4 \times 1.875/2 - 2V_3 \times 1.875$$

$$= -P_1 + 89.06 - 3.75V_3 \tag{4.16}$$

For element EBGJ, taking moments about BG, vector moment JE

$$= P_1 \times 4 + (4 \times 1.875) \times 4 \times 2 - 4V_4 \times 2$$

$$= 4P_1 + 60 - 8V_4 \tag{4.17}$$

For element EBGJ, taking moments about BE, vector moment GJ

$$= P_1 \times 1 - 4V_4 \times 1.875 + (1.875 \times 4) \times 4 \times 1.875/2$$

$$= P_1 - 7.5V_4 + 28.13 \tag{4.18}$$

For element FJHD, taking moments about FD, vector moment JH

$$= 2V_3 \times 1 + (2 \times 3.125) \times 4 \times 1 + 3.125V_2 \times 2$$

$$= 2V_3 + 6.25V_2 + 25 \tag{4.19}$$

For element FJHD, taking moments about DH, vector moment FJ

$$= 3.125V_2 \times 3.125/2 + 2V_3 \times 3.125$$
$$+ (3.125 \times 2) \times 4 \times 3.125/2$$

$$= 4.883V_2 + 6.25V_3 + 39.06 \tag{4.20}$$

For element JGCH, taking moments about GC, vector moment HJ

$$= (4 \times 3.125) \times 4 \times 4/2 - 3.125V_2 \times 4 + 4V_4 \times 2$$

$$= 100 - 12.5V_2 + 8V_4 \tag{4.21}$$

For element JGCH, taking moments about HC, vector moment JG

$$= -3.125V_2 \times 3.125/2 + 4V_4 \times 3.125$$
$$+ (4 \times 3.125) \times 4 \times 3.125/2$$

$$= -4.883V_2 + 12.5V_4 + 78.13 \tag{4.22}$$

Solving equations (4.15) to (4.22), either by the classical schoolboy method, which is quite easy, or with a computer, gives

$P_1 = 15.79$ kN, $V_2 = 3.442$ kN,

$V_3 = 1.74$ kN, $V_4 = -0.8714$ kN

For element AEJF, vector moment EJ, from equation (4.15)

$= 129.9$ kN m, or per unit length

$= 129.9/1.875 = 69.30$ kN m/m

For element AEJF, vector moment JF, from equation (4.16)

$= 66.75$ kN m, or per unit length

$= 66.75/2 = 33.37$ kN m/m

For element EBGJ, vector moment GJ, from equation (4.18)

$= 50.46$ kN m, or per unit length

$= 50.46/4 = 12.61$ kN m/m

For element FJHD, vector moment JH from equation (4.19)

$= 49.99$ kN m, or per unit length

$= 49.99/3.125 = 16.00$ kN m/m

It is desirable to have a strong band of slab beneath the point load and also at the edge AB, so the moment will be kept the same along EJ. Along the other edges the slab will be made stronger so the distribution of moments is chosen as shown in Figure 4.28(e).

In the above example, if there had been no point load, then the boundary line FG would have traversed through the point load and for each of the four elements, moments could have been taken about EJ, JH, FJ and JG, without involving the point load and therefore not having to assess the proportion of it carried by each element. These equations give the support reactions and then the moments can be calculated for EJ, JH, FJ and JG. Also the shear forces of this example, along FG and EH, are all zero. To explain this statement, for a simply supported beam, when there is a change in the sign of the shear force at a point load, then the shear force at the point load is zero. Thus the necessity for the shear forces in example 4.17 is because of the distributed loading causing the line of zero average shear FG to not pass through the position of the point load.

4.10 Traditional U.K. design office methods

From the earliest of days when slabs of difficult shapes had to be designed without suitable published guidance and even up to the present day, the practice in specialist reinforced concrete U.K. design offices has been essentially to carry loads and portions of loadings by main, secondary etc. beams formed with reinforcement within the slab depth and all designed simply for adequate strength against bending moments and shear forces. This kind of design is inconsistent with the satisfying of deflection considerations of the various internal beam and slab members. The individual sections of the conceived constituent members are designed by code methods based formerly on elastic theory but now on plastic theory.

The problem of holes is dealt with by pushing aside the reinforcement, which would have traversed the holes, to form narrow beams at the sides of the holes. Then nominal corner bars are placed at the corners to reduce cracking there because from photo-elastic tests high stresses are known to occur at these corners. Away from the hole this impairs the reinforcement system locally. This is checked by calculations making simplifying assumptions and extra reinforcement placed locally accordingly, although often no such reinforcement is necessary. Example 4.15 would be designed by code tables and then the treatment for the hole would be as just described.

4.11 General discussion of design methods for two-way and flat slabs

This discussion excludes a design method which consists of assessing the distribution of bending moments and shear forces by elastic analysis and then designing the reinforcement by code methods either elastic (CP 114) or plastic (CP 114, CP 110 and BS 8110). This is the best method in the writer's opinion, but limited to those slabs of shapes and loadings covered by tables and computer programs.

Johansen's[8,9] yield-line method is attractive in that it considers the way in which slabs collapse. It is upper bound. But if the most sensible modes of failure are considered the design should not be very significantly upper bound. It usually, however, commences with a most uneconomic reinforcement layout. Reinforcement can be curtailed but this involves extra mechanisms being considered and the process of curtailment is not particularly systematic. The method is of course most suitable for mesh reinforcement.

Kemp's[21] method can be used for many of the problems, except for example those dealing with triangular and trapezoidal shapes of slabs, which can be solved with Hillerborg's strip and advanced methods. It is easy to understand but laborious in practice. Just a single point load is easy to deal with but distributed loads, practically, have to be considered as individual loads, one on each element of a slab as though a great number of point loads, and an analysis has to be effected for each of these 'point loads' and then all analyses finally added together. Sensible engineering choice enters into the method so it is not very suitable for computer programming.

The method of Fernando and Kemp[22] has to make use of a computer.[23,24] It is similar to a grillage elastic analysis excluding torsional resistance of members. It is more limited than Hillerborg's methods in that it cannot deal with the sort of problem found in Example 4.14.

Hillerborg produced a considerable treatise in his book[39] justifying his methods. After this justification he gives many different examples but one has to read, digest and understand the considerable treatise before one can understand the examples. The examples also assume that the reader has considerable other background knowledge and experience. The justification is difficult and very time-consuming to follow and thoroughly understand, yet the method is easier to use in practice than the Johansen, Kemp and Fernando and Kemp methods already discussed. It is more versatile than them; seemingly it can be used to design any shape of slab with any loading. It has advantages over Johansen's method in being lower bound, giving sensible practical arrangements of reinforcement, and allowing easy curtailment of reinforcement. It is much more economic than the traditional U.K. design office methods (see Section 4.10).

Wood[17] and Armer[19,20] have studied and made tests to justify Hillerborg's methods. Several U.K. and U.S.A. textbooks have included Hillerborg's strip method. A very few have, seemingly reproduced Hillerborg's advanced method from the scientific works just quoted of Wood and Armer. They are certainly not written for students.

The author has given examples, covering most types of loading and a very difficult shape in Example 4.14, completely explained as they are pursued. They are best attempted in the sequence given and some syllabi may exclude Example 4.14, whch the author found difficult to explain. After understanding these examples the reader may find he can understand many of the further examples given in Hillerborg's book.[39]

If the elastic analyses described in the first paragraph of this section are not available from tables or computer packages then Hillerborg's methods or the method of Section 4.9 seem to offer the best solution to any problem.

References

1. Jones, L. L. and Wood, R. H., *Yield-Line Analysis of Slabs*, Thames and Hudson and Chatto and Windus, London (1967).
2. Wood, R. H. and Armer, G. S. T., The theory of the strip method for design of slabs. *Proceedings of the Institute of Civil Engineers*, Oct. (1968).
2a. Wilby, C. B., Design of reinforced concrete slabs. *Civil Engineering*, August (1980).
3. Scott, W. L., Glanville, W. H. and Thomas, F. G., *Handbook on B.S. Code CP 114 (1957–65)*, Concrete Publications Ltd.
4. Westergaard, H. M. and Slater, W. A., Moments and stresses in slabs. *A.C.I. Proceedings*, **17**, p. 415 (1921).
5. Westergaard, H. M., Formulas for the design of rectangular floor slabs and the supporting girders. *A.C.I. Proceedings*, **22**, p. 26 (1926).
6. Marcus, H., *Die Theorie Elastischer Gewebe und ihre Anwendung auf die Berchnung beigsaner Platten*, 2nd edn, p. 362, Julius Springer, Berlin.
7. Loser, B., *Bemessungsverfahren*, W. Ernst und Sohn, Berlin.
8. Johansen, K. W., *Pladeformler*, Polyteknish Forening, Copenhagen, 2nd edn (1949).
9. Johansen, K. W., *Pladeformler: Formelsamling*, Polyteknish Forening, Copenhagen, 2nd edn (1954).
10. Di Stassio, J. and Van Buren, M. P., Slabs supported on four sides. *A.C.I.*, 7, No. 3, Jan.–Feb. (1936); *Proceedings*, **32**, p. 350.
11. Bertin, R. L., Di Stassio, J. and Van Buren, M. P., Slabs supported on four sides. *A.C.I.*, 16, No. 6, June (1945); *Proceedings*, **41**, p. 537.
12. European Committee for Concrete, Information Bulletin No. 35, English Translation issued by Cement and Concrete Association, London (1962).
13. Bares, R., *Tables for the Analysis of Plates, Slabs and Diaphragms Based on the Elastic Theory*, Bauverlag, Germany (1971).
14. Hendry, A. W. and Jaeger, L. G., *The Analysis of Grid Frameworks and Related Structures*, Chatto and Windus, London (1958).
15. Bares, R. and Massonnet, C. *Analysis of Beam Grids and Orthotropic Plates by the Guyon–Massonnet–Bares Method*, Crosby Lockwood, London (1968). Earlier non-English edition, Prague (1966).
16. Taylor, R., Hayes, B. and Mohamedbhai, G. T. G., Coefficients for the design of slabs by the yield-line theory. *concrete*, **3**, No. 5, p. 171, May (1969).
17. Wood, R. H. and Armer, G. S. T., The theory of the strip method for design of slabs. *Proceedings of the Institute of Civil Engineers*, **41**, p. 287, Oct. (1968).
18. Armer, G. S. T., Ultimate load tests of slabs designed by the strip method. *Proceedings of the Institute of Civil Engineers*, **41**, p. 313, Oct. (1968).
19. Armer, G. S. T., The strip method: a new approach to the design of slabs. *Concrete*, **2**, No. 9, p. 358, Sept. (1968).
20. Crawford, R. L., Limit design of reinforced concrete slabs. *Proceedings of the American Society of Civil Engineers, Engineering Mechanics Division*, EM5, p. 321, Oct. (1964).
21. Kemp, K. O., A strip method of slab design with concentrated loads or supports. *Structural Engineer*, **49**, No. 12, Dec. (1971).

22. Fernando, J. S. and Kemp, K. O., A generalized strip deflexion method of reinforced concrete slab design. *Proceedings of the Institute of Civil Engineers*, Part 2, **65**, Mar. (1978).
23. Wilby, C. B., Computation of the strip deflection method. *Proceedings of the Institute of Civil Engineers*, Part 2, TN244, June (1980).
24. Wilby, C. B., The strip deflection method of reinforced concrete slab design. B. Eng. (Tech.) major project, University of Wales Institute of Science and Technology (1979).
25. Wilby, C. B., Design tables for two-way spanning slabs. *The Structural Engineer*, Oct. (1981) (Synopsis on p. 337).
26. Thakkar, M. C. and Rao, J. K. S., Design of two-way reinforced concrete rectangular slabs by modified Hillerborg's Strip Method. *Indian Concrete Journal*, April (1970).
27. Wilby, C. B., Design tables for slabs using strip deflection method. *Indian Concrete Journal* (1982).
28. Dunham, C. W., *Advanced Reinforced Concrete*, McGraw–Hill, U.S.A. (1964).
29. Ferguson, P. M., *Reinforced Concrete Fundamentals*, John Wiley, U.S.A. (1973).
30. Wilby, C. A., The plastic and elastic design and analysis of reinforced concrete containers. *M.Sc. Thesis, University of Oxford* (1976).
31. Wilby, C. A., Structural analysis of reinforced concrete tanks. *Journal of Structural Division, Proceedings of the American Society of Civil Engineers*, May (1977).
32. Wilby, C. A., General optimisation of the design of rectangular containers. *Indian Concrete Journal*, May (1977).
33. Wilby, C. A., Optimization of design of circular tanks. *Proceedings of the Institute of Civil Engineers*, Part 2, TN170, Dec. (1977).
34. Wilby, C. A., Optimization of design of tanks of regular polygonal shape. *Indian Concrete Journal*, May–June (1978).
35. Wilby, C. A., Application of the strip method to the design of tanks. *Indian Concrete Journal*, April (1979).
36. Wilby, C. A., Rectangular tanks. *Civil Engineering*, July (1979).
37. Hughes, B. P., *Limit State Theory for Reinforced Concrete*, Pitman, London (1971).
38. Wilby, C. B. and Naqvi, M. M., Deflections of rectangular plates supported elastically. *Proceedings of the Institute of Civil Engineers*, Paper 7411, August (1971).
39. Hillerborg, A., Strip method of design, A Viewpoint Publication, London (1975).

5

Columns and walls

5.1 General

BS 8110 and CP 110 recommend ultimate load design using plastic theories and not elastic theory as was allowed by CP 114.

5.2 Slender and short columns

Slender and *short* columns are ones affected and not affected by buckling, respectively. BS 8110 defines a column as short when both the ratios l_{ex}/h and l_{ey}/b are less than 15 for a braced and 10 for an unbraced column. A slender column is designed as a short column required to withstand an additional bending moment due to buckling.

5.3 Axially loaded short columns

The assumptions of this analysis are given in Section 3.7.1. Figure 5.1(a) shows the cross section of a column and Figure 5.1(b) the distribution of stress across the cross section. Basically the concrete strains in a plastic fashion until the reinforcement yields (or, for high yield steel, its strain is so great as) to realise the maximum strain which can be tolerated by the concrete. The latter occurs when the stress in the concrete is about $0.67f_{cu}$, that is the 0.67 is based on experimental evidence; the 150 mm cube strength is affected by nearness of loading platens because

Figure 5.1

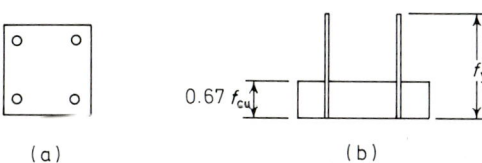

(a) (b)

of friction between concrete and platens restricting movement associated with Poisson's ratio. Hence resolving vertically

$$\text{Ultimate axial load} = 0.67 f_{cu} A_c + f'_y A_{sc} \qquad (5.1)$$

where A_c = area of concrete, A_{sc} = area of compression steel, f_{cu} = characteristic strength of concrete and f'_y = characteristic strength of steel in compression. $f'_y/\gamma_m = f_y/1.15$. f_{cu} is divided by a γ_m of 1.5 so the ultimate axial load for design purposes

$$= (0.67/1.5) f_{cu} A_c + 0.87 f_y A_{sc}$$

$$= 0.45 f_{cu} A_c + 0.87 f_y A_{sc}$$

where f_y is the characteristic tensile strength of the steel. As loads in practice are rarely axial, BS 8110 recommends for design an ultimate axial load

$$= 0.4 f_{cu} A_c + 0.75 A_{sc} f_y \qquad (5.2)$$

for columns that cannot be subjected to significant moments. Or

$$= 0.35 f_{cu} A_c + 0.67 f_y A_{sc}$$

for short braced columns (i.e. for example columns held from buckling laterally by substantial walls) supporting an approximately symmetrical arrangement of beams, and where

(a) the beams are designed for uniformly distributed imposed loads; and
(b) the beam spans do not differ by more than 15% of the longer.

Example 5.1. Design a short reinforced concrete column, which is not subjected to significant moments, for an ultimate axial load of 2900 kN.

Suppose $f_{cu} = 25$ N/mm² and $f_y = 250$ N/mm² and assume say 2% of reinforcement, then

$$A_{sc} = 0.02(A_c + A_{sc}), \therefore A_{sc} = 0.02041 A_c$$

From equation 5.2,

$$2900 = 0.4 \times 25000 \times A_c + 0.75 \times 250000 \times 0.02041 A_c$$

Therefore $A_c = 0.2097$ m² and

$$A_{sc} = 0.02041 \times 0.2097 = 0.004281 \text{ m}^2$$

Gross cross sectional area of column = 0.2097 + 0.0043 = 0.2140 m². Use a column 470 mm square with (see Table 3.2) four 32 mm diameter and four 20 mm diameter bars.

5.4 Plastic analysis for eccentrically loaded short columns

This is a column required to be designed for an ultimate axial load N combined with an ultimate bending moment M where $M = Ne$. Figure 5.2(a) shows the cross-section of a column of any shape, Figure 5.2(b) the distribution of stress assumed by BS 8110 for design purposes,

Figure 5.2

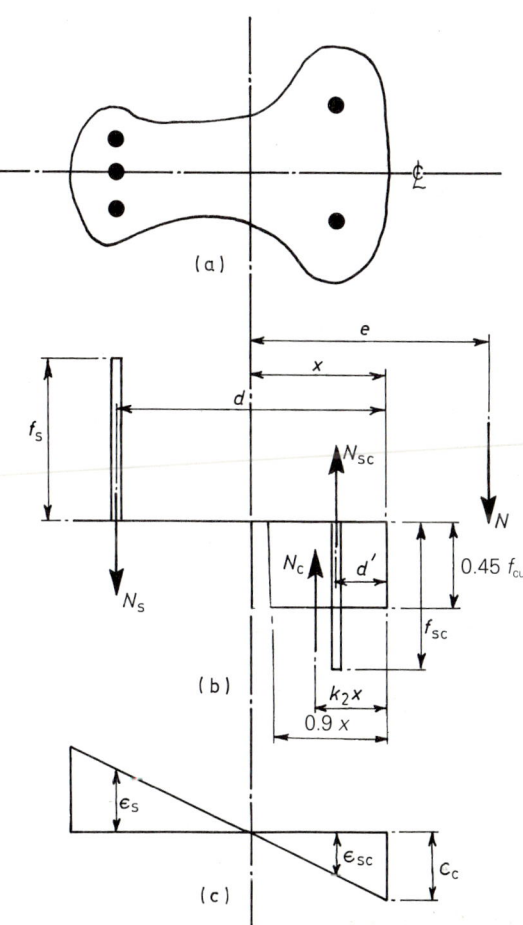

and Figure 5.2(c) the distribution of strain. The $0.45f_{cu}$ comes from $(0.67/1.5)f_{cu} = 0.45f_{cu}$ where $\gamma_m = 1.5$.

Resolving vertical forces $N = N_c + N_{sc} - N_s$ where N_c is the force in the concrete over the gross area in compression, and N_{sc} and N_s are forces in the steel in compression and tension, respectively. Therefore

$$N = N_c + N_{sc} - N_s = 0.45f_{cu}A_c + A_{sc}f_{sc} - A_s f_s \tag{5.3}$$

where A_c = area of concrete in compression, A_{sc} and A_s = areas of steel in compression and tension respectively, $k_2 x$, where $k_2 = 0.45$, is the distance to the line of action of N_c (that is to the centroid of A_c), and f_{sc} and f_s = design strengths (stresses) of compression and tension steels, respectively. Taking moments for convenience about the line of action of N_s

$$N(e + d - x) = N_c(d - k_2 x) + N_{sc}(d - d') \tag{5.4}$$

For large eccentricities, failure is initiated by the tension steel yielding or straining excessively (for high-yield steel) causing the value of x to be reduced, see Figure 5.2(c), until eventually the concrete crushes. For small eccentricities the concrete may crush to cause failure when the steel remote from N is in compression or only modestly strained in tension. Between these two types of failure we have what is called a *balanced condition* where the failure is caused by simultaneous crushing of the concrete and yielding or excessive straining of the tension steel. For this condition let $N = N_b$ and $e = e_b$. Then from Figure 5.2(c), taking $\varepsilon_c = 0.0035$ as recommended by BS 8110 because tests show that this is approximately the maximum strain which is experienced at crushing of the concrete, and taking $\varepsilon_s = f_s$

$$\frac{x}{d} = \frac{\varepsilon_c}{\varepsilon_c + \varepsilon_s} = \frac{0.0035}{0.0035 + f_s/E_s} \tag{5.5}$$

where f_s is $f_y/\gamma_m = f_y/1.15$, and $E_s = 200 \text{ kN/mm}^2$ from BS 8110: Part 1, Fig. 2.2.

For this condition, for a defined section, x is given by equation 5.5, then A_c calculated and then equation 5.3 gives N_b if we know f_{sc}. From Figure 5.2(c) ε_{sc} can be found for this balanced condition and, from the stress–strain curve for this steel, f_{sc} can be determined. Normally ε_{sc} is large enough to develop the maximum stress in the steel (it might not do so if this steel were unusually near to the neutral axis). Then e_b can be determined from equation 5.4.

Loads with eccentricities less than e_b cause primary compression failures at ultimate loads greater than N_b, whereas loads with eccentricities greater than e_b cause primary tension failures at loads smaller than N_b.

Thus if $e > e_b$, $f_s = f_y/1.15$ and $\varepsilon_c = 0.0035$. $f_{sc} = f'_y/1.15$ where f'_y is the yield strength of the steel in compression should it be a different steel to that in tension (the yield stresses in tension and compression are equal for any particular steel according to BS 8110: Part 1, Fig. 2.2. Then for a defined section and a known e, equations 5.3 and 5.4 can be solved for the two unknowns N and x.

But if $e < e_b$, $\varepsilon_c = 0.0035$, so equation 5.5 has two unknowns x and f_s. If this f_s is substituted in equations 5.3 and 5.4 and $f_{sc} = f_y/1.15$ and then N eliminated between these two equations, a cubic equation for x results. It may be solved by trial and error (a computer can help), estimating sensible values of x, or by using a computer program for solving a cubic equation. When x is obtained N can then be obtained from either of the equations from which it was eliminated.

In the first case ε_{sc} and in the second case ε_s and ε_{sc} can be determined finally from Figure 5.2(c) to see if they are great enough to correspond to the values of f_{sc} and f_s assumed, using the stress–strain curve of CP 110, Fig. 2. If not, then values of f_{sc} and f_s are estimated and the above calculations repeated until the values assumed for f_{sc} and f_s have value ε_{sc} and ε_s which agree with their values on the stress–strain curves.

In the following examples, eccentricity is specified from the centre line of a column, as this is a more practical case for the reasons given in Section 5.6.

Example 5.2. The cross section of a column is rectangular of width 250 mm ($= b$) by depth 450 mm ($= h$), and $A_s = A_{sc} = 1473$ mm² (three 25 mm diameter bars – see Table 3.2), $d' = 50$ mm and $d = 450 - 50 = 400$ mm. If $f_{cu} = 25$ N/mm², f_y for A_s and A_{sc} is 250 N/mm², $E_s = 200$ kN/mm² and the eccentricity of the line of action of the load from the centre line of the column $= e_1 = 420$ mm, determine the BS 8110 ultimate load for the column.

For balanced design condition $f_s = 250/1.15 = 217.4$ N/mm². From equation 5.5

$$\frac{x}{0.4} = \frac{0.0035}{0.0035 + 0.2174/200} \quad \therefore x = 0.3052 \text{ m}$$

From Figure 5.2(c),

$$\varepsilon_{sc} = 0.0035 \times (305.2 - 50)/305.2 = 0.002\,927$$

Now the design yield stress corresponds to a strain = $250/(200\,000 \times 1.15) = 0.001\,087$. As $0.002\,927$ is greater than this, $f_{sc} = 250/1.15 = 217.4\ \text{N/mm}^2$.

From equation 5.3,

$$N_b = 0.45 \times 25\,000 \times 0.25 \times 0.3052 \times 0.9 + 0.001\,473$$
$$\times 217\,400 - 0.001\,473 \times 217\,400$$
$$= 772.5 + 320.2 - 320.2 = 772.5\ \text{kN}$$

From equation 5.4,

$$N_b(e_b + 0.4 - 0.3052) = 772.5 \times (0.4 - 0.45 \times 0.3052)$$
$$+ 320.2 \times 0.35 = 315.0$$

$$\therefore e_b = 0.4077\ \text{m}$$

Therefore value of e_1 for balanced design

$$= e_b - x + h/2 = 0.4077 - 0.3052 + 0.225 = 0.3275\ \text{m}$$

This is less than 420 mm, hence failure is by yielding of tension steel. Equation 5.3 gives (assuming yielding of compression steel)

$$N = 0.45 \times 25\,000 \times 0.25 \times 0.9x + 320.2 - 320.2 = 2531x$$
$$e = e_1 - h/2 + x$$
$$\therefore e + d - x = e_1 - h/2 + d = 0.42 - 0.225 + 0.4 = 0.595\ \text{m}$$

From equation 5.4

$$0.595N = 0.45 \times 25\,000 \times 0.25 \times 0.9x(0.4 - 0.45x)$$
$$+ 320.2 \times 0.35$$

From the above two equations in N and x, $x = 0.1646$ and $N = 416.5$ kN.

Now as $\varepsilon_c = 0.0035$, from Figure 5.2(c)

$$\varepsilon_{sc} = 0.0035 \times (164.6 - 50)/164.6 = 0.002\,437$$

Now from BS 8110:Part 1, Fig. 2.2, yield occurs at a stress of $f_y/\gamma_m = 250/1.15 = 217.4\ \text{N/mm}$ and as $E_c = 200\,000\ \text{N/mm}^2$, this stress corresponds to a strain of $217.4/200\,000 = 0.001\,087$. As $0.002\,437$ is greater than this $0.001\,087$ the compression steel must have yielded as assumed. (Had this not been so, it would be necessary to obtain f_{sc} from the strain ε_{sc}, i.e. $f_{sc} = E_c \cdot \varepsilon_{sc}$. Then repeat the above calculations. Then the ε_{sc} calculated would correspond to a slightly different value of f_{sc} to the one

Eccentrically loaded short columns 211

taken. The whole process is repeated as many times as necessary to obtain the required accuracy of N).

Example 5.3. Repeat Example 5.2, only using $e_1 = 180$ mm.

As before, the value of e_1 for the balanced design condition = 0.3275 m. This is greater than 0.18 m, hence failure is by compression of concrete. Equations 5.3, 5.4 and 5.5 become

$$N = 0.45 \times 25000 \times 0.25 \times 0.9x + 288.9 - 0.001\,473f_s \tag{5.6}$$

$$e + d - x = e_1 - h/2 + d = 0.18 - 0.225 + 0.4 = 0.355 \text{ m}$$

$$0.355N = 0.45 \times 25000 \times 0.25 \times 0.9x(0.4 - 0.45x) + 320.2 \times 0.35 \tag{5.7}$$

Then from Figure 5.2(c) and taking $E_c = 200$ kN/mm² = 200 000 000 kN/m².

$$\varepsilon_s = f_s/200\,000\,000$$

then

$$\frac{x}{0.4} = \frac{0.0035}{0.0035 + f_s/200\,000\,000} \tag{5.8}$$

One way of solving these equations is to assume that f_s is say 217 400 kN/m² (that is 250/1.15 N/mm²), then calculate x from the last equation. With these values calculate values of N from the previous two equations. These will normally differ. Adjust the value of f_s and start again. Repeat until the values of N from the two equations are sufficiently in agreement. Alternatively the equations can be algebraically reduced to a cubic equation. It is then very easy and rapid to solve this with an electronic hand programmable calculator and guessing values of x, or with a computer library program. The former method gave $x = 0.3248m$, $f_s = 162\,097$ kN/m², and $N = 903.6$ kN. From Figure 5.2(c)

$$\varepsilon_{sc} = 0.0035 \times (324.8 - 50)/324.8 = 0.002\,961$$

Now from BS 8110:Part 1, Fig. 2.2, yield occurs at a stress of $f_y/\gamma_m = 250/1.15 = 217.4$ N/mm² and as $E_c = 200\,000$ N/mm², this stress corresponds to a strain of $217.4/200\,000 = 0.001\,087$. As 0.002 961 is greater than this 0.001 087 the compression steel must have yielded, as assumed. Also from Figure 5.2(c)

$$\varepsilon_s = \frac{0.0035 \times (400 - 324.8)}{324.8} = 0.000\,810\,3$$

Now from BS 8110:Part 1, Fig. 2.2, yield occurs at a stress of $f_y/\gamma_m = 250/1.15 = 217.4$ N/mm^2 and as $E_c = 200\,000$ N/mm^2, this stress corresponds to a strain of $217.4/200\,000 = 0.001\,087$. As $0.000\,810\,3$ is less than this $0.001\,087$ the tensile steel has not yielded, confirming that failure is by compression of the concrete and therefore

$$f_s = 200\,000 \times 0.000\,810\,3 = 162.1 \text{ N/mm}^2$$

5.4.1 Design of eccentrically loaded columns

To be in accordance with BS 8110 it is probably easiest to choose columns from the Charts of Part 3 of BS 8110. If a column section cannot be obtained in these Charts then it has to be checked as in Section 5.4.

5.5 Reinforced concrete walls

Load-bearing reinforced concrete walls are designed as columns, but if any structural reliance is made on the reinforcement, such reinforcement needs to have ties across the wall to prevent the bars buckling outwards. Such ties are highly undesirable in practice, causing much trouble to both the steelfixer and concretor. It is therefore usually more economical to design the wall as though it contained no reinforcement. It would not, however, be built without any reinforcement because differential settlement, shrinkage and temperature expansion or contraction could all cause cracking, which would be most noticeable on a concrete surface. Such small movements also cause hair cracks between the bricks of brickwork walls, but even if occasional bricks are cracked the cracks blend with the pattern of the wall and are not noticeable to the layman. Cracks in concrete surfaces tend to concentrate into a few of large size, rather than many of a small size, and ramble in various directions in an unsightly way. Consequently for good practice horizontal and vertical reinforcement is placed in both faces of a reinforced concrete wall, whether the wall is load-bearing or not, the horizontal reinforcement usually being nearer the surface than the vertical reinforcement. In practice, the vertical bars are usually made of at least 12 mm diameter, except in the case of very thin walls, as these have to support the horizontal reinforcement. The construction of walls can be very difficult if light reinforcement fabrics are used. The difficulty is in maintaining the desired cover of concrete accurately to the concrete.

5.6 Design of columns to frameworks

In accordance with BS 8110, frameworks are analysed using elastic theory for forces and bending moments, assuming the members to be concentrated at their centre lines (see Chapter 7). The designer may then choose to redistribute these bending moments as described in Chapter 6. Each column section then needs to be designed for a bending moment about its centre line and an axial force whose line of action is through this centre line. It will be appreciated from Section 5.4 and its examples that a direct design calculation is difficult because of decisions as to whether primary compression or tension failures or balanced design conditions are relevant. The designer will often desire the column to be as large as possible to aid detailing of column and interconnecting beam reinforcement, to avoid long column instability, and for economy as the concrete is a more economic material than steel with regard to the carrying of compression forces. However, the larger the column the more it restricts circulation space in the building, and for this reason and sometimes aesthetic considerations the architect will often want columns to be as few and as slender as possible. The designer often chooses the size of a column using these considerations, and an assessment of strength. To assess the size of the column and its reinforcement to carry the load required a very approximate design is usually made. This can then be checked more accurately by using the design charts of BS 8110 (or computer programs based on the charts), or analytically, similar to the method of Section 5.4 if the section is not included in the design charts. If the approximate design is inadequate or uneconomic then this design is altered accordingly and the above procedure repeated until the designer is satisfied. This is a long process if charts, or a computer program, are not used. For the initial approximate design the gross cross-sectional area can be obtained by dividing the ultimate axial load by $0.42f_{cu}$ if the line of action of the eccentric load is outside and $0.45f_{cu}$ if within the section. These figures are for rectangular or square cross sections. For circular cross sections the figures would be $0.39f_{cu}$ and $0.42f_{cu}$, respectively. In all cases the amount of longitudinal reinforcement, $f_y = 250$ N/mm², can be taken to be 2.0% of the gross cross-sectional area.

Example 5.4. Make an approximate initial design for a circular column required to withstand a design ultimate moment of 153 kN m and an axial load of 2400 kN for $f_{cu} = 50$ N/mm² and $f_y = 425$ N/mm².

Eccentricity of load = 153/2400 = 0.0638 m.

214   Columns and walls

The size of the column is not yet known. Assume that the line of action of the axial load is inside the section, and check this later.
Cross-sectional area required

$$= 2400/(0.42 \times 50\,000) = 0.1143 \text{ m}^2$$

Diameter of column

$$= \sqrt{(0.1143/0.7854)} = 0.3815 \text{ m, say } 400 \text{ mm}$$

The line of action of the axial load is within the section. Total area of steel reinforcement

$$= 0.02 \times 0.1143 \times (425/250) \text{ m}^2 = 3886 \text{ mm}^2$$

This checks reasonably well with design charts/computer programs.

Example 5.5. Make an approximate initial design for a rectangular column required to withstand a design ultimate moment of 91 kN m and an axial load of 2460 kN for $f_{cu} = 50 \text{ N/mm}^2$ and $f_y = 425 \text{ N/mm}^2$.

Eccentricity of load $= 91/2460 = 0.037$ m.
Assume that the line of action of the axial load is inside the section and check this later.
Cross-sectional area required $= 2460/(0.45 \times 50\,000) = 0.1093 \text{ m}^2$.
If one dimension is 450 mm, the other needs to be

$$0.1093/0.45 \text{ m} = 243 \text{ mm, say } 250 \text{ mm}$$

Thus the line of action of the axial load is within the section, as assumed.
Total area of steel reinforcement

$$= 0.02 \times 0.1093 \times (250/425) \text{ m}^2 = 1286 \text{ mm}^2$$

Use four 20 mm diameter bars. Using 30 mm cover to these bars.
This checks reasonably well with design charts/computer programs.

5.7   Very slender columns

Since 1968 the author and Dr V. R. Pancholi have conducted research for the S.R.C. (now S.E.R.C.) at the University of Bradford, with the assistance of A. Dracos and D. H. Schofield, into very slender columns with length to least lateral dimension ratios of 30 to 79, mainly 30 to 60, see Refs. 1 and 2. The columns were loaded axially and eccentrically and with combinations of axial load and end bending moments. Occasionally very slender columns have been used for important structures, for example supports to bridge decks for (a) the bridge across the River Derwent at Hobart, Tasmania, Australia, and (b) the approaches to the main span of the Almo bridge, Sweden.

Our tests show that the columns fail by instability when the maximum concrete strain is well below the 0.0035 used by BS 8110 (and a similar figure used by the A.C.I. code), for example often no more than 0.001. Thus the codes mentioned are incorrect in using theories based on material failure for very slender columns.

The instability failure experienced manifests itself in that the column collapses due to excessive lateral deflection. In a practical structure, if the column ends were still secure and the collapse loading was not redistributed to other members, the column would eventually fold up under a lower load than caused the initial instability failure. This 'disintegration' after initial failure is irrelevant but at this stage the basic code theories (that is maximum strain of 0.0035) should apply.

References

1. Pancholi, V. R., 'The Instability of Slender Reinforced Concrete Columns', Ph.D. Thesis, University of Bradford (1978).
2. Wilby, C. B. and Pancholi, V. R., Design of very slender reinforced concrete columns. *Civil Engineering*, London (1978).

6

Reinforced concrete frames and continuous beams and slabs

6.1 Introduction

Although this chapter is really part of, and its contents are used in, Chapter 7, it is useful for it to be separated for clarity and easy reference as it contains Tables 6.1 and 6.2 which designers refer to considerably, even though it makes a very short chapter. The author wrote Chapters 6 and 7 because a lecturer complained that many books considered elements, for example beams, slabs, columns, in isolation and not as a complete structure, and of course one great advantage of reinforced concrete has always been its use in monolithic constructions, which have many advantages – stability, inherent strength, economy, etc.

6.2 Frames

BS 8110 accepts frames being designed for bending moments and shear forces obtained by elastic analysis. The second moments of areas are not usually varied according to the disposition of reinforcement. It is common practice to calculate the second moments of areas of the gross concrete cross sections only, ignoring reinforcement. The individual sections are then designed for ultimate limit states of bending moment and shear force. The disposition of this reinforcement influences the distribution of bending moments towards plastic collapse of a frame.

Much research has been done (for example the author has supervised the work of Refs. 1, 2, 3) with regard to the plastic redistribution of bending moments towards collapse. The fear is that if a designer chooses to make the resistance moment of a section excessively weak, then the section might fail by the extreme concrete fibre strain trying to exceed 0.0035 (the maximum amount experienced before concrete crushing), or fail in shear, before the other sections of the collapse mechanism have

Frames 217

realised their full resistance moments. To allow reasonable plastic redistribution of moments but to safeguard against the above, BS 8110 recommends that 'The ultimate resistance moment provided at any section of a member must not be less than 70% of the moment at that section obtained from an elastic maximum moments diagram covering all appropriate combinations of ultimate loads'. Then BS 8110 is concerned that the sections should be reasonably under-reinforced (because of the fear of concrete compression failure, which occurs suddenly). Where, as a result of redistribution, the ultimate resistance moment at a section is reduced, it therefore restricts the neutral axis depth to be not greater than $(\beta_b - 0.4)d$, where d is the effective depth and β_b is the ratio of the resistance moment of section after redistribution to that before redistribution. Also, for buildings of more than four storeys BS 8110 more cautiously allows elastic moments to be reduced by only 10%, not 30% as mentioned previously.

Design in accordance with the first paragraph of this chapter is commendable, in that it automatically gives good control of crack widths and deflections (limit states of serviceability). Design in accordance with the second paragraph endeavours to give increased economy and reinforcement systems which are easier to detail and assist in concreting. For example, the steel required over the supports is often reduced to help detailing, particularly when there are two continuous beams at right-angles to one another joining a column at the same place and the architect has requested that the column should have a small cross-section.

Table 6.1.

Dead load	Live load
0.125	0.125
0.071 0.071	0.096 0.096
0.100 0.100	0.117 0.117
0.080 0.025 0.080	0.101 0.075 0.101
0.107 0.072 0.107	0.121 0.107 0.121
0.077 0.036 0.036 0.077	0.099 0.081 0.081 0.099
0.105 0.080 0.080 0.105	0.120 0.111 0.111 0.120
0.078 0.033 0.046 0.033 0.078	0.100 0.080 0.086 0.080 0.100

Table 6.2.

Dead load	Live load
0.38 0.62 0.62 0.38	0.44 0.62 0.62 0.44
0.40 0.50 0.60 0.60 0.50 0.40	0.45 0.58 0.62 0.62 0.58 0.45
0.39 0.54 0.46 0.61 0.61 0.46 0.54 0.39	0.45 0.60 0.57 0.62 0.62 0.57 0.60 0.45
0.40 0.53 0.50 0.47 0.60 0.60 0.47 0.50 0.53 0.40	0.45 0.60 0.59 0.58 0.62 0.62 0.58 0.59 0.60 0.45

6.3 Continuous beams and slabs

The previous section deals with frames, but applies similarly to continuous beams and slabs. Tables 6.1 and 6.2 are most useful for designers. Table 6.1 gives bending moment coefficients for continuous beams or slabs whose spans are equal, or do not vary by more than say 10%, carrying uniformly distributed loads. For live loads the coefficients are for complete spans loaded in the worst possible arrangement. The elastic bending moment at either support or span = Coefficient × Total load on span × Span. Similarly, Table 6.2 gives coefficients for shear forces. The elastic shear force at a support = Coefficient × Total load on span.

References
1. Wilby, C. B. and Pandit, T., Inelastic behaviour of reinforced concrete single-bay portal frames. *Civil Engineering and Public Works Review*, Mar. (1967).
2. Noor, F. A., 'Elastic and Inelastic Behaviour of Reinforced Concrete Frames', Ph.D. Thesis, University of Bradford (1970).
3. Chapman, B. C., 'Flexural Behaviour of Redundant Reinforced Concrete Frames', Ph.D. Thesis, University of Bradford (1973).

7

Design of structures

7.1 Design of an *in situ* R.C. framed building

Some other books give designs which check the adequacy and design of, and design the reinforcement for, beams, slabs and columns, whose sizes and layouts are given without any explanation of derivation. In other words the essential speedy creation (which a designer has to perform) of the design, giving layout and sizes of members, is not done. The beginner following such designs naturally asks, 'How were this layout and these sizes chosen?' These books might be thought useful for designers of sufficient experience as not to need guidance on determination of layout and sizes, but then such designers do not need the information which the books give.

The beginner needs to be able to create/design suitable structural systems and the sizes of the beams, slabs and columns involved, with the knowledge that the reinforcement will properly fit in the sections upon subsequent detailing and that more comprehensive or accurate design will not require revision of the outline drawings. The self-weight of reinforced concrete members is very significant in their structural design and is unknown until layout and sizes, which it affects, have been determined. As speed is important for economy in design, it is therefore necessary to determine the adequacy of the layout and outlines from simple (approximate), basic, reliable and rapid calculations considering the most influential design requirements first. For example, it is certainly not unknown for a beginner to have inadequate guidance and to design a continuous T-beam by firstly concentrating on making full use of the flange in flexural compression at mid span and then finding that this needs to be revised radically several times because of other design requirements (such as concerning considerations of shear and flexural compression at the supports, and the practicality of detailing reinforcement etc.) which, for efficiency and speed of design, should have been considered previously.

This present chapter, therefore, takes the beginner through the system of creating/designing a beam and slab layout (from merely a column layout required by the architect planning the client's requirements), obtaining the sizes and checking the practicality of the main practical design problems (for example, that reinforcement can be detailed in sections) in the sequence in which a professional designer has to rapidly perform this operation. Continuous T-beams, mentioned previously, are designed for layout, size and reinforcement in a speedy practical sequence. The reader is able to follow the mind of the professional designer through this present Section 7.1 and its sub-sections 7.1.1 to 7.1.5 inclusive. Following the designer's mind in creating a structure is not the same as submitting tidy calculations justifying one's creations, to checkers of the designs, for structural adequacy. In Section 7.1.6, therefore, the designs are set out in a suitable way for submission to others who wish to check general structural adequacy. These could be: one's supervisor in the design concern or an outside supervisory authority (local, national or consultant working for one or other of these). This setting out of the calculations also acts as a summary to the designs in the previous sub-sections.

Chapter 6 belongs to Chapter 7 but has been separated for clarity of the very useful design tables it contains – these are used where required in Chapter 7.

Figure 7.1 shows a layout of columns, which has been determined to be sympathetic to the arrangement of the windows and layout of internal requirements (for example, partition walls, equipment, machinery). The building is four 7 m bays wide and ten 5 m bays long. Table 7.1 gives a

Figure 7.1

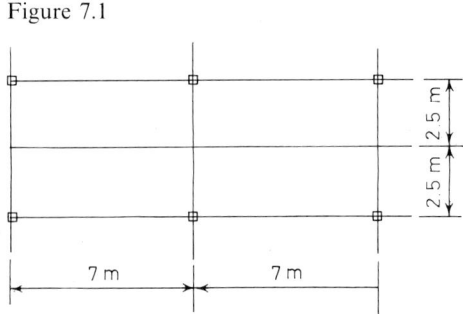

very approximate guide for preliminary design proportioning. If there were no intermediate beams and the floor slabs were designed as 7 m × 5 m two-way spanning they would be, from Table 7.1, say about 7/40 m = 175 mm thick. This is a rather thick slab. Intermediate beams reduce it considerably so that the total amount of concrete and reinforcement is less, and the load on the supporting beams, columns and foundations is less. Also the shuttering does not need to be as strong. The intermediate beams can be as shown in Figure 7.1 and from Table 7.1 the slab is about 2.5/35 m = 71 mm say 125 mm thick as this is about a minimum floor thickness for practical reasons, and for deflection in this example, see later. Otherwise they could have been at right-angles to these, giving two-way spanning slabs 5 m × 3.5 m, of thickness, from Table 7.1, approximately 5/40 m = 125 mm say 150 mm for practical reasons. If these two schemes are compared the first is favoured as the shorter beams carry a greater proportion of the load on each 7 m × 5 m panel.

7.1.1   Floor slab

This is therefore a 125 mm thick one-way slab, continuous for 20 bays, each of 2.5 m span. Suppose the floor carries bedrooms for a hotel or hospital. CP 3 requires the floor to be designed for a uniformly distributed load of 2 kN/m². As the slab was made thicker than required for practical reasons the concrete will only need to be weak but as it is a slab and not very thick we do not want the cover to be very great. Considering mild exposure in BS 8110 Clause 3.3.3 we choose concrete with $f_{cu} = 35$ N/mm² and a minimum cement content of 300 kg/m³, so that we can have 20 mm cover – not 25 mm as with Grade 30 concrete – so as not to reduce the effective depth of the reinforcement. Referring to Table 7.1.

Ratios of span to overall depth	
Simply supported beams	20
Continuous beams	25
Cantilever beams	7
Slabs spanning in one direction, simply supported	30
Slabs spanning in one direction, continuous	35
Slabs spanning in two directions, simply supported	35
Slabs spanning in two directions, continuous	40
Cantilever slabs	7

BS 8110, Pt. 2, Cl. 4.3.1 the slab will have a one hour fire resistance and we assume that this is satisfactory. The floor will carry lightweight partitions and there will be floor finishes, perhaps tiles on the floor, and either plaster or a suspended ceiling and minor services below; assume all this weighs 1.5 kN/m². The self-weight of the floor (taking the weight density of reinforced concrete as 23.6 kN/m³) = 0.125 × 23.6 = 2.95 kN/m².

Thus characteristic dead load = 2.95 + 1.5 = 4.45 kN/m². The building is wide and long enough compared to its height for wind forces to be neglected (see CP 3). From Table 1.1, design load = 1.4 × 4.45 + 1.6 × 2 = 9.43 kN/m².

From BS 8110 Cl. 3.4.3, maximum bending moment is at the first interior support and

$$= 1.1 \times 9.43 \times 2.5^2 = 6.48 \text{ kN m per metre width of slab}$$

Using $f_y = 460$ N/mm² and BS 8110 Part 3 Design Chart No. 2, and (guessing 8 mm diameter bars and thus $d = 125 - 24 = 101$ mm)

$$M(bd^2) = 0.006\,48/0.101^2 \text{ MN/m}^2 = 0.6352 \text{ N/mm}^2$$

then $100 A_s/(bd) = 0.17$

$$A_s \therefore A_s = 0.17 \times 0.101/100 = 0.000\,171\,7 \text{ m}^2/\text{m}$$

From Table 3.2 use 8 mm diameter bars at 290 mm centres. This is reasonable for detailing. The steel at other locations can be obtained *pro rata* to the bending moments of BS 8110 Cl. 3.4.3. For example the smallest of these is at the middle of the interior spans and as Chart No. 2 is linear for smaller values of $M(bd^2)$ than 0.6352 N/mm² (namely $0.6352 \times 0.07/0.11 = 0.404$)

$$A_s = 0.1717 \times 0.07/0.11 = 0.1093 \text{ mm}^2/\text{m}$$

To check that the deflection is not excessive BS 8110 Table 3.10 allows a span-to-effective-depth ratio of 26 and from BS 8110 Table 3.11 the modification factor can be taken as 1.6. Therefore allowable maximum span

$$= 1.6 \times 26 \times 0.101 = 4.20 \text{ m} > 2.5 \text{ m}$$

We need some steel at right-angles to the above steel. This is usually called *distribution steel*; it helps to distribute point loads across the width of a slab, to resist shrinkage and temperature stresses, and to help fix the main steel. Using high-yield distribution steel, reference to BS 8110 Table 3.27 gives the area of this steel as

$$0.0013 \times 0.101 = 0.000\,162\,5 \text{ m}^2/\text{m}$$

From Table 3.2 use 8 mm diameter bars at 300 mm centres. This is the minimum area of reinforcement recommended in each direction.

With regard to the limit state of cracking, BS 8110 clause 3.5.8 says that for normal conditions of exposure no special check is required if our slab thickness is less than 200 mm thick which it is; and the spacing of the reinforcement bars must not exceed $3d$, or 300 mm for main steel and $3d$ or 400 mm for distribution steel. For the 8 mm diameter bars $= 3 \times 101 = 303$ mm, and this agrees with the above.

It is very unlikely that shear reinforcement will be required (in this eventuality we would normally avoid having to use it by making the slab thicker). From BS 8110 Table 3.6 maximum shear force

$$= 0.6 \times 9.43 \times 2.5 = 14.15 \text{ kN per metre width of slab}$$

Referring to Section 3.4

$$V/bd = 14.15/0.101 \text{ kN/m}^2 = 0.1401 \text{ N/mm}^2$$

which is obviously satisfactory from Section 3.4.

7.1.2 Beams of 7 m span

BS 8110 Cl. 3.3.3 gives a minimum cover of 20 mm for mild exposure and concrete with $f_{cu} = 35$ N/mm^2 and a minimum cement content $= 300$ kg/m^3. Using a fire resistance of 1 hour, as for the slab, BS 8110 Pt. 2 Cl. 4.3.1 requires a minimum concrete *cover to main reinforcement of 20 mm* and requires a beam width of 80 mm. To allow for 20 mm cover for durability to say 5 mm dia. stirrups (links) use 25 mm cover to the main reinforcement.

The continuous beam supporting the heaviest loading is the penultimate continuous beam of the building. From BS 8110 Table 3.6 the reaction on this beam from the slab $= (0.6 \times 0.55) \times$ load. Hence the characteristic dead load from the slab

$$= 1.15 \times 4.45 \times 2.5 = 12.79 \text{ kN/m}$$

and the characteristic live load from slab

$$= 1.15 \times 2 \times 2.5 = 5.75 \text{ kN/m}$$

From Table 7.1 the overall depth of the beam required is approximately 7.25 = 0.28 m. Within reason the greater the depth the more economic and easy the design, detailing and fixing of the reinforcement. A small amount of extra vertical shuttering (which does not alter scaffolding costs) and of concrete can save expensive reinforcement and its fixing and reduce concreting costs of placing concrete around high percentages of reinforcement. Architects often require the overall depths of beams to be a reasonable minimum for reasons of aesthetics. Deeper beams increase the heights of buildings where strict use is made of minimum headrooms, but we are only talking about altering the depth of beams by perhaps about 0.1 m or so to increase the economy and speed of the reinforced concrete construction. In this example suppose the architect for aesthetic reasons does not desire a beam with overall depth deeper than 0.4 m. The breadth of the rib of a beam will often be about $\frac{1}{3}$ to $\frac{1}{2}$ of the overall depth with a minimum sufficient to accommodate three 25 mm diameter bars. Using 19 mm down coarse aggregate the horizontal distance between bars, from BS 8110, must be greater than $19 + 5 = 24$ mm, say, 25 mm. Hence width of rib to accommodate three 25 mm diameter bars = $5 \times 25 + 2 \times 25$ (that is covers) = 175 mm. Hence use a beam of overall depth 0.4 m and breadth of rib of 0.2 m. The effective depth, assuming 25 mm diameter bars with 25 mm cover at mid span, will be approximately $400 - 25 - 25/2 = 362$ mm and then the span-to-effective-depth ratio = $7000/362 = 19.3$. This is less than 26 from BS 8110 Table 3.10 so limit state of deflection is satisfied.

Then characteristic self-weight of rib

$$= (0.4 - 0.125) \times 0.2 \times 23.6 = 1.30 \text{ kN/m}$$

and the total characteristic dead load is

$$12.79 + 1.3 = 14.09 \text{ kN/m}$$

The design ultimate load is then

$$1.4 \times 14.09 + 1.6 \times 5.75 = 28.93 \text{ kN/m}$$

If the support moments can be carried, and the reinforcement will practically fit in the sections, then the spans should be adequate to resist flexural compression; there is a considerable area of the T-flange available, whereas there is only the rib to take compression at the supports, and the bending moments at mid spans and supports are similar in magnitude.

It is also important to know if the maximum shear force can be carried by the rib with suitable reinforcement if necessary – sometimes being able to practically detail the shear reinforcement can be the critical problem to overcome with the design, necessitating a larger rib.

From BS 8110 Cl 3.4.3, the maximum bending moment at a support (and anywhere)

$$= 0.11 \times 28.93 \times 7^2 = 155.9 \text{ kN/m}$$

and the maximum shear force (adjacent to the inner support of either end span)

$$= 0.6 \times 28.93 \times 7 = 121.5 \text{ kN}$$

At this support the beam is cracked in flexure in the top and acts as a rectangular beam. The overall depth of the beam is 0.4 m. The slab has top main steel of 8 mm diameter with 20 mm cover. The main beam steel over the support must be beneath this slab steel; hence, assuming it is of 25 mm diameter bars its effective depth

$$= 400 - 28 - 12.5 = 359 \text{ mm}$$

Using the same type of reinforcement as for the slab, $f_y = 460 \text{ N/mm}^2$, then using BS 8110 Part 3 Design Chart No. 2

$$M/bd^2 = 155.9/(0.2 \times 0.359^2) \text{ kN/m}^2 = 6.05 \text{ N/mm}^2$$

This is beyond the range of the Chart, so compression steel is required. So using Chart No. 7, ($d' = 25 + 12 = 37$ mm, $d'/d = 37/359 = 0.10$ say 0.15)

$$A_s = 1.77 \times 200 \times 359/100 = 1271 \text{ mm}^2$$

and

$$A'_s = 1.0 \times 200 \times 359/100 = 718 \text{ mm}^2$$

From Table 3.2 use two 25 mm diameter bars and one 20 mm dia. bar as tension steel and two 25 mm diameter bars as compression steel.

$$V/bd = 121.5/(0.2 \times 0.359) \text{ kN/m}^2 = 1.692 \text{ N/mm}^2$$

which is satisfactory as it is $<0.8\sqrt{35} = 5.9$ and 5 N/mm². Then see Example 3.8. At support, A_s provided $= 982 + 314 = 1296$ mm² and

$$\therefore 100 A_s/(bd) = 100 \times 1296/(200 \times 359) = 1.805$$

From BS 8110 Table 3.9 shear resistance provided by concrete alone

$$= 0.7934 \times 200 \times 359 \text{ N} = 57.0 \text{ kN}$$

Hence shear reinforcement is required and has to resist $121.5 - 57.0 =$

64.5 kN. Using stirrups the V/d required is $64.5/0.359 = 180$ N/mm. From Table 3.5, using steel with $f_{yv} = 250$ N/mm², use 8 mm diameter two-arm stirrups at 120 mm centres.

There are significant bending moments and the shear forces because of the ends not being pin-jointed to the external columns. This reduces the maximum shear force and bending moments used above. Hence this beam is capable of being designed and detailed with regard to ultimate limit state from the above.

The maximum span bending moment is in the end span and from BS 8110 Cl. 3.4.3

$$= 0.09 \times 28.93 \times 7^2 = 127.6 \text{ kN/m}$$

Assuming 25 mm diameter bars will be used (two of these need to carry through the supports to provide compression steel there) and using 25 mm cover to them, the effective depth

$$= 400 - 25 - 12.5 = 362 \text{ mm}$$

The whole of the large slab portion of the T-beam is unlikely to be required in compression, so the centre of the compression force is unlikely to be lower than half the slab depth. Therefore the moment arm can be taken as

$$= 362 - 125/2 = 299 \text{ mm}$$

$$\therefore A_s = \frac{127\,600\,000}{299 \times 460} = 927.7 \text{ mm}^2$$

Use two 25 mm diameter bars (Table 3.2).

The continuous T-beam of this type will normally be adequately strong in flexural compression. This can be checked, and a more accurate calculation for the reinforcement made, as in Example 3.18, using BS 8110 clause 3.4.1.5 to obtain the effective width of the flange.

The limit state of cracking is easy to comply with in the detailing, see BS 8110.

As mentioned before, bending moments due to the beams framing into the external columns cause moments and shear forces along the continuous beams, mainly advantageously. In any case the bending moment in the beam at this junction must be assessed as given at the end of Section 7.1.3 for detailing the beam at this location. If a more accurate design is to be produced, use can be made of Table 7.3.

7.1.3 External columns between ground and first floor

Figure 7.2 shows an external column. The base shown rests on a cohesive soil (clay) and is designed for uniform soil pressure. That is, the base is assumed to rotate because of the inelastic or plastic action (or creep) of the soil. So that the shutters can be unaltered for economy, the external column BG is designed for BC to be as small in girth as possible and then the upper portions of the column BG are kept the same size, their reinforcement being reduced. The greatest vertical load is little greater at B than at C, yet it is combined with a substantial bending monent at C, which is therefore the critical section for design.

Considering durability (mild exposure) and fire resistance (1 hour) BS 8110 Clauses 3.3.3 and 4.3.1 mean that the cover of concrete with $f_{cu} = 30$ N/mm² and min. cement content $= 275$ kg/m³, to the links needs to be not less than 25 mm and the minimum dimension of the concrete needs to be 200 mm.

The vertical loads can be accurately obtained from the shear forces of the beams framing into the columns and estimating the self-weight of the columns. The 5 m long beams should therefore be designed in a similar way (NB BS 8110 Cl. 3.4.3 cannot be used for point loads) to that given in Section 7.1.2 before the columns are designed. Assume that the vertical load at C comprises a characteristic dead load of 665 kN and a maximum characteristic live load of 166 kN. To estimate the size of the column

Figure 7.2

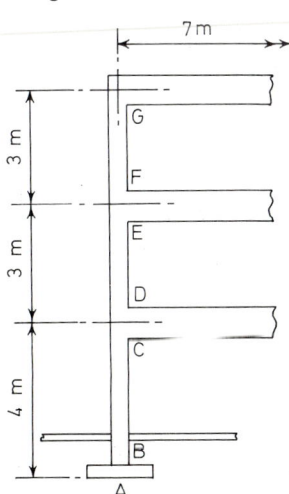

assume it is axially loaded, ignore the strength of its reinforcement, and increase the cross-sectional area by about 30% to allow for neglecting bending moment. The design ultimate load

$$= 1.4 \times 665 + 1.6 \times 166 = 1197 \text{ kN}$$

Then the cross-sectional area of column required (see equation 5.2)

$$= 1.3 \times 1\,197\,000/(0.4 \times 30) = 129\,700 \text{ mm}^2,$$
$$\text{say } 350 \text{ mm by } 450 \text{ mm}$$

This is rather generous but bending moments have not been considered (admittedly the reinforcement has not either). Also if there are no particular architectural restrictions upon size, larger columns within reason have less steel and are more economic in that concrete is a less expensive material for resisting compression than steel. In addition there are often severe difficulties in detailing reinforcement from beams into columns of small size.

Second Moment of Area for the column

$$= 350 \times 450^3/12 = 2658 \times 10^6 \text{ mm}^4$$

The stiffnesses of columns DE and CB, respectively, are $2658 \times 10^6/3000 = 886\,000 \text{ mm}^3$ and $2658 \times 10^6/4000 = 664\,500 \text{ mm}^3$, respectively. The Second Moment of Area of the beam poses a problem as the beam is a T-beam at mid span but a rectangular beam in effect at the location of cracks near the supports. The writer considers the former as the more accurate assumption, as did Scott and Glanville, and many structures have been designed on this basis in the past. However, the latter assumption is easier for calculation, gives higher moments in the columns, and is favoured in books by Allen (1974) and Higgins and Hollington (1973) of the Cement and Concrete Association which has greatly influenced BS 8110. Using this latter assumption the Second Moment of Area for the beam

$$= 200 \times 400^3/12 = 1067 \times 10^6 \text{ mm}^4$$

and the stiffness

$$= 1067 \times 10^6/7000 = 152\,400 \text{ mm}^3$$

For this beam the total design ultimate load = 28.93 kN/m (see Section 7.1.2) and if it were fixed the end moment (from Table 7.2)

$$= 28.93 \times 7^2/12 = 118.1 \text{ kN m}$$

The bending moment at C, from CP 110 clause 3.5.2, is

$$118.1 \times \frac{664.5}{664.5 + 886 + 152.4/2} = 48.24 \text{ kNm}$$

This simple type of assessment of bending moment based on Hardy Cross's moment distribution method was recommended by CP 110 and previous codes. BS 8110 no longer gives these formulae perhaps because so many computer programs are available for frame analyses these days. Still a frame analysis cannot be commenced until sensible sizes of the members have been created, so the formulae for assessing moments in columns still give a rapid way of assessing column sizes. A subsequent computer frame analysis, should this be desired if the job went ahead, can keep these sizes and economise in the reinforcement, for example.

As A in Figure 7.2 is assumed to be in effect a hinge, it would be more accurate to reduce the stiffness of CB, but CP 110 does not suggest this, and it thus gives a higher moment at C.

There are walls between the external columns, various internal walls and the overall height of the building compared to its horizontal dimensions is sufficiently low for lateral wind forces to be ignored. It can be assumed therefore that the beam column junctions will not move laterally, that is the columns can be considered as 'braced' as defined by BS 8110 clause 3.8.1.5. From BS 8110 Cl. 3.8.1.6 take the effective height of column CB as the length CB = $0.95 \times 3.7 = 3.515$ m (guessing AB as 0.3 m). Then $3.515/0.35 = 10.04$ which is less than 15, hence column can be treated as a short column (see Section 5.2).

From BS 8110 clause 3.8.2.4 the minimum design ultimate bending moment uses an eccentricity of $0.05 \times 450 = 22.5$ mm but not more than 20 mm and therefore

$$= 1197 \times 0.020 = 23.94 \text{ kN m} < 48.24 \text{ kN m},$$

hence design for this latter. If the links are 8 mm diameter this means that the cover to the main steel is $25 + 8 = 33$, say, 35 mm. Suppose 25 mm bars are to be used in a single layer at each side of the column then effective depth $= 450 - 48 = 402$ mm and $d/h = 402/450 = 0.8933$. Thus for $f_{cu} = 30$ N/mm^2 and $f_y = 460$ N/mm^2 use Design Chart 29 of BS 8110 Part 3

$$N/bh = 1.197/(0.35 \times 0.45) \text{ MN/m}^2 = 7.6 \text{ N/mm}^2$$

$$M/bh = 0.048\,24/(0.35 \times 0.45^2) \text{ MN/m}^2 = 0.6806 \text{ N/mm}^2$$

therefore no reinforcement is required. But the percentage of reinforcement should not be less than 0.4 (see BS 8110 clause 3.12.5), hence the area of reinforcement required

$$= \tfrac{0.4}{100} \times 350 \times 450 = 630 \text{ mm}^2$$

Then from Table 3.2, four 16 mm diameter bars can be used.

For junctions with beams higher up the external column, bending moments will be similar but the vertical loadings much less. The column is kept the same size all the way to economise on shuttering. A greater predominance of bending moment relative to axial load could require greater reinforcement so it would be precipitous to redesign this portion CB of the column to be smaller with more reinforcement just because the reinforcement required is fairly modest, although four 16 mm diameter bars are quite good for detailing purposes particularly at the junctions with the beams.

With the present design the bending moment in the beam framing into the column at its support is the sum of the bending moments in the columns at C and D, that is

$$= 118.1 \times \frac{664.5 + 886}{664.5 + 886 + 152.4/2} = 112.6 \text{ kN m}$$

It was mentioned at the end of Section 7.1.2 that this bending moment would be calculated here. Its effect can be assessed along the continuous beam using Table 7.3.

7.1.4 Bases

From Section 7.1.3 the design ultimate vertical load at C was 1197 kN and AB the thickness of the base was guessed to be 0.3 m. The characteristic self-weight of the column BC can be taken as

$$0.450 \times 0.350 \times (4.0 - 0.3) \times 23.6 = 13.75 \text{ kN}$$

The design ultimate vertical load at B is therefore

$$13.75 \times 1.4 + 1197 = 1215 \text{ kN}$$

The weight of the base gives a characteristic pressure on the soil of $0.3 \times 23.6 = 7.08 \text{ kN/m}^2$ and an ultimate design pressure of $1.4 \times 7.08 = 9.91 \text{ kN/m}^2$. As mentioned before, the soil is cohesive and is considered to give a uniform pressure beneath the base. Assume the soil beneath the base can safely withstand a pressure of 217 kN/m². Using a load factor of say 1.8 the ultimate pressure on the soil can be $217 \times 1.8 = 390$ kN/m². Then the area of the base needs to be

$$1215/(390-9.91) = 3.2 \text{ m}^2$$

Making it square to save shuttering it needs to be 1.8 m × 1.8 m.

From BS 8110 Cl. 3.3.3 and $f_{cu} = 35$ n/mm² and minimum cement content = 300 kg/m³, cover to reinforcement for moderate exposure (buried concrete) = 35 mm. If 16 mm diameter bars are to be used to form a square mesh, then the effective depth for the bars in the upper layer = $300 - 35 - 24 = 231$ mm.

For shear using BS 8110:

$$\text{shear stress at the column face} = 1197/(2 \times 0.35 \times 0.45) \text{ kN/m}^2$$

$$= 3.8 \text{ N/mm}^2$$

this is satisfactory as it is less than $0.8\sqrt{35} = 4.73$ and 5 N/mm² the critical section for punching shear is $1.5d$ ($= 1.5 \times 231 = 347$ mm) from the faces of the column therefore the critical perimeter is a rectangle $0.35 + 2 \times 0.347 = 1.044$ m by $0.45 + 2 \times 0.347 = 1.144$ m and has total length $= 2 \times 1.044 + 2 \times 1.144 = 4.376$ m

area within critical perimeter

$$= 1.044 \times 1.114 = 1.194 \text{ m}^2$$

area outside critical perimeter

$$= 1.8^2 - 1.194 = 2.046 \text{ m}^2$$

ultimate design uniform pressure on base, excluding self-weight of base

$$= 1197/1.8^2 = 369.4 \text{ kN/m}^2$$

∴ punching shear force on critical perimeter

$$= 369.4 \times 2.046 = 755.9 \text{ kN}$$

∴ punching shear stress on critical perimeter

$$= 755.9/(bd) = 755.9/4.376/0.231 \text{ kN/m}^3 = 0.7478 \text{ N/mm}^2$$

From BS 8110 clause 3.4.3.4 this will perhaps be in order with the help of the main steel for resisting bending moments.

This is to be so for bending considerations later.

Using BS 8110 the maximum bending moment is at a section passing completely across the base of the face of the column and

$$= \frac{369.4}{2} \times \left(\frac{1.8-0.35}{2}\right)^2 = 97.08 \text{ kN m/m}$$

For $f_{cu} = 35$ N/mm^2 and $f_y = 250$ N/mm^2 use BS 8110, Part 3, Design Chart 1

$$\frac{M}{bd^2} = \frac{0.09708}{0.231^2} \text{ MN/m}^2 = 1.819 \text{ N/mm}^2$$

$$\therefore A_s = 0.89\% \text{ of } bd = \frac{0.89}{100} \times 1000 \times 231 = 2056 \text{ mm}^2/\text{m}$$

From Table 3.2 use, say, 16 mm diameter bars at 90 mm centres.

With regard to limit state of cracking, the clear distance between bars should not exceed 300 mm so the 90 mm centres suggested are satisfactory.

The limit state of deflection can be ignored for bases. Deflections would not be unsightly underground; also the greater proportion of the loading is dead not live so the deflection due to the latter would not normally result in undesirable springiness of the base.

Traditionally the local bond stresses associated with the shear forces above are checked for adequacy, but BS 8110 has decided that these do not now need checking. Local bond stresses must not be excessive and no doubt codes other than BS 8110 require these to be checked. The BS 8110 general detailing recommendations plus the design requirements due to shear and bending and anchorage of the reinforcement bars usually mean that local bond stresses are satisfactory.

The reinforcement must be able to develop adequate anchorage length, within the size of the base, from the position of the maximum bending moment. The distance from the face of the column, where the maximum bending moment was calculated to the periphery of the base = $(1.8-0.35)/2 = 0.725$ m. Using an end cover of 40 mm this gives a possible overall bar length of $0.725-0.040 = 0.685$ m available for anchorage. Referring to Table 2.9 (and see Example 2.5)

$$l_b = 39d_b = 39 \times 16 = 624 \text{ mm}$$

If hooks are used as anchors to the bars, from Table 2.12 a hook is equivalent to an anchorage length of 256 mm. Hence the overall anchorage length required $= 624-256 = 368$ mm, so 685 mm is satisfactory.

It is interesting that the BS 8110 calculation for bending moment gives the same result as using Hillerborg's methods for strips and elements as shown in Figure 7.3(a) and as using Johansen's method with a yield-line pattern as shown in Figure 7.3(b). Applying Johansen's method to the yield-line pattern of Figure 7.3(b) gives an upper-bound solution to that using Figure 7.3(a) as a yield-line pattern.

7.1.5 Anchorage of column bars into bases (see Sections 2.6–2.6.9)

In this example the column bars are in compression. There will be 'starter bars' projecting from each base, as shown in Figure 7.4. These are lapped with a 'compression lap' with the column bars. Distance a_1 is this lap plus a tolerance of, say, 20 mm (that is one aims at having a gap of 20 mm between the column bars and pad A). A is a 'kicker pad' of concrete, say 50 mm deep, for holding the column shutters apart and to hold them in position at this location. The base must be adequately thick to accommodate the distance a_2, which needs to be the 'compression anchorage length'.

Extra length such as a_3 cannot be counted in the compression lap for similar reasons to those given in Section 2.6.8. If the base is too thin to

Figure 7.3

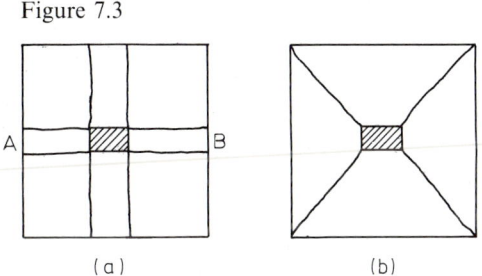

(a)   (b)

Figure 7.4

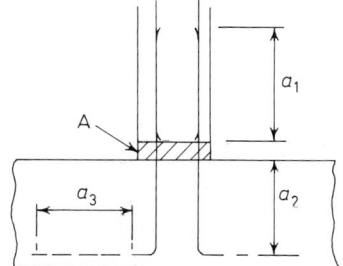

234 *Design of structures*

accommodate a_2 then it may need to be thickened, giving economies in the reinforcement for bending moments in the base. Alternatively, larger diameter or more starter bars, or both, may be used.

7.1.6 Design calculations

As mentioned in Section 7.1, this present sub-section sets out the previous calculations, in a style used in design offices for ease of checking by one's supervisor, or an outside supervisory authority, wishing to check general structural adequacy. Two margins are used, one for titles and the other to give information required later or when detailing reinforcement. The following also acts as a summary of the designs of the previous sub-sections:

Slab	span 2.5 m (continuous 20 bays) loading (char.): live 2 kN/m² finishes 1.5 self-weight, SW 2.95 $\overline{4.45}$ kN/m²	125 mm thick
	ultimate design load: $1.6 \times 2 = 3.2$ $1.4 \times 4.45 = 6.23$ $\overline{9.43}$ kN/m²	mild exposure $f_{cu} = 35$ N/mm² min. cement content $= 300$ kg/m³ fire resistance 1 hr
	support $M = 0.11 \times 9.43 \times 2.5^2 = 6.48$ kN m/m $d = 125 - 20 - 4 = 101$ mm	cover $= 20$ mm $d = 101$ mm
	$\dfrac{M}{bd^2} = \dfrac{0.00648}{0.101^2} = 0.6352$ BS 8110 Part 3 Chart 2	$f_y = 460$ N/m²
	$\dfrac{100 A_s}{bd} = 0.17$ $A_s = 0.17 \times 0.101/100 = 0.0001717$ m²/m Interior span $A_s = 0.1667 \times 0.07/0.11 = 0.1072$ mm²/m so use distribution steel below $V = 0.6 \times 9.43 \times 2.5 = 14.15$ kN/m $V/(bd) = 14.15/0.101 = 0.1401$ N/mm² Distribution steel $A_s = 0.0013 \times 0.125$ $= 0.0001625$ m²/m	8 mm φ at 290 mm centres 8 mm φ at 300 mm centres
Beam	span 7 m (4 bays) 7 m span loading (char.) live from slab $1.15 \times 2 \times 2.5 = $ 5.75 kN/m dead from slab $1.15 \times 4.45 \times 2.5 = $ 12.79 SW rib $= (0.4 - 0.125) \times 0.2 \times 23.6 = $ 1.30 Total dead 14.09 kN/m Design load $= 1.4 \times 14.09 + 1.6 \times 5.75$ $= 28.93$ kN/m support $M = 0.11 \times 28.93 \times 7^2 = 155.9$ kN m	(T-beam section sketch: flange 0.125 m, web 0.2 m, total depth 0.4 m) mild exposure $f_{cu} = 35$ N/mm² min. cement content $= 300$ kg/m³ fire resistance $= 1$ hr

Design of an in-situ R.C. framed building

max. $V = 0.6 \times 28.93 \times 7 = 121.5$ kN
$d = 400 - 20 - 8 - 12.5 = 359$ mm
$d' = 25 + 12 = 7$ mm

$$\frac{d'}{d} = \frac{7}{359} = 0.131$$

Use Chart No. 7 of BS 8110 Pt. 3
for $d'/d = 0.15$

$$\frac{M}{bd^2} = \frac{155.9}{0.2 \times 0.359^2} \text{ kN/m}^2 = 6.05 \text{ N/mm}^2$$

cover to main reinforcement 25 mm
bars 25 mm (provisional)

$d = 359$ mm

$f_y = 460$ N/mm²

$$\frac{100 A_s}{bd} = 1.77$$

$A_s = 1271$ mm²

$$A_s = \frac{1.77 \times 200 \times 359}{100} = 1271 \text{ mm}^2$$

2–25 mm φ and 1–20 mm φ

$$A'_s = \frac{1.0 \times 200 \times 359}{100} = 718 \text{ mm}^2$$

2–25 mm φ

$$\frac{V}{bd} = \frac{121.5}{0.2 \times 0.359} \text{ kN/m}^2 = 1.692 \text{ N/mm}^2$$

$< 0.8\sqrt{35} = 5.9$ and 5 N/mm²

$$\frac{100 A_s}{bd} = \frac{100 \times 1296}{200 \times 359} = 1.805$$

shear resistance, concrete only
$= 0.7934 \times 200 \times 359 = 57.0$ kN
shear steel required for
$V = 121.5 - 57.0 = 64.5$ kN

$$\frac{V}{d} = \frac{64.5}{0.359} = 180 \text{ N/mm}$$

$f_{yv} = 250$ N/mm²

{ 8 mm φ 2-arm stirrups
  at 12 mm centres

span M (max.) $= 0.09 \times 28.93 \times 7^2 = 127.6$ kN m
$d = 400 - 25 - 12.5 = 362$ mm
$z = 362 - 125/2 = 299$ mm

cover 25 mm
$d = 362$ mm
$f_y = 460$ N/mm²

$$A_s = \frac{127\,600\,000}{299 \times 460} = 927 \text{ mm}^2$$

2–25 mm φ

Column

At C
char. dead load $= 665$ kN
char. live load $= 166$ kN
design ult. load $= 1.4 \times 665 + 1.6 \times 166$
$\qquad\qquad\qquad\quad = 1197$ kN

$$\text{cross-sectional area} \simeq \frac{1.3 \times 1\,197\,000}{0.4 \times 30}$$
$$= 129\,700 \text{ mm}^2$$

350×450

$(157\,500 \text{ mm}^2)$

Second mt. area $= \dfrac{350 \times 450^3}{12} = 2658 \times 10^6$ mm⁴

mild exposure $f_{cu} = 30$ N/mm²

Stiffness of cols:

min. cement content $= 275$ kg/m³

$$\text{DE}: \frac{2658 \times 10^6}{3000} = 886\,000 \text{ mm}^3$$

fire resistance $= 1$ hr

236 *Design of structures*

$$CB: \frac{2658 \times 10^6}{4000} = 664\,500 \text{ mm}^3$$

Beam:
Second moment of area:

$$\frac{200 \times 400^3}{12} = 1067 \times 10^6 \text{ mm}^4$$

$$\text{Stiffness} = \frac{1067 \times 10^6}{7000} = 152\,400 \text{ mm}^3$$

Fixed end M

$$= \frac{29.93 \times 7^2}{12} = 118.1 \text{ kN m}$$

Col. M at C

$$= \frac{118.1 \times 664.5}{664.5 + 886 + 152.4/2}$$

$$= 48.24 \text{ kN m}$$

effective height/lateral dimension

$$= \frac{0.95 \times 3.7}{0.35} = 10.04 < 15$$

min. design ult. $M = 1197 \times 0.020$
$= 23.94$ kN m < 48.24
cover to main steel $= 25 + 8 = 33$ say 35

effective depth $= 450 - 48 = 402$

$$\frac{d}{h} = \frac{0.402}{0.45} = 0.893$$

$$\frac{N}{bh} = \frac{1.197}{0.35 \times 0.45} \text{ MN/m}^2 = 7.6 \text{ N/mm}^2$$

$$\frac{M}{bh^2} = \frac{0.048\,24}{0.35 \times 0.45^2} \text{ MN/m}^2 = 0.6806 \text{ N/mm}^2$$

BS 8110 Part 3 Design Chart 29
$A_{sc} = \frac{0.4}{100} \times 350 \times 450 = 630 \text{ mm}^2$

M in beam at junction

$$= 118.1 \times \left(\frac{664.5 + 886}{664.5 + 886 + 152.4/2}\right)$$

$$= 112.6 \text{ kN m}$$

Bases

design ult. load at C $= 1197$ kN
SW col. CB $= 0.45 \times 0.35 \times 3.7 \times 23.6$
$= 13.75$ kN
design ult. load at B
$= 1197 + 1.4 \times 13.75 = 1215$ kN
SW base $= 0.3 \times 23.6 = 7.08$ kN/m^2
SW base: ultimate design pressure
$= 1.4 \times 7.08 = 9.91$ kN/m^2
Safe soil pressure $=$
Using load factor for soil $= 1.8$
ultimate pressure on soil can be
$= 217 \times 1.8 = 390$ kN/m^2

min. cover to links $= 25$ mm

∴ short col.

35 mm
25 mm φ bars
402 mm

$f_{cu} = 30$ N/mm^2
$f_y = 460$ N/mm^2
4–16 mm φ
(804 mm^2)

217 kN/m^2
moderate exposure $f_{cu} = 35$ N/mm^2
min. cement content $= 300$ kg/m^3

area of base $= \dfrac{1215}{390-9.91} = 3.197$ m^2

make base —
effective depth of bars (in upper layer)
  using 16 mm φ bars and 35 mm cover
  $= 300-35-24 =$

ult. design pressure $= \dfrac{1215}{3.24} = 375$ kN/m^2

Shear stress at the column face
  $= 1197/(2 \times 0.35 \times 0.45)$ kN/m^2
  $= 3.8$ N/mm^2
satisfactory as $< 0.8\sqrt{35} = 4.73$ and 5 N/mm^2
Critical shear section $1.5 \times 231 = 347$ mm
from faces of col.
Critical perimeter is rectangle
  $0.35 + 2 \times 0.347 = 1.044$ m by
  $0.45 + 2 \times 0.347 = 1.144$ m
and has total length
  $= 2 \times 1.044 + 2 \times 1.144 = 4.376$ m
area within critical perimeter
  $= 1.044 \times 1.144 = 1.194$ m^2
area outside critical perimeter
  $= 1.8^2 - 1.194 = 2.046$ m^2
ultimate design uniform pressure
on base, excluding self-weight of base
  $= 1197/1.8^2 = 369.4$ kN/m^2
∴ shear force on critical perimeter
  $= 369.4 \times 2.046 = 755.9$ kN
∴ shear stress on critical perimeter
  $= 755.9/4.376/0.231$ kN/m^2
  $= 0.7478$ N/mm^2
From BS 8110 clause 3.4.3.4 this will
perhaps be in order later.
max. M

$= \dfrac{369.4}{2} \times \left(\dfrac{1.8-0.35}{2}\right)^2 = 97.08$ kN m/m

BS 8110, Part 3, Design Chart 1

$\dfrac{M}{bd^2} = \dfrac{0.097\,08}{0.231^2} = 1.819$ N/mm^2

$A_s = \dfrac{0.89}{100} \times 1000 \times 231 = 2056$ mm^2/m

overall anchorage length available
  $= (1.8-0.35)/2 - 0.040 = 0.685$ m
From Table 2.9
  $l_b = 39d_b = 39 \times 16 = 624$
From Table 2.12
  hook $\equiv 256$ mm
  $a_h = 624 - 256 = 368$ mm
  < 685 mm ∴ satisfactory

cover $= 35$ mm

1.8 m × 1.8 m
(3.24 m^2)

231 mm

16 mm φ bars
at 90 mm centres
40 mm end cover

7.1.7 Student design office exercise

Each member of the class can be given a different column grid layout similar to that of Figure 7.1, that is 7×5 m, 6.9×5 m, 6.8×5 m,

238   *Design of structures*

7 × 4.9 m etc. A student can check calculations at all stages, with his colleagues working on grids immediately on either side of his own. This helps supervision enormously.

The exercise can be as in Sections 7.1–7.1.6 and can be more accurately designed using bending moment envelopes.

Groups of students can design structures of the same geometry and exposure, each student with a different fire resistance. Groups of students can also design structures of the same geometry and fire resistance but using different exposure conditions. Similarly two students can use the same exposure and fire resistance but use different strength steels, namely with $f_y = 250$ and 460 N/mm² respectively.

7.1.8   Floor of building (two-way and flat slabs)

Suppose that the floor designed in Section 7.1.1 is supported by a 5 m square (as opposed to the rectangular) system of columns. It then seems natural, because of symmetry, to choose two-way spanning slabs or a flat slab, rather than a system of one-way spanning slabs with subsidiary and main beams. Using Table 7.1 the two-way continuous slabs would need to be approximately $5000/40 = 125$ mm thick. This is reasonably thin; hence an intermediate system of crucifix beams, making the slabs 2.5 m square, is not required. The beams between the columns supporting the slab, from Table 7.1, will perhaps need an overall depth of about $5000/25 = 200$ mm, and breadth say about half of this, namely 100 mm, say 125 mm as 100 mm is rather too small to accommodate beam reinforcement.

Ignoring shear, a flat slab needs an overall depth of about $125/0.9 = 139$ mm, say 150 mm. With drops the slab would need to be about 125 mm thick and the drops about $1.4 \times 125 = 175$ mm thick. In either case it would be normal to avoid the need for shear reinforcement and this would usually necessitate the slab being thicker even if column heads are used.

7.2   Design tables

Table 7.2 is useful for the design of the beams shown and also for giving fixed end moments for commencing moment distribution analysis. In this table, as regards the end restraints, F denotes free to rotate and C denotes constrained (i.e. fixed or encastré). The bending moments at A, B and C, respectively, are $\alpha_A Ql$, $\alpha_B Ql$, $\alpha_C Ql$ respectively, where Q is the total load on span l, and C is the position of maximum positive bending

Design tables

moment in the span. The maximum deflection along the span is $\beta Q l^3/(EI)$. The reaction at A is $P = \gamma Q$. Also $\alpha_1 = 1-\alpha$, $\alpha_2 = 2-\alpha$, $\alpha_3 = 3-\alpha$ and $\alpha' = 1+\alpha-\alpha^2/2$.

Table 7.3 is very useful in conjunction with Tables 7.2, 6.1 and 6.2. It is for continuous beams of spans AB, BC, CD, etc. The first section is for a unit bending moment applied to A, whilst the second section is for unit bending moments applied simultaneously at A and the other end of the continuous beam. The bending moment at any support is the applied bending moment M at the end (or ends) times the coefficient. The shear force next to any support is $M \times$ Shear force coefficient divided by the

Table 7.2

Loading	End restraint A B	Coefficients bending moment α_A	α_C	α_B	Deflection β	P γ
(UDL)	F F C F C C	— $-1/8$ $-1/12$	$1/8$ $1/14.2$ $1/24$	— — $-1/12$	$1/76.8$ $1/185$ $1/384$	$1/2$ 0.625 $1/2$
(Point load at $l/2$)	F F C F C C	— $-1/5.33$ $-1/8$	$1/4$ $1/6.4$ $1/8$	— — $-1/8$	$1/48$ $1/107.3$ $1/192$	$1/2$ 0.688 $1/2$
(Triangular)	F F C F C C	— $-1/6.4$ $-1/9.6$	$1/6$ $1/9.51$ $1/16$	— — $-1/9.6$	$1/60$ $1/139.5$ $1/274.3$	$1/2$ 0.656 $1/2$
(Point load at αl)	F F C F C C	— $\alpha\alpha_2/2$ $-\alpha\alpha_1^2$	$\alpha\alpha_1$ $\alpha^2\alpha_1\alpha_3/2$ $2\alpha^2\alpha_1^2$	— — $-\alpha^2\alpha_1$	$\alpha^2\alpha_1^2/3$ — 	α_1 $\alpha_1\alpha'$ $\alpha_1^2(1+2\alpha)$

240   Design of structures

Table 7.3

No. of spans	Bending moment coefficients							Shear force coefficients										
	A	B	C	D	E	F		AB	BA	BC	CB	CD	DC	DE	ED	EF	FE
2	−1.00	+0.25	—	—	+0.25	—		+1.250	−1.250	—	—	—	—	—	—	−0.250	+0.250
3	−1.00	+0.267	—	—	−0.067	—		+1.267	−1.267	−0.333	+0.333	—	—	—	—	+0.067	−0.067
4	−1.00	+0.268	−0.071	—	+0.018	—		+1.268	−1.268	−0.339	+0.339	+0.089	−0.089	—	—	−0.018	+0.018
5	−1.00	+0.268	−0.072	+0.019	−0.005	—		+1.268	−1.268	−0.340	+0.340	+0.091	−0.091	−0.024	+0.024	+0.005	−0.005
2	−1.00	+0.500	—	—	+0.500	−1.00		+1.500	−1.500	—	—	—	—	—	—	−1.500	+1.500
3	−1.00	+0.200	—	—	+0.200	−1.00		+1.200	−1.200	0	0	—	—	—	—	−1.200	+1.200
4	−1.00	+0.286	−0.143	—	+0.286	−1.00		+1.286	−1.286	−0.429	+0.429	+0.429	−0.429	—	—	−1.286	+1.286
5	−1.00	+0.263	−0.053	−0.053	+0.263	−1.00		+1.263	−1.263	−0.316	+0.316	0	0	+0.316	−0.316	−1.263	+1.263

Table 7.4. *Weights of materials*

	kN/m³		kN/m²
Aluminium	27.0	Concrete hollow tile slabs	
Ashes (dry)	6.3	125 mm thick	2.14
Asphalt	20.4	150 mm thick	2.38
Brickwork, cement mortar		190 mm thick	2.68
common brick	19	Corrugated sheeting	
pressed brick	23	galvanised iron	0.144
Cement		asbestos-cement	0.156
loose	11.8–13.3	Doors	0.384
bags	11.0–12.6	N-light roof glazing	0.264
bulk	12.6–14.1	Roofing felt (two-layer built up)	0.048
Coal		Windows	0.240
solid	12.8		
crushed washed	9.0		
crushed unwashed	9.3		
Concrete			
plain or reinforced	23.6		
granolithic or terrazzo	23.6		
foamed slag non-structural	13–15		
foamed slag structural	21		
aerated	8.5–9.4		
Cork	2.4		
Copper	85.9		
Fibreboard	2.9		
Fibreboard, compressed	5.0		
Glass	24–27		
Iron	70.6		
Lead	112		
Lime plaster	18.8		
Macadam	21		
Mortar (set)			
cement screeds	22.6		
lime screeds	15.7–17.3		
Plasterboard	9.3		
Rubber	9.6		
Steel (cast or mild)	77		
Tarmacadam	23		
Vermiculite/cement screed	5.8		
Wood paving	8.7		
Wood wool/cement slabs	5.8–7.2		
Woodwork			
red pine	4.8–7.2		
teak	6.4–8.8		
pitch pine	6.6–7.2		
greenheart	10–12		

span; EF is always the end span, otherwise the spans read consecutively from left to right (that is AB, BC, CD, etc.). The use of Table 7.3 is described in Sections 7.1.3 and 7.1.2.

Table 7.4 gives the weights (for $g = 9.807$ m/s^2) of various building materials.

7.3   Creation (design or selection) of structural system

The designer has to initially decide which type of construction to use. Of course he could design and obtain prices from contractors for many different alternative schemes. Generally the time available for, and the cost of, design and pricing or estimating mitigate against this procedure. The structural designer in conjunction with the architect has therefore generally to decide upon the structural system before obtaining tenders from contractors. The better their experience, the better the selection for the structural system should be, and the selection may have minimum cost as its only objective or may be a compromise between cost, aesthetics and quality.

The author has not mentioned the type of firm where he gained considerable experience, namely the designer–contractor. This is because over the past twenty or more years these types of firms have become fairly insignificant with regard to the total amount of design work effected in the U.K. These firms had a great advantage in designing structures to suit the economics of their own construction organisation, that is making full economic use of exactly the type of plant, works and personnel possessed by the company. Also when in competition with other designer–contractors the client was assured of obtaining the most economic construction. The disadvantage of the system was that the client and architect did not have the advantage of a consultant structural engineer independent of the contractor and this is probably the reason for consultants mainly being used in preference to designer–contractors because the architect advises the client and professionally he will probably prefer to have the services of a consultant from the beginning before having to tangle with contracting costs.

For a building the column layout will be determined from the use of the building and will be as regular a system (or systems) as possible to give repetition for keeping down the contractors' costs (for example, of shuttering, or formwork).

In this book we are considering flat roofs. For those interested in shell and folded plate roofs the author has produced many publications and of these would recommend to beginners Refs 1 to 7; all but Ref. 5 concern cylindrical shells, Ref. 4 also concerns hyperbolic paraboloidal shells (or hypars), Ref. 5 concerns conoidal shells (or conoids) and Ref. 6 also concerns folded plates. For further reading the author has produced Refs 8 to 11.

For a flat roof the superimposed loads used in the U.K. are light in weight relative to the self-weight of the concrete; for example, in Sections 7.1 to 7.1.2 the self-weight of the slab and beams is considerably greater than the superimposed loads they are designed to carry. Whatever type of construction is used to support the superimposed loads between the columns therefore needs to be as light in weight as possible as regards cubic metres of concrete used. Lightness in weight reduces the amount of reinforcement required, but this can also be effected by using a greater overall construction depth. The area of shuttering required for the soffit is the same for all types of *in-situ* concrete roofs. The sides of beams require shuttering and the deeper these beams to reduce reinforcement requirements the greater the amount and therefore cost of this shuttering. Architects often do not like deep, heavy looking beams. Also if the overall construction depth is excessive say at the roof and every other floor of a tall building, then the building will require extra wall cladding and will end up taller than necessary. The complexity of reinforcement bending and fixing may be borne in mind by an estimator as slowing down the construction programme, yet the actual difference in cost between normal and complex bending and fixing per tonne will generally be fairly insignificant with regard to the total cost of the construction.

A flat slab will not therefore usually provide a very economic roof because although there are no beam sides to shutter, the construction is heavy (meaning large quantities of concrete and stronger soffit shutters) and shallow in overall depth. Thus large quantities of reinforcement are required because of both the heavy self-weight and the small overall depth. For example the roof slab may be 200 mm thick for a flat slab roof whereas the slab of an alternative design with beams and slab might well be 125 mm thick. The flat slab can be made lighter by having 'dropped-panels', that is the area around each column is made thicker than the remainder of the slab. This involves the expense of shuttering the vertical periphery of each drop panel. Hollow tiles have also been incorporated in flat slabs to reduce weight but this will generally be found to be uneconomic.

Apart from systems with either or both thicker slabs and/or larger

244 *Design of structures*

columns, flat slabs are usually supported by columns with heads flared out each in the shape of an inverted pyramid or cone to reduce the high shear stress in the slab around the periphery of each column. The term 'punching shear stress' used to be used in this connection, the scenario being that of columns punching through a flat slab. The term was used and a check was made of the slab tearing in shear on a plane vertically above the periphery of the column. Tests show that this never happens and that the failure in shear, characterised by an inclined crack basically due to diagonal tension but also affected by bending moment etc., occurs a short distance away from the periphery of the column.

Flat slabs can be made more economic by reducing the amount and thus the weight of the concrete by introducing voids of minimum shuttering cost. For example a 'waffle slab' uses say standard glass fibre moulds (for lightness and ease of stripping) as shown in position in Figure 7.5 with the idea that the tremendous repetition of use which each of these moulds can sustain will make this shuttering very inexpensive. More recently this idea has been extended using larger voids, see Figures 7.6 and

Figure 7.5 (courtesy the Cement and Concrete Association).

Creation of structural system

7.7, these giving more of a standardised two-way spanning slab supported by beams to the column arrangement.

Continuing our discussion of roofs (the flat slab discussion having led to regular systems of two-way spanning slab arrangements just mentioned) after flat slabs the next *in-situ* arrangements would be those of beams supporting either one-way or two-way spanning slabs. If the column arrangement is square in plan or of up to say 1.5 to 1.0 length-to-width ratio then two-way spanning slabs may well be useful. Again for a roof it is desirable to keep the thickness of the slab to a minimum of say 125 mm. Panel sizes can be designed on this basis. If they are too large for a suitable

Figure 7.6 (courtesy the Cement and Concrete Association).

246 *Design of structures*

division of the distances between columns with a suitable beam system, then the thickness can be kept the same for smaller panels and economies made in the amount of reinforcement required, because the depth is then greater than the minimum requirement.

For more rectangular column layouts one-way spanning slabs of minimum thickness, say, 125 mm would be used in preference to two-way spanning slabs.

The beam layouts for these beam/slab systems sometimes involve main beams supporting secondary beams. This can cause enormous weight on the main beams which may need to be of shorter span and greater depth than the secondary beams.

Figure 7.7 (courtesy the Cement and Concrete Association).

Figure 7.8 shows one of many types of precast concrete roofs. The type shown incorporates voids to keep the weight down. Generally the beam units of these types of roofs (and floors) can be of ordinary or prestressed reinforced concrete. An alternative type commonly used is such that each unit is in effect a hollow beam. The author has been concerned with units which comprise halves of these, and patented for his employer a system using such halves either together or as singles or in pairs between hollow blocks making three types of floor from a machine-produced block and a machine-produced half hollow beam unit. Generally it has been the author's experience that precast slabs are less expensive than *in-situ* slabs mainly because of the shuttering cost. If a slab is at a height then there is also an advantage with precast units in the saving of considerable scaffolding costs for the shuttering.

Generally *in-situ* slabs are of better quality than precast slabs. For example they are more robust, that is they do not have thin (for example 30 mm thick) unreinforced members supporting the top surfaces, although the top surfaces of precast floors are often strengthened, for example, with

Figure 7.8 (courtesy the Cement and Concrete Association).

25 mm thick screeds. Still an *in-situ* slab is more robust against, say, a blow from a sledge hammer, a load accidentally dropped on the floor, say, on one of its corners (that is an impact point load), and so on.

Precast units have to sit on beams and be fastened to them. This causes clumsy detailing problems. If the units simply sit on the beams then the overall depth is unnecessarily high. Beams sometimes have their sides provided with seatings for the precast units; the beams are then rectangular and cannot benefit from being T-beams as in *in-situ* construction. Supporting beams sometimes have their top part cast after the precast units are in place so that the beams can be T-beams but then the beams have to be propped whilst the units are placed, unless the T action is only for subsequent live load.

The units are sometimes filled in with *in-situ* reinforced concrete over the supports to gain the advantage which *in-situ* floors automatically have, that is of continuity.

The type of roof shown in Figure 7.8 can be made with the beam unit supporting the blocks incomplete and requiring supporting until finally completed with *in-situ* concrete. Apart from the disadvantages of this system one advantage is being able to make a very standard and lightweight beam unit. One such system used burnt clay tiles with grooves in them for accommodating prestressing wires for fabricating the beam units.

Previously in this section the discussion has concentrated on roofs. Similar considerations apply to floors, particularly for those carrying lightweight superimposed loads. For floors supporting heavy superimposed loads, the self-weight of the concrete is a smaller proportion of the total weight of the construction than for roofs and it is therefore not so important to try and reduce the self-weight as described previously for roofs.

Flat slabs supported by columns with flared heads will not be economic for small spans because of the size and therefore cost of these heads.

Precast frames[12,13] are economic for single-storey buildings commonly of column layouts 9 m by 4.5 m to 6 m. The author[12,13] has designed and constructed many of these and exceptionally designed for a column layout of 15 m by 9.3 m – the doubly-pitched portal frames in this case carried overhead cranes and needed to be post-tensioned.

With precast frames, joints are the weakness and the problem. The author in Ref. 14 describes a joint suitable for use in multi-storey buildings.

7.4 Creation of structures generally

Previously in this chapter concrete structures have been considered. Generally if very large spans for bridges are required, very lightweight construction is necessary and steel suspension bridges are used. As these are very flexible the decks have to be stiffened with for example steel lattice girders or steel plate box girders (the Humber Bridge) or prestressed concrete box girders; suitably shaped on top to provide carriageways and footpaths as desired. Services can be carried within the box sections.

For slightly shorter spans cable stayed bridges are used. The continuous beam which spans between the cable support points can be of steel lattice girder or steel plate box girder or prestressed concrete box girder construction.

Bridges of slightly shorter span are constructed of steel lattice girders or R.C. arches.

R.C. arches can be very useful for steep sided ravines (e.g. in the Alps) where the arch supports can be on rock and the arch can be high. An arch can be shaped so that it is principally in compression and concrete is a very economic material for use in compression.

Then for similar or slightly shorter spans steel girder bridges or prestressed concrete beams can be used. For shorter spans still R.C. beams can be used.

In buildings, for light roof loads and long spans, doubly pitched steel portal frames (warehouses, station halls, etc.), and steel trusses or castellated beams (supermarkets) are useful. For spans not quite as long, concrete shell and folded plate roofs, doubly pitched precast or *in-situ* portal frames with precast purlins of inverted L- or rectangular cross section, prestressed concrete members and steel beams are useful. For shorter spans R.C. beams can be used.

The steel or R.C. portal frames, or cylindrical shell roofs can carry overhead cranes.

Multi-storey buildings can have structural steel or *in-situ* or precast concrete frames. The floor slabs can be as follows:

(a) For long slender spans, prestressed concrete floor units can be used (e.g. see Fig. 7.8). Deflection due to live load needs consideration, and finishes can crack over supports. The units usually have a substantial camber and this means that if a granolithic topping of a certain thickness is specified – a sig-

nificant amount of extra thickness has to be provided over and near the supports.
(b) For similar and slightly shorter spans, continuous *in-situ* hollow tile floor slabs can be used.
(c) For slightly shorter spans, precast concrete units can be used. In the author's experience these are usually the least expensive of the various floor constructions.
(d) For similar spans, continuous *in-situ* floor slabs can be used, two-way or one-way spanning, according to support conditions.
(e) If no supporting beams are provided then flat slab construction can be used. Sometimes these are used for aesthetic reasons, but the appearance of a clean flat soffit with mushroom headed circular columns can be ruined if services have to be fastened to the outside of the columns. For the longer spans the slab tends to become rather thick and heavy so drop panels can be used. But this sometimes means double shuttering or the use of hollow tiles for the central shallower areas. The waffle slab has been very useful in reducing the self-weight of large span flat slabs and makes use of standard plastic moulds which can be used a great number of times, see Fig. 7.5. See also Sections 4.3 and 7.3.

This section should be read in conjunction with Sections 2.5 and 7.3 and the part on shell and folded plate roofs.

References

1. Evans, R. H. and Wilby, C. B., *Concrete – Plain, Reinforced, Prestressed and Shell*, Edward Arnold, London (1963).
2. Wilby, C. B. and Bellamy, N. W., *Elastic Analysis of Shells by Electronic Analogy*, Edward Arnold, London (1962).
3. Wilby, C. B., A proposed 'exact' theory for analysing shells, and its solution with an analogue computer. *Proceedings of the Institute of Civil Engineers*, July (1962).
4. Wilby, C. B. and Khwaja, I., *Concrete Shell Roofs*, Applied Science Publishers Ltd., Amsterdam, London, New York (1977).
5. Wilby, C. B. and Naqvi, M. M., *Reinforced Concrete Conoidal Shell Roofs – Flexural Theory, Design Tables*, Cement and Concrete Association, London (1973).
6. Wilby, C. B., *Concrete for Structural Engineers*, Newnes–Butterworths, London (1977).
7. Wilby, C. B., *Design Graphs for Concrete Shell Roofs*, Applied Science Publishers, London (1980).
8. Wilby, C. B. and Bellamy, N. W., *Analysis Elastico de Cascarones por la Analogia Electronica*, Compania Editorial Continental, S.A., Mexico (1963).

References

9. Wilby, C. B., A method of designing north-light shell roofs. *Indian Concrete Journal*, Jan. (1961).
10. Wilby, C. B., *Elastic Stability of Post-tensioned Prestressed Concrete Members*, Edward Arnold, London (1964).
11. Wilby, C. B., 'Shell Roofs', *Handbook of Structural Concrete*, Editors Kong and Evans, Chapter 32, Pitman Books Ltd., London (1982).
12. Wilby, C. B., A warehouse with continuous precast frames. *Concrete & Constructional Engineering*, Jan. (1958).
13. Wilby, C. B., Precast concrete framed roofs – design of joints and post-tensioning. *Indian Concrete Journal*, Feb. (1960).
14. Wilby, C. B., Structural behaviour of a special type of joint for connecting precast concrete members of industrialised buildings. *Proceedings RILEM-CEB-CIB Symposium*, University of Athens (1978).

8

Prestressed concrete

8.1 Prestressing

Prestressing consists of initially applying loads to a member to counteract the effects of the working loads to which it will eventually be subjected. Concrete is relatively weak in tension compared with compression, so the prestressing forces are used to compress zones which will subsequently be required to carry tension. Prestressing forces are usually applied in one of the following ways:

1. Stretching wires, cables or bars on a bed, concreting the member around such wires, and then releasing the wires when the concrete is sufficiently hard. When the wires are released, they shorten, and compress the concrete member, the line of action of such compression for each wire being the profile of the wire in the beam. This procedure is known as *pretensioning*.

2. A member is concreted and a duct is formed in the member either with a metal sheath or an inflatable tube. A tendon, consisting of either a bar, cable (for example Strand) or group of wires, is threaded through the duct and tensioned when the concrete is sufficiently hard, and anchored to the concrete member, so that the concrete member is compressed by this tendon. The procedure is known as *post-tensioning*, and it is usual subsequently to fill the duct surrounding the cable with grout. A grout of cement, with no more than sufficient water for the workability required, is suitable. Sand is not recommended.[1,2] Special plasticisers[3,4] are recommended to give better quality grouting. Air entrainers[4] can be used instead of or in addition to plasticisers. It is vitally important not to trap pockets of water in ducts, as they have frozen in winter and caused trouble. Soroka and Geddes[5] report 'the ultimate moment and pattern of cracking are hardly influenced by grout quality'. Szilard[6] reports particular concern with regard to the adequacy of the strength of the

grout and its corrosion resisting properties. Refs. 5 and 6 list 75 and 103 references respectively on this subject. The author has experience of a special polyester material which would seem to be excellent for strength and workability for use in even damp ducts, though its rapid setting time would be its greatest disadvantage in use and some development in this respect would be necessary. Epoxy resins[3] are affected by water, as in the experience of the author the hardeners react chemically with water.

3. A variation on method 2 is to place the tendon in the sheath before concreting. It is usually easier to thread the tendon in the sheath before concreting than in the duct after concreting. This does not, of course, allow inflatable tubes to be used for forming the duct. The latter method appears to be cheaper from the point of view of forming the duct, but on the whole, in the U.K., when the extra cost of positioning the inflatable tubes and threading the ducts they form is considered, it is usually more economical to place the tendon in the expendable tubing before casting.

4. Another variation on method 2 is to make the concrete member in precast portions which are placed together on the site, the joints between such members being dry packed with cement:sand mortar, usually after the tendons have been threaded through the blocks. An alternative material for jointing is polyester resin. The author devotes Chapter 16 of Ref. 7 to 'Beams Consisting of Segments – Joint Efficiency'.

5. A variation[7] on method 4 is to cast each portion against the previous portion, sometimes post-tensioning each portion to the previous one, and the finally post-tensioning all portions together.

6. Prestressing forces can be exerted on structures in suitable places by jacks. For example, hydraulic jacks have been used in the abutments of dams arched in plan, to exert known forces in favourable directions and achieve economies in the amounts of concrete required in the dams.

8.1.1 Advantages and disadvantages of prestressing

The chief advantages of prestressed concrete are in reducing the quantities of steel and concrete required and in eliminating or reducing the widths of cracks. The disadvantages are the extra labour costs connected with the stressing of the tendons, and with other items.

Prestressing strengthens a beam in shear and can give a useful saving in shear reinforcement, useful with regard to cost and sometimes especially with regard to facility of detailing. The author has on occasions post-tensioned jointed precast structures solely because of the weakness of the joints in shear.

In the U.K., if a member can be equally well constructed in prestressed or ordinary reinforced concrete, then the latter is usually more economical. When, however, large spans are required with shallow depths, for example for bridges, precast floors and so on, and the ordinary reinforced concrete is structurally unacceptable, then prestressed concrete is the only answer in concrete, and, if there is a reasonable repetition in the making of members (to reduce shuttering costs), in the U.K. it is sometimes more economical than structural steelwork. If a factory is highly organised in the manufacture of prestressed flooring units it is sometimes found that units which could be of ordinary reinforced concrete can be made shallower in prestressed concrete and can thus be less expensive overall by making savings in transportation, handling and stacking. When the spans of bridges are sufficiently short to make prestressing cheaper than steelwork, prestressed concrete has the great advantages over steelwork of relative freedom from maintenance, and fire resistance.

Prestressed concrete construction is often more expensive to design than ordinary reinforced concrete work. In post-tensioned *in-situ* structures, prestressing procedures have to be carefully planned because tensioning one cable makes previously tensioned cables deficient in stress and can cause undesirable stresses to develop due to the eccentricity of the prestressing force; this eccentricity will usually be eliminated when the prestressing is satisfactorily completed. Sometimes this planning involves larger amounts of structures to be shuttered or alternatively supported before prestressing than would be necessary if the structure were of ordinary reinforced concrete. In such circumstances prestressing sometimes slows down the speed of construction and increases the shuttering required for a contract.

Members designed with prestressed concrete can be very flexible and the designer must be particularly careful that deflections, cambers and flexibilities are satisfactory.

It is conceivable even to pay more for prestressed concrete structures than for ordinary reinforced concrete structures when resistance to corrosion is important; the life of the prestressed structure can be greater because of the absence of cracks. Structures such as docks, wharfs and jetties which are exposed to sea water, exposed structures at gas works, bridges exposed to pollution, structural work in dairies exposed to lactic acid, are common examples of concrete structures exposed to corrosive elements and can benefit from prestressing.

8.2 Materials

Prestressed concrete uses highly stressed steel and concrete, and good materials and workmanship are most important. Failures have occurred due to corrosion of tendons. The concrete and grouting materials must be non-corrosive to the steel and dense for strength and for resistance against water or corrosive liquids endeavouring to come into contact with the steel. Calcium chloride is detrimental in concrete that allows water to contact the steel (see Chapter 2). Generally the quantities of chlorides and sulphates should be strictly limited in the concrete materials. The corrosion of tendons can also be due to pitting and hydrogen embrittlement as well as stress corrosion; Ref. 6 is useful.

If the use of high alumina cement is contemplated, refer to Section 2.1.

The writer has seen a system of shell roofs with post-tensioned prestressed concrete tie members of rectangular cross section 152 mm wide by a depth varying between 252 mm at mid-span to 155 mm at supports, each with a centrally disposed eight wire cable without sheathing, where many of these cables have had seven wires completely corroded through, in about twenty seven years, even though the ties were completely wrapped with roofing felt which had no apparent defects. Calcium chloride had been used to accelerate hardening of the concrete and it was agreed that this was the trouble.

8.2.1 Stress corrosion

This is a very important problem. There have been failures due to stress corrosion, and there has been much research, particularly in connection with prestressed concrete bridges, and also because the strands used for tendons are also used for cables of suspension and cable-stayed bridges and funicular mountain railways. The author has seen considerable research in progress in stress corrosion in Paris and Zurich. Leonhardt[3] establishes conditions which must exist for stress corrosion of prestressing wires. The author understands that British steel has always been manufactured not to experience this problem.

8.3 Losses of prestress

The stress initially effected in the tendons is reduced by the following losses. (For examples of how BS 8110 deals with these losses, see Example 8.5.)

1. *Relaxation (creep) of steel.* The high stresses used in the tendons mean that the steel is sometimes stressed slightly beyond its limit of

proportionality. Hence, after anchorage the strain in the steel can correspond to a lower stress as creep occurs. With pretensioned members, this loss can be greatly reduced by tensioning say in the afternoon and then suitably increasing the strain in the tendons next morning before casting. This is an operation which interferes with progress and increases labour costs, and for overall economy it is usually better not to try to eliminate creep but to consider it as a loss in prestress. A rise of temperature helps the steel to creep and can thus increase creep loss. The relaxation loss depends on the type of tendon and the magnitude of the stress it experiences.

2. *Elastic deformation (strain) of concrete.* When pretensioned wires are released they compress the concrete, the concrete strains, and thus reduces the strain and hence the stress in the wires. This is known as loss of prestress due to strain (or elastic deformation). A post-tensioned member with only one tendon does not, in theory, experience a strain loss because as the jack strains the tendon it compresses the concrete. When more than one tendon is used, then as each tendon is strained the jack increases the strain in the concrete; this reduces the strain in the tendons already anchored; that is strain losses occur in all but the last tendon to be stressed. All these losses total less than those experienced with pretensioned concrete. When pretensioned wires are released they will shorten owing to the concrete becoming strained and stressed (that is prestressed). The shortening of the wires divided by their length is the loss in strain of the wires, say ε_1. This shortening must be the same for the concrete immediately in contact with the wires and this is unstressed before the shortening and then stressed due to the shortening. Hence the strain in the concrete is also ε_1. Applying Hooke's law, the loss of stress in the wires is $E_s \varepsilon_1$ and the gain of stress in the concrete is $E_c \varepsilon_1$. Hence loss of stress in wires $= E_s \varepsilon_1 = E_s \cdot ($Stress in concrete$/E_c) = \alpha_e \cdot ($Stress in concrete$)$. This is most important and will be summarised as follows:

For pretensioning:

$$\text{Loss of strain in wires} = \frac{\text{shortening of wires}}{\text{length}}$$

$$= \text{gain of strain in concrete} = \varepsilon_1$$

$$\text{Gain in strain in concrete} = \frac{\text{stress in concrete}}{E_c} = \varepsilon_1$$

Losses of prestress

$$\text{Loss of stress in wires} = E_s \varepsilon_1$$

$$= E_s \times \left(\frac{\text{stress in concrete}}{E_c} \right)$$

$$= \alpha_e \times (\text{stress in concrete})$$

3. *Shrinkage of concrete.* Shrinkage is discussed in Chapter 2. As concrete shrinks after the tendons have been anchored to the concrete, the concrete member shortens and hence so does the tendon, thus releasing some stress in the tendon. In the case of pretensioned concrete, the shrinkage effect begins as soon as the concrete is cast, but with post-tensioned concrete the concrete is able to shrink before the tendon is stressed. If there were no longitudinal reinforcement the shrinkage would be restricted only by friction with moulds, etc., and most of the shrinkage would occur before stressing. Humidity and temperature also affect shrinkage. For practical design BS 8110 gives suitable recommendations for calculating the loss of prestress due to shrinkage of the concrete (see Example 8.5).

4. *Creep in concrete.* Creep has already been explained in Chapter 2. As the concrete creeps it reduces the strain and hence the stress in the prestressing tendons. With prestressed concrete, creep is not under a constant stress, as considered in Chapter 2, because the stress in the concrete is reduced as the concrete creeps. The creep loss may be estimated by reference to BS 8110. The loss is greater for pretensioned than for post-tensioned members. Pretensioned tendons rely upon their bond to the concrete for anchorage and in time this releases (or creeps) slightly; this *creep of bond stress* is not counted as a separate loss, so is accounted for as an increase in creep loss.

5. *Slip of anchorage.* This refers to the tendons losing stress after anchorage due to the anchorage device slipping; for example wedges are pulled forward in their jaws as the stress is taken up by the anchorage. This should be assessed for the particular system used. For prestressing over short distances it is preferable that this allowance should be as small as possible, as the greater the allowance the greater the probable error in the reliability of this quantity. For this reason the author[8] found certain bars useful for prestressing over short lengths; the relative movement between the nuts and threads of the system caused only very little loss of stress.

6. *Friction in jack and anchorage system.* In pretensioning, if the extension of the wire is measured directly then the friction in the jack is not a loss to be deducted from this prestress measurement. If, however, in pretensioning, and mostly in post-tensioning, the prestress is measured say on the body of the jack with a vernier recording the movement of the movable part relative to the stationary part, then as there will be friction in the jack, this frictional force will be included in our measurement of the force in the tendon using the oil pressure gauge on the jack. The difference between the forces measured in a tendon by these two methods depends upon the type of jack used. This difference should be reasonable for the particular type of jack as excessive oil pressure would indicate that the jack had seized up and further stressing would damage it. Certain jacking equipment can incorporate a strain gauge load cell. In post-tensioning systems where the tendons are deviated just before the final anchorage, indeed deviated to effect the final anchorage because of the space required by jacks and anchorages, there is a frictional loss related to the pressure of the tendons on the sides of the deviating device used by the particular system. This must be allowed for in determining the prestressing force in the tendons and it depends upon the particular system.

7. *Friction along duct in post-tensioning.* With regard to a straight tendon it will normally be detailed not to touch its duct, but in practice, because of lack of straightness of tendon and duct, there will be some contact between a tendon and its duct. The duct will tend to deviate the tendon, perhaps in some kind of wobble along the duct. Because of this deviation there must be pressure between tendon and duct and thus frictional forces between the two when the tendon is being stressed. When a tendon is taken round a bend, the tendon exerts pressure on the duct, or concrete tank wall, etc., and there is friction associated with this pressure on tensioning the tendon.

BS 8110 recommends the following formula which has been justified by tests:

$$P_x = P_0 \exp[-Kx - (\mu x/r_{ps})] \tag{8.1}$$

which for small values of Kx and $\mu x/r_{ps}$ can be approximated to

$$P_x = P_0[1 - Kx - (\mu x/r_{ps})] \tag{8.2}$$

where

P_0 = Prestressing force in a tendon at the jacking end.
P_x = Prestressing force at any distance x from the jack.

Limit state design of members

K is a constant depending on how much the duct is likely to deviate, that is how rigid the sheath is, how often it is supported, how much vibration is used for the concrete.

μ is the coefficient of friction between the tendon and the duct surface, or surface of concrete tank wall, etc.

r_{ps} = Radius of curvature.

e = 2.718.

In the case of tendons which are not to be finally bonded to the member, they can be lubricated, and some tendons can be purchased enclosed in polythene sheaths packed with suitable grease. The author has used the latter (drawn Strand in lubricated sheaths), and was very impressed with it, on the outside of a dome which required strengthening as an emergency measure – the polythene and grease gave good weather resistance.

The author in Ref. 7 devotes Chapter 5 to friction describing methods of assessment, giving derivation of formulae (such as equations 8.1 and 8.2) and suggesting a different approach. It gives K and μ values for normal work and, in a Table 5.1, for both mastic coated and pregreased tendons which are used to reduce friction in the U.S.A. Then Appendix 1.6 in the book gives K and μ values used in many different countries for many various types of tendons and ducts.

8. *Steam curing.* This can interfere with the losses due to creep and shrinkage of the concrete, and relaxation of the steel.

8.4 Limit state design of members

Members must be designed for the following:

1. Limit state of cracking, due to flexure.
2. Ultimate limit state, due to flexure.
3. Prestressing requirements; losses; maximum initial prestress; end-block design or transmission length requirements.
4. Ultimate limit state; shear.
5. Limit state of deflection.
6. Considerations affecting design details.
7. Torsion.

For an exposed structure it might be required to eliminate cracks completely, whatever the particular loading being experienced by the member. In the U.K. absence of cracks was originally considered to be one of the prime advantages of prestressed concrete. Generally speaking the greater the amount of flexural tension which can be allowed the more economic will be the construction, for example less tendons required, but the greater the danger of cracking.

The ultimate strength of a prestressed concrete beam is generally not greatly different whether the prestressing is applied or not. The greater the amount of prestressing applied to such a beam the more it will be possible to reduce the size of the cracks (they can even be eliminated), the amount of the deflection, and its rigidity (proportional to second moment of area).

For design purposes (with regard to the limit state for cracking) BS 8110 suggests three classes of structures thus:

Class 1. No tension is allowed to be taken by the concrete.

Class 2. Tension is allowed to be taken by the concrete but the amount is limited to preclude noticeable cracking.

Class 3. Refers to *partial prestressing*, where large theoretical tensile stresses are allowed which cannot exist because of exceeding the modulus of rupture of the concrete, but these theoretical tensile stresses are limited so that the cracks which will occur are not likely to allow rainwater to penetrate to the reinforcement, etc.

The designer has to choose his limit state of cracking according to the conditions of exposure, and the quality required, of the structure. If the designer is, say, concerned about temperature stresses due to the member not being perfectly free to move, then he may require no tensile stresses, and he may require the minimum compressive stress at any stage of loading experienced under working conditions to be slightly in excess of the maximum tensile temperature stresses. This would be as Class 1.

Class 1 would be used say for exposed structures (exposed to polluted atmosphere, sea water, etc.). Class 2 would be used for more economy than Class 1, when durability is not so important. Class 3 would be used for greater economy, but of course one of the advantages of prestressed concrete, namely absence of cracking, is sacrificed. Class 3 could be suitable where there would be no tensile stresses under most working loads, but yet for the infrequent maximum working load of short duration tensile stresses would be induced in the member. It has been used for some railway bridges.

The sequence of design for Classes 1 and 2 is suggested to be in the order 1–6 (see the beginning of this section). For Class 3 the sequence is suggested to be 2, 5, 1, 3, 4, 6.

To design for the various limit states of BS 8110 it is necessary to be able

Limit state design of members

to calculate stresses and deflections at working loads. This is done with the elastic theory. It is also necessary to assess ultimate strength, and this is done using the plastic theory. Also, transmission lengths for prestressing wires and end-block designs, for post-tensioned members, must all be adequate. All these methods will now be discussed.

In design it is always necessary to find a suitable section and its reinforcement, before all the checks of the adequacy of 1–6 (previously in this section) are ascertained with adequate precision. With experience the original estimate of the section may need little or no alteration as a result of these checks. For optimisation one would program the procedure so that the computer can keep modifying the original estimate of the section to satisfy the various checks as economically as possible. This is simply a matter of programming the procedures of design which follow in this chapter.

8.4.1   Simple assessment of size of prestressed members

As previously mentioned, experience helps this procedure. One can be guided by observing sizes of members of similar jobs from publications, etc. Alternatively, or in addition, one can choose the type of concrete to be used – one which is not too difficult to achieve with the methods to be used and standard of product required – and proceed as follows.

Example 8.1. An initial estimate is required of a suitable I-shaped cross section for a prestressed concrete beam which has to resist a total bending moment at mid span at working loads of 870 kN m (inclusive of its self-weight), and is to be designed for a limit state of cracking of Class 1.

It is fairly easy for the manufacturer to obtain a concrete of characteristic cube strength at 28 days (when we assume the structure may need to withstand its working load) of 40 N/mm², and this concrete can be made of early enough strength for the requirements at transfer.

Referring to BS 8110 Cl. 4.3.4.2, the allowable compressive stress is

$$0.33 \times 40 = 13.2 \text{ N/mm}^2 = 13\,200 \text{ kN m}^2$$

If the tendons are to be straight then the bending moment due to the weight of the member will reduce the prestressing at mid span, but not at the supports, and thus the supports are the critical sections for deciding the amount of prestressing. At these sections at working loads the prestressing could be as Figure 8.1(a). This would also be the prestressing at the mid-span section, because of the straight tendons. The total bending moment at mid span can therefore give a stress distribution as Figure 8.1(b), which is superimposed upon (a) to give (c) in Figure 8.1, assuming the section to have its neutral axis at mid depth of the section. The section can hence be designed as for Figure 8.1(b); thus the section modulus needs to be

$$870/13\,200 = 0.065\,91 \text{ m}^2$$

Try the sections shown in Figure 8.2. Its second moment of area is

$$[(0.45 \times 1.1^3)/12] - [(0.3 \times 0.8^3)/12] = 0.037\,11 \text{ m}^4$$

and its section modulus is therefore

$$0.037\,11/0.55 = 0.067\,48 \text{ m}^3.$$

This section will therefore be suitable. The bending moment used included an allowance for the self-weight of the member. This had to be estimated and should be checked against the section now obtained. If the estimate is found to be wrong then the section we have just designed gives a good clue to a revised estimate of the self-weight of the beam for use in a revised design.

Figure 8.1

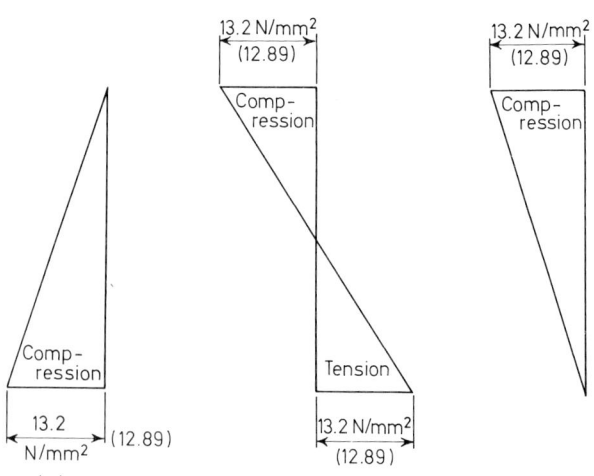

Limit state design of members

Example 8.2. It might be useful to continue the design of Example 8.1 to assess approximately the tendons required.

Suppose the beam is pretensioned and 7 mm diameter wires are to be used. From BS 8110 to specified characteristic strength of these wires is 60.4 kN and the cross-sectional area of each wire is 38.5 mm². According to BS 8110 the maximum initial prestressing force in a wire would normally be 70% of this 60.4 kN = 42.28 kN. The ACI-ASCE 323 Report suggests that for approximate purposes total losses can be taken as 245 N/mm² for pretensioning and 175 N/mm² for post-tensioning, but loss due to friction between tendon and duct must be added to the 175. Hence total loss of prestressing force per wire

$$= 0.245 \times 38.5 = 9.433 \text{ kN}$$

and prestressing force per wire after losses

$$= 42.28 - 9.43 = 32.85 \text{ kN}$$

The prestressing force required after losses (see Figure 8.1(a)) will be the average prestress multiplied by the area of the cross section.

$$\text{Area of cross section} = 0.45 \times 1.1 - 0.3 \times 0.8 = 0.255 \text{ m}^2$$

Now the section was slightly larger than required. We might as well allow for this, so the stress in Figure 8.1(a) now becomes

$$870/0.067\,48 = 12\,890 \text{ kN/m}^2 = 12.89 \text{ N/mm}^2$$

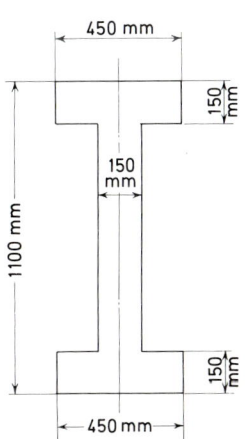

Figure 8.2

This figure will therefore be used instead of 13.2 in Figure 8.1. It is shown in brackets in the figure.

Prestressing force required after losses when member finally in use, from Figure 8.1(a) $= 0.5 \times 12890 \times 0.255 = 1643$ kN. Therefore

$$\text{Number of wires required} = 1643/32.85 = 50.02 = 51 \text{ wires}$$

The designer has to check whether or not these can be placed in the section, with the distances between wires and covers specified by BS 8110; and the centroid of the wires should coincide with the centroid of the force calculated from the stress distribution of Figure 8.1(a) and the cross-sectional areas of Figure 8.2. If this is not possible, then larger tendons may be satisfactory, but if unsatisfactory the designer starts again with another size of section.

In this example the wires can be accommodated in the section. They will mostly be placed in the bottom flange, perhaps two or more in the top flange and perhaps a few in the web.

From Figure 8.1(a) the bending moment due to the prestressing force

$$= 1643 \times e = (12890/2) \times 0.06748$$

where e is the depth of the resultant prestressing force below the centre of the depth of this symmetrical section. Therefore $e = 0.2647$ m.

As mentioned before, the wires have to be disposed so that their centroid is at this depth.

8.4.2 Assumptions for elastic design

Of the following assumptions, 1–4 are the same as those described in Chapter 3:

1. Plane sections subjected to bending remain plane after bending.
2. Stress is proportional to strain for both the steel and the concrete.
3. Perfect bond is assumed between the steel and the concrete. In the case of post-tensioning this theoretically applies after the tendon has been grouted.
4. Depths of reinforcements relative to the depth of the concrete member are considered to be negligible.
5. Allowances must be made for shrinkage and creep losses.
6. Young's modulus for concrete is the same in tension as compression; this is reasonably true.

8.4.3 Limit states of stresses and deflections

During the life of a prestressed concrete beam there are many changes in the stresses and deflections it experiences, and all the worst possibilities should be investigated. When all this has been evaluated, if anything is wrong then one has to return to the beginning and re-estimate the size of the section. Hence one has to concentrate on the most likely worst cases first, so that if re-design is necessary one finds this out as soon as possible.

Essentially a member has to be designed for stresses at *transfer* of prestress from tendons to concrete. This is an important limit state, as the concrete is often not very old and hence not as strong as it will be when in the final structure; also, the tendons have not experienced losses as great as they will experience in the final structure.

Then the member, if not *in situ*, will be handled, stacked, loaded, transported, unloaded, perhaps stacked and then lifted into position. All these operations, if not skilfully performed, could impose many adverse stresses. It is usually best, for prestressed concrete, to have spreaders for slings of cranes and to use lorries with long backs so that beams are always supported at their ends as they have been at transfer, and will be in the final structure. Then adverse stresses can be eliminated, and there is no need to design for this limit state of handling, transportation, erection, etc.

A member must also, of course, be designed for its limit states of stresses and deflection when in its final position in the structure.

8.4.4 Simplified elastic design of prestressed concrete beams

The simplification is by way of ignoring the steel reinforcement in calculating the cross-sectional area, depth of neutral axis and second moment of area of the concrete section. This reduces the work of the calculations considerably, as for very accurate calculations various different sectional properties are required. For example, when pre-tensioning, the areas of the wires and the concrete they displace should be included in the calculations of the sectional properties and different modular ratios should be used for transfer and final serviceability. For post-tensioning, at transfer, the tendon and duct should be excluded, but any other steel included, in calculations of sectional properties. On the other hand, for limit state of serviceability (that is after the duct has been grouted) all the concrete (including grout), tendons and any other reinforcements should be included in the calculations of sectional properties, these reinforcements having different modular ratios to those used at transfer.

The simplified method is adequate for many purposes, as the percentage of steel in the cross section is generally low enough to cause little error, and this error tends to cause excess safety.

The losses are firstly taken as percentages of the initial prestressing tendon forces. This enables the concrete stresses to be obtained and then the losses can be obtained more accurately from these stresses. If it is then found that the original estimate of losses was not good enough, an adjustment is made and the design repeated. The process can be repeated until the designer is satisfied – it can quite simply be programmed for a computer. However, with experience a designer often does not need to alter his first estimate, as he will have slightly overestimated so as not to have the trouble of re-design; the computer is of course useful for optimisation here. The method is illustrated in the following examples.

Example 8.3. Continue the design of the beam of Example 8.1. Having approximately checked the stresses it might now be best approximately to check the limit states of deflection in case we have to alter the section on this count.

When we are interested in the maximum deflection in service, the concrete then has a characteristic strength of 40 N/mm², and γ_m for concrete is unity, so from BS 8110:Part 2, Table 7.2, $E_c = 28$ kN/mm². Shrinkage and creep have been allowed for in the losses assumed. When we consider deflection at transfer we will assume that the concrete has a characteristic strength of 30 N/mm², and γ_m for concrete is unity, so from BS 8110:Part 2, Table 7.2,

$$E_c = 26 \text{ kN/mm}^2$$

Assuming the beam is simply supported over a span of 22 m and that all loading is uniformly distributed (q), then

$$(q/8) \times 22^2 = 870, \therefore q = 14.38 \text{ kN/m}$$

From BS 8110, the limit states of deflection are as follows:

1. If finishes are to be applied the span-to-total-upward-deflection ratio should exceed 300. This refers chiefly to floor and roof units which can have varying cambers, often because of releasing the wires when the concrete is not strong enough on the prestressing beds – the indicative cubes are sometimes compacted very much more thoroughly and

sometimes cured more favourably than most of the concrete in a member and, under these bad circumstances, are misleading. The less the upward deflection, the less the problem and hence this limitation suggestion of BS 8110.

At transfer the losses will not be as great as finally. For pretensioning they can be very approximately 10–15% (assuming the relaxation losses of the steel are kept modest). Supposing we take 10% to be on the safe side. At transfer the smaller losses give greater concrete stresses, which are usually the most limiting consideration at transfer. (Note that a safe and not excessively conservative figure for post-tensioned concrete would be just the steel relaxation loss if the tendons are stressed simultaneously and there are no excessive losses due to severe curvature, such as for a circular tank or dome.)

Prestressing force of 1643 kN was based on losses of

$$(9.43/42.28) \times 100 = 22.3\%$$

Then at transfer prestressing force after losses

$$= \frac{1643 \times (100-10)}{100-22.3} = 1903 \text{ kN}$$

The bending moment due to this prestressing force

$$= 1903 \times 0.2647 = 503.7 \text{ kN m}$$

The deflection upwards due to this constant bending moment (the span on the prestressing bed is the overall length of the beam, say 23 m)

$$= \frac{503.7 \times 23^2}{8 \times 26 \times 10^6 \times 0.03711} = 0.03457 \text{ m}$$

This is reduced by the downwards deflection due to the self-weight of the beam which is

$$\frac{5 \times 6.018 \times 23^4}{384 \times 26 \times 10^6 \times 0.03711} = 0.2272 \text{ m}$$

where the self-weight of the beam, assuming the weight density of prestressed concrete is 23.6 kN/m³ (mass density of prestressed concrete = 2400 kg/m³), is

$$0.255 \times 23.6 = 6.018 \text{ kN/m}$$

At transfer, therefore, the total upward deflection

$$= 34.6 - 22.7 = 11.9 \text{ mm}$$

This gives a span-to-deflection ratio of 1849, which is greater than 300 and therefore satisfactory should the member be used in this way. (In this particular example it was not necessary to calculate the 21.2, as the 32.1 without the reduction of 21.1 would still have been satisfactory for the span-to-deflection ratio of 300, but this will not always be the case.)

2. The final span-to-deflection ratio should exceed 250, the deflection being measured below the level of the supports. In the present example the deflection downwards

$$= \frac{5 \times 14.38 \times 22^4}{384 \times 28 \times 10^6 \times 0.03711} = 0.04221 \text{ m}$$

The bending moment due to the prestressing force

$$= 1643 \times 0.2647 = 434.9 \text{ kN m}$$

The deflection upwards due to this constant bending moment

$$= \frac{434.9 \times 22^2}{8 \times 28 \times 10^6 \times 0.03711} = 0.02532 \text{ m}$$

Hence the deflection below the supports is

$$42.2 - 25.3 = 16.9 \text{ mm}$$

This gives a span-to-deflection ratio of 1302, which is satisfactory as it exceeds 250.

3. Partitions and finishes, either above or below, if the beam is in a building, can be damaged by excessive deflections. BS 8110 generally suggests limiting the deflection to 20 mm and to a span-to-deflection ratio greater than 350. These deflection calculations are for deflections after the fixing of the partitions and the applications of the finishes. We can therefore assume the concrete to be at least 28 days old and we are essentially interested in the subsequent deflection due to live load. However, if say glass partitions are built up to the soffit of the beam, then no live load deflection is tolerable and details have to be devised to, for example, allow a beam to slide past rather than bear on to a partition.

In the present example, the self-weight of the beam, from 1 above, is 6.018 kN/m, hence the live load is

$$14.38 - 6.018 = 8.362 \text{ kN/m}$$

Limit state design of members

and the deflection due to this

$$= (8.362/14.38) \times 42.21 = 24.55 \text{ mm}$$

The span-to-deflection ratio is 896.1, which is greater than 350 and therefore satisfactory.

The deflection is, however, greater than 20 mm and therefore not as recommended by BS 8110. If this is acceptable practically then the design does not need revision to reduce this deflection. Thus in the present example the beam might not be suitable in a building.

Example 8.4. Before we examine in more detail the preliminary design given in Examples 8.1 and 8.2, it would be advisable to determine approximately the adequacy of stresses at transfer of this design.

At transfer the losses will not be as great as finally. They can be 10–15%. Supposing we take 10% to be on the safe side. At transfer less losses give greater concrete stresses, which are usually the most limiting consideration at transfer.

Prestressing force of 1643 kN was based on losses of 22.3%. Hence, referring to Figure 8.1(a), the stress of 12.89 N/mm², which is directly proportional to the prestressing force, will be altered for conditions at transfer pro rata to the different prestressing forces at transfer and finally, and it thus becomes

$$\frac{12.89 \times (100 - 10)}{100 - 22.3} = 14.93 \text{ N/mm}^2$$

Figure 8.3(a) therefore shows the distribution of prestress at transfer at the supports. At transfer the member will usually hog upwards and hence the mid-span section withstands the maximum bending moment due to the self-weight of the member superimposed upon the prestress at this section. For calculating this bending moment we should use the overall length (23 m) of the beam. The maximum bending moment due to self-weight

$$= (6.018 \times 23^2)/8 = 397.9 \text{ kN m}$$

and the extreme fibre stresses due to this bending moment

$$= 397.9/0.067\,48 \text{ kN/m}^2 = 5.897 \text{ N/mm}^2$$

Figure 8.3(b) therefore shows the distribution of stress at mid span due to the self-weight loading. Algebraically adding these stresses to the prestress

270   *Prestressed concrete*

shown in Figure 8.3(a) we obtain Figure 8.3(c), which gives the resultant distribution of stress at mid span at transfer.

Referring to Clause 4.3.5.1 of BS 8110:Part 1, the concrete strength at transfer will need to be the greater of $14.93/0.5 = 29.86$ N/mm² or $9.033/0.4 = 22.58$ N/mm². This agrees with our assumption of 30 N/mm² in Example 8.3.

Example 8.5. In Examples 8.1, 8.2, 8.3 and 8.4 we have made an approximate design of a prestressed concrete beam. We shall now check for this beam the limit states determined by elastic theory and concerning stresses and losses.

For these limit states BS 8110 gives $\gamma_m = 1$ for steel and 1.3 for concrete.

Considering the losses:

1. *Relaxation of steel.* BS 8110 recommends reference to the manufacturer's U.K. Certificate of Approval and adapting this with a Relaxation Factor from its Table 4.6. Choosing a cold drawn and prestraightened low relaxation wire a relaxation loss of 2% is obtained. Otherwise refer to Clause 4.8.2 of BS 8110:Part1 and BS 5896:1980.

2. *Elastic deformation of concrete*

(a) At transfer

Figure 8.3

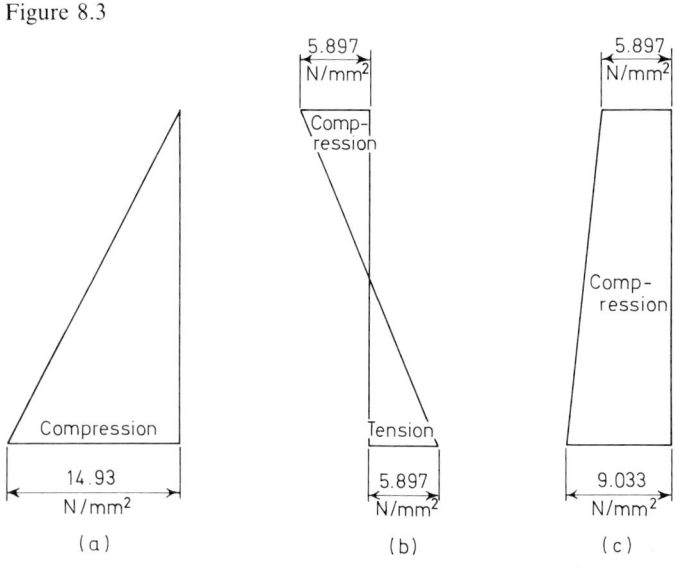

Limit state design of members

Support: stress in concrete at level of centroid of wires (from Figure 8.3(a))

$$= \frac{550 + 264.7}{1100} \times 14.93 = 11.06 \text{ N/mm}^2$$

From Example 8.3, $E_c = 26$ kN/mm². BS 8110 for the wires, gives $E_s = 200$ kN/mm². Hence

$$\alpha_e = 200/26 = 7.692$$

Therefore, as explained earlier, loss of prestress

$$= 7.692 \times 11.06 = 85.08 \text{ N/mm}^2$$

Using cross-sectional area of 38.5 mm², the loss of force per wire

$$= 85.08 \times 38.5 \text{ N} = 3.275 \text{ kN}$$

Hence the percentage loss of initial prestressing force

$$= (3.275/42.28) \times 100 = 7.747\%$$

Mid span: at level of centroid of wires, stress due to self-weight

$$= 5.897 \times (264.7/550) = 2.838 \text{ N/mm}^2$$

Therefore resultant stress at this level due to self-weight and prestress

$$= 11.06 - 2.838 = 8.22 \text{ N/mm}^2$$

Therefore, as previously, percentage loss of initial prestressing force

$$= 7.747 \times (8.22/11.06) = 5.758\%$$

(b) In service

Support: stress in concrete at level of centroid of wires (from Figure 8.1(a))

$$= \{(550 + 264.7)/1100\} \times 12.89 = 9.55 \text{ N/mm}^2$$

From Example 8.3, $E_c = 28$ kN/mm². BS 8110 for the wires gives $E_s = 200$ kN/mm². Hence

$$\alpha_e = 200/28 = 7.143$$

Therefore loss of prestress

$$= 7.143 \times 9.55 = 68.21 \text{ N/mm}^2$$

Therefore percentage loss of initial prestressing force

$$= 7.747 \times (68.21/85.08) = 6.211\%$$

Mid span: from Figure 8.1(c) stress at level of wires

$$= 12.89 \times (550 - 264.6)/1100 = 3.344 \text{ N/mm}^2$$

Therefore percentage loss of prestress

$$= 6.211 \times (3.344/9.55) = 2.175\%$$

3. *Shrinkage of concrete.* Supposing the beam is cured in effect in water – say by covering with wet hessian cloth which is covered with polythene sheet; there will then be no shrinkage loss at transfer. Let us assume that we cure after transfer at normal exposure until it is in use, then, guided by Figure 2.9, take the maximum shrinkage per unit length as 0.04%. Taking the shortening movement of the tendon as the same as the concrete shrinkage, then strain loss in tendon due to shrinkage

$$= 400 \times 10^{-6}$$

Hence corresponding loss of stress

$$= 400 \times 10^{-6} \times 200 \times 10^3 = 80 \text{ N/mm}^2$$

Therefore percentage loss of prestress (finally)

$$= (80/85.08) \times 7.747 = 7.284\%$$

4. *Creep of concrete.* At transfer, creep has had negligible time to take place, hence we take this loss as zero. The stress in the concrete at transfer will cause subsequent creep, and it is upon this stress that the final creep is based. BS 8110 is not as prescriptive with its creep loss recommendations as CP 110 so CP 110 will be used to illustrate a calculation of this loss. Referring to clause 4.8.2.5 of CP 110, as cube strength at transfer is 30 N/mm², creep per unit length is

$$48 \times 10^{-6} \times (40./30) = 64 \times 10^{-6} \text{ per N/mm}^2$$

According to CP 110, if the maximum stress at transfer exceeds $\frac{1}{3} \times$ Cube strength at transfer

$$= \tfrac{1}{3} \times 30 = 10 \text{ N/mm}^2$$

then the creep loss should be increased. At transfer Figure 8.3(a) gives the stresses at each support and Figure 8.3(c) gives the stresses at mid span.

Limit state design of members

At mid span the stresses do not exceed 10 N/mm² so the creep loss is satisfactory. At each support, as the maximum stress is approximately half the cube strength, the creep per unit length from CP 110 is

$$1.25 \times 64 \times 10^{-6} = 80 \times 10^{-6} \text{ per N/mm}^2$$

The stress causing creep will depend upon whether the beam is supporting its own weight only or its full load most of its life. At the level of the centroid of the wires the former gives a stress of 11.06 N/mm² at support and 8.22 N/mm² at the mid span, whilst the latter, from Figure 8.3(a) and Figure 8.1(b), gives 11.06 N/mm² at support and

$$11.06 - (264.7/1100) \times 12.89 = 7.96 \text{ N/mm}^2$$

at mid span. Supposing the imposed load is rarely applied, so that we take the worst of the cases just mentioned. When in use therefore the creep per unit length is

(a) $11.06 \times 80 \times 10^{-6} = 885 \times 10^{-6}$ at the support

and (b) $8.22 \times 64 \times 10^{-6} = 526 \times 10^{-6}$ at mid span

As the movement of the concrete is assumed to be the same as that of the tendon, then the loss of stress in the tendon is

(a) $885 \times 10^{-6} \times 200 \times 10^3 = 177 \text{ N/mm}^2$ at the support

and (b) $526 \times 10^{-6} \times 200 \times 10^3 = 105.2 \text{ N/mm}^2$ at mid span

These can be expressed as

(a) $(177/61.61) \times 5.61 = 16.12\%$ at support

and (b) $(105.2/61.61) \times 5.61 = 9.58\%$ at mid span

5. *Slip of anchorage.* Suppose the wedges at each end pull in 3 mm and our system is one where we jack the movable wire anchorage block away from the prestressing bed, which has a length of say 75 m; then this loss, if not allowed for when stressing, would be

$$(6/75\,000) \times 200 \times 10^3 = 16 \text{ N/mm}^2$$

But we will allow for this when stressing and extend the movable anchorage block 6 mm more than its required amount.

6. *Friction in jack and anchorage system.* This is nil because of the way we are pretensioning; see 5 above and also Section 8.3, para. 6.

Summarising the losses (the numbers in brackets refer to the number of the loss heading, previously used in this present Section, for example 1. refers to relaxation of steel):
at transfer, total loss at mid span

$$= 2(1) + 5.76(2) = 7.76\%$$

and at a support

$$= 2(1) + 7.75(2) = 9.75\%$$

Finally, in use, total loss at mid span

$$= 2(1) + 2.18(2) + 7.28(3) + 9.58(4) = 21.04\%$$

and at a support

$$= 2(1) + 6.21(2) + 7.28(3) + 16.12(4) = 31.61\%$$

At transfer we took the losses as 10%, so for greater accuracy we could now try 9.7% for support sections and 7.7% for mid-span sections and repeat the above design. Further such repetitions can then be made until the desired degree of accuracy is achieved. When in use we took the losses as 22.3%. For greater accuracy we would repeat the above design and use losses of 21% for mid span and 31% for support sections. Also for more accurate design, we would repeat the example, considering losses for the wires at their respective levels. We have considered them all as though concentrated at their centroid and this is slightly erroneous. For normal purposes our present accuracy in this problem could be considered satisfactory and hence our design is justified.

8.4.5 Ultimate limit state due to flexure (bonded tendons)

If a member has been designed as shown previously, then if the tendons are arranged so that most have a reasonably generous effective depth, checking the ultimate limit state is almost a formality. The exception to this is in the case of BS 8110, Class 3, structures (partially prestressed) – see Section 8.4. These could be designed for ultimate limit state first, then deflections at working loads checked before checking the stress systems at working loads and transfer.

In the case of a rectangular beam with one tendon, this is required at about $\frac{1}{3}$ of the height of the beam. Hence when cracking occurs due to overloading, the effective depth of this tendon is small, so the tendon does not control the crack widths very well at the soffit. In a case like this the ultimate limit state might not be satisfactory so additional non-prestressed reinforcement might be used and placed as near to the soffit as possible. Likewise in the case of a pole of circular cross section, when the bending moment can be in any direction; if the tendons are arranged around the periphery then the ultimate limit state will most probably be all right, but not if there is say just one tendon down the centre.

As in Chapter 3 the equivalent rectangular stress block due to C. S. Whitney is favoured[9] for predicting actual ultimate resistance moments,

that is, $f_{cm} = (\alpha/2\beta)f'_c$ and $x_1 = 2\beta x$

where (taking $f'_c = 0.84 f_{cu}$)

$$\alpha = 0.72 \text{ for } f_{cu} \leqslant 33 \text{ N/mm}^2$$

and decreases by 0.04 for every 8.21 N/mm² above 33 N/mm²

$$\text{and } \beta = 0.425 \text{ for } f_{cu} \leqslant 33 \text{ N/mm}^2$$

and decreases by 0.025 for every 8.21 N/mm² above 33 N/mm²
Thus for $f_{cu} \leqslant 33$ N/mm²,

$$f_{cm} = 0.85 f'_c = 0.714 f_{cu} \text{ and } x_1 = 0.85x$$

It is generally used in the U.S.A. and is the basis of the BS 8110 simplified method and many other codes internationally. Towards failure in bending a prestressed concrete beam cracks and behaves like a non-prestressed reinforced concrete beam apart from:

1. The strain in the tendon was not zero at zero loading, as in the reinforcement of the reinforced concrete beam. At zero loading the strain ε_p in a tendon corresponds to the force in the tendon after losses (that is, the losses which have occurred up to the time of loading to failure) divided by the cross-sectional area of the tendon and its Young's modulus. Strain in the tendon caused by the loading adds to ε_p. The strain due to the prestress in the concrete can be ignored, as it is negligible compared to ε_p and the strains at failure.

2. The stress–strain relationships for tendons are different to those for a reinforcement bar (see Figure 8.4). The ultimate resistance moment for an under-reinforced prestressed beam can be obtained as in Example 3.15 for under-reinforced sections, provided the ultimate tensile strength (stress) of the tendons is used for f_s in equations 3.60 and 3.62. To determine if the section is under-reinforced we need to calculate the maximum concrete strain corresponding to the tensile reinforcement strain when ultimate steel stress (or a suitable proof stress) is reached (see

Figure 8.4

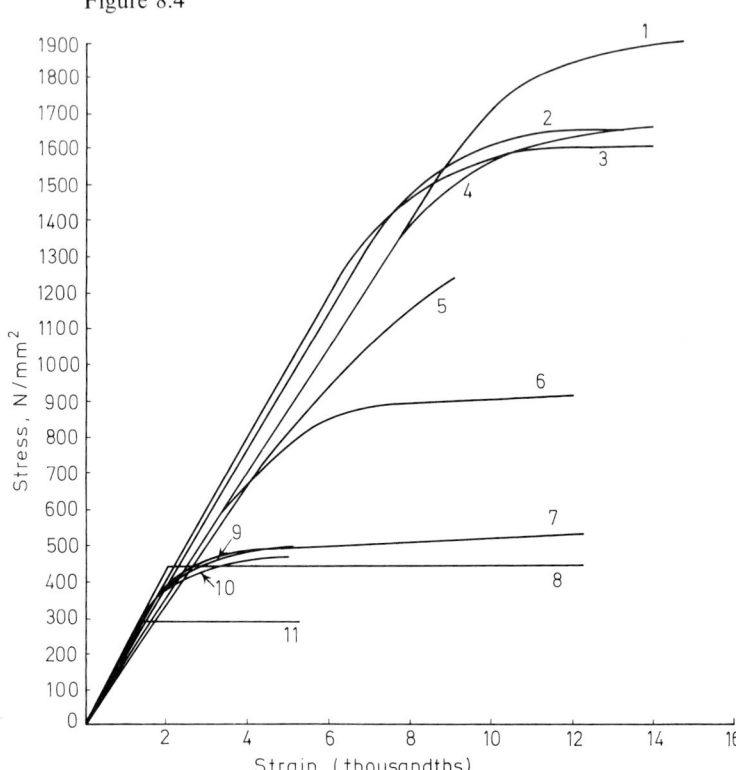

1. 12.7 mm dia. super quality strand, $E_s = 176$ kN/mm²
2. 15.2 mm dia. drawn strand, $E_s = 192$ kN/mm²
3. 5 mm dia. crimped prestressing wire, $E_s = 200$ kN/mm²
4. 15.2 mm dia. French strand, $E_s = 178$ kN/mm²
5. 28.6 mm dia. strand, $E_s = 169$ kN/mm²
6. 32 mm dia. prestressing alloy bar, $E_s = 175$ kN/mm²
7. 16 mm dia. round cold worked high yield reinforcing bar, $E_s = 200$ kN/mm²
8. 20 mm dia. round hot rolled high yield reinforcing bar, $E_s = 213$ kN/mm²
9. 9.5 mm square twisted high yield reinforcing bar, $E_s = 198$ kN/mm²
10. 11 mm square twisted with chamfered edges high yield reinforcing bar, $E_s = 208$ kN/mm²
11. 20 mm hot rolled mild steel reinforcing bar, $E_s = 208$ kN/mm²

Apart from 4, all the above are British products.

Figures 8.4 and 8.6) to check that this is less than the maximum known to be possible from experiments, namely 0.003 according to Whitney (0.0035 is used by BS 8110). Figures 8.5(a) and (b) show the distribution of stress across the cross section and the corresponding distribution of strain, respectively.

From similar triangles in Figure 8.5(b), the maximum concrete strain

$$= \varepsilon_c = \frac{(\varepsilon_{su} - \varepsilon_p) x}{d - x} \tag{8.3}$$

If this is greater than 0.003 (Whitney) then it is an over-reinforced prestressed concrete beam and its ultimate resistance moment cannot be assessed as above, as the concrete will disintegrate before the steel reaches its ultimate tensile strength (and corresponding strain ε_{su}).

For an over-reinforced section a simple direct calculation as above cannot be made because the stress in the steel at failure is less than its ultimate tensile strength and is not known initially, hence x_1 cannot be immediately obtained etc.; also, the stress–strain curve for the steel cannot be represented by a simple mathematical expression. A simple solution is by successive approximations (a method suitable for the digital computer).

Figure 8.5

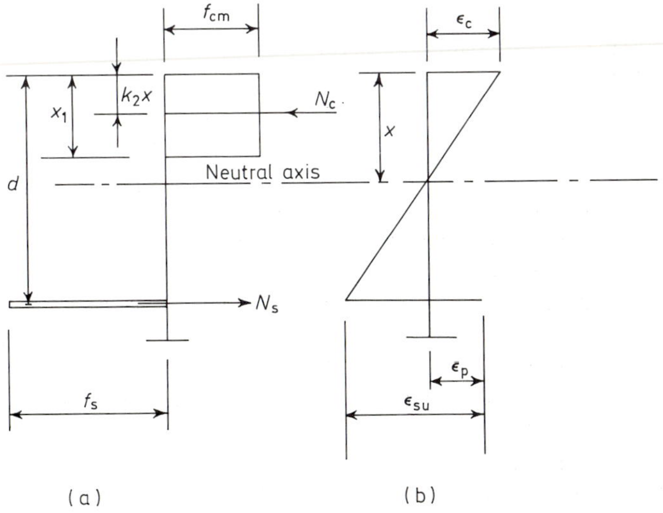

A value of x is assumed and x_1 is obtained from x as before (Whitney). Then equating longitudinal forces

$$f_{cm} A_c = A_s f_s \tag{8.4}$$

where A_s is the cross-sectional area of the tendons, f_s the stress in the tendons at failure, f_{cm} the mean concrete stress of the equivalent stress block, and A_c the area of concrete (cross section can be of any shape) subjected to f_{cm}. This gives f_s and the corresponding strain ε_s is obtained from the stress–strain curve for the tendon. Then from equation 8.3 but substituting $\varepsilon_c = 0.003$ (Whitney) and $\varepsilon_{su} = \varepsilon_s$

$$0.003 = \frac{(\varepsilon_s - \varepsilon_p) x}{d - x} \tag{8.5}$$

which now gives x. Now if this disagrees with the value assumed, the calculation is repeated until it is correct. When we are satisfied, taking moments about the line of action of N_c

$$M_u = A_s f_s (d - k_2 x) \tag{8.6}$$

In the case of a rectangular beam, using Whitney's theory, $k_2 x = 0.5 x_1$.

Example 8.6. A beam of rectangular cross section 0.25 m wide by 0.45 m deep is post-tensioned by one 25 mm diameter bar at 75 mm above its soffit. The duct enclosing the bar is grouted. Determine the ultimate resistance moment of the section using the simplified BS 8110 method and assuming $f_{cu} = 40$ N/mm², γ_m for steel $= 1.15$, and stress in 25 mm diameter tendon after losses $= 570$ N/mm².

From BS 4486:1980, Table 1: for tendon of nominal size 25 mm, the nominal cross-sectional area $= 491$ mm², and the characteristic breaking load $= 505$ kN.

Equating longitudinal forces

$$0.25 \times 0.9 \times x \times 0.45 \times 40\,000 = 505/1.15, \therefore x = 0.1084 \text{ m}$$

For the simplified BS 8110 method, $x_1 = 0.9 x = 0.097\,56$ m.

Figure 8.6 is prepared from BS 8110, Fig. 2.3

$$f_{pu}/\gamma_m = 505/491/1.15 = 0.8944 \text{ kN/mm}^2$$

and 80% of this value $= 0.7155$ kN/mm² and the minimum strain for the maximum stress to be realised

Limit state design of members

$$= \varepsilon_{su} = 0.005 + 0.8944/206 = 0.009\,342$$

Also $\varepsilon_p = 0.57/206 = 0.002\,767$

Hence from equation 8.3

$$\varepsilon_c = \frac{0.009\,342 - 0.002\,767 \times 108.4}{450 - 75 - 108.4} = 0.002\,673$$

This is less than 0.0035, so the section is under-reinforced. Hence using equation 8.6

$$M_u = (505/1.15)(0.45 - 0.075 - 0.5 \times 0.097\,56) = 143.3 \text{ kN m}$$

Example 8.7. Repeat Example 8.6, only with two bars instead of one, both at the same level.

Equating longitudinal forces gives $x = 2 \times 0.1084 = 0.2168$ m. From equation 8.3

$$\varepsilon_c = \frac{(0.009\,342 - 0.002\,767) \times 216.8}{450 - 75 - 216.8} = 0.009\,010$$

This is greater than 0.0035, hence section is over-reinforced. Assume

$$x = 0.1953 \text{ m, i.e. } x_1 = 0.9 \times 0.1953 = 0.1758$$

Figure 8.6

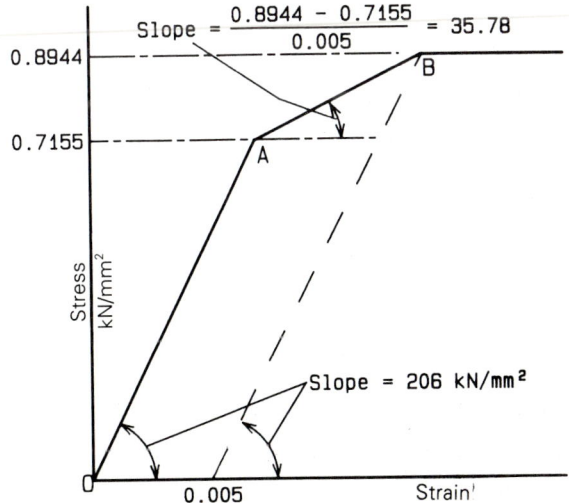

Equating longitudinal forces

$$0.25 \times 0.1758 \times 0.45 \times 40\,000 = 2 \times 491 \times f_s$$

$$\therefore f_s = 0.8055 \text{ kN/mm}^2$$

As before $f_{pu}/\gamma_m = 0.8944 \text{ kN/mm}^2$

so $0.8 f_{pu}/\gamma_m = 0.7155 \text{ kN/mm}^2$

Hence from Figure 8.6,

$$\varepsilon_s = (0.7155/206) + (0.8055 - 0.7155)/35.78 = 0.005\,988$$

Then from equation 8.3 but substituting $\varepsilon_c = 0.0035$ (BS 8110) and $\varepsilon_{su} = 0.005\,988$

$$0.0035 = \frac{(0.005\,988 - 0.002\,767)\,x}{0.45 - 0.075 - x} \quad \therefore x = 0.1953 \text{ m}$$

This is in order, the estimate of x being correct. Normally several attempts would be required. Although this trial and error method is favoured by others, and by the writer when using real stress–strain curves for the tendons, the writer prefers direct calculation when using the simplified stress–strain curves of BS 8110. To illustrate this, instead of assuming x as before, assume that the strain in the tendons is in the range AB of Figure 8.6. Then equating longitudinal forces

$$0.25 \times 0.9 \times x \times 0.45 \times 40\,000 = 2 \times 491 f_s \quad \therefore f_s = 4.124x$$

From Figure 8.6,

$$\varepsilon_s = \frac{(0.7155/206)}{206} + \frac{(4.124x - 0.7155)}{35.78} = 0.1153x - 0.016\,52$$

Using this for ε_{su} and $\varepsilon_c = 0.0035$ in equation 8.3

$$0.0035 = \frac{(0.1153x - 0.016\,52 - 0.002\,767)\,x}{0.375 - x}$$

$$\therefore x = 0.1952 \text{ m}, \ x_1 = 0.9 \times 0.1952 = 0.1757 \text{ m}$$
and $f_s = 4.124 = 0.8050 \text{ kN/mm}^2$

Hence f_s does lie in range AB and the calculation is satisfactory. If f_s had been in range AO then one would assume it in this range and make a similar but simpler calculation to the above. Then from equation 8.6

$$M_u = (0.7833 \times 2 \times 491)(0.375 - 0.5 \times 0.1757) = 220.9 \text{ kN m}$$

8.4.6 Additional untensioned steel (bonded tendons)

If the ultimate resistance moment is inadequate, and the other limit states satisfactory, sometimes extra untensioned steel is added. This has negligible effect on the other limit states, and thus saves re-design. This extra steel might be extra prestressing tendons which are not stressed, or reinforcement bar. This steel is placed with maximum effective depth.

Additional untensioned steel is sometimes necessary for crack control when post-tensioned bonded tendons are located at some distance from the tensile face of the concrete.

Example 8.8. Repeat Example 8.7 but add two non-prestressed 12 mm diameter bars, with 25 mm concrete cover, in the bottom of the beam. Assume $f_y = 460$ N/mm² for these bars.

Equating longitudinal forces,

$$0.25 \times 0.9 \times x \times 0.45 \times 40\,000 = 2 \times 505/1.15 + 226 \times 0.46/1.15$$

$$\therefore x = 0.2392 \text{ m}$$

For tendons, from Example 8.6, $\varepsilon_{su} = 0.009\,342$ and $\varepsilon_p = 0.002\,767$. Hence from equation 8.3,

$$\varepsilon_c = (0.009\,342 - 0.002\,767) \times 0.2392/(0.375 - 0.2392) = 0.011\,58$$

Depth of centroid of 12 mm diameter bars $= 450 - 25 - 6 = 419$ mm

Using distribution of strain diagram and similar triangles, strain in 12 mm diameter bars

$$= 0.011\,68 \times (419 - 239.2)/239.2 = 0.008\,705$$

so that maximum stress can be realised in these bars (see Appendix 4 which shows Fig. 2.2 of BS 8110), that is strain greater than $\dfrac{f_y}{\gamma_m \times 200\,000}$

$$= \dfrac{460}{1.15 \times 200\,000} = 0.002.$$

As ε_c is greater than 0.0035, section is over-reinforced. Equating longitudinal forces

$$0.25 \times 0.9 \times x \times 0.45 \times 40\,000 = 2 \times 491 \times f_s + 226 \times 0.46/1.15$$

$$\therefore f_s = 4.124x - 0.092\,06$$

From Figure 8.6, assuming f_o is between A and B

$$\varepsilon_s = (0.7155/206) + (f_s - 0.7155)/35.78 = 0.1153x - 0.0191$$

Using this for ε_{su} and $\varepsilon_c = 0.0035$ in equation 8.3

$$0.0035 = \frac{(0.1153x - 0.0191 - 0.002767)x}{0.375 - x}$$

$\therefore x = 0.2128$ m, and $f_s = 0.7855$ kN/mm²

As f_s lies in the range AB in Figure 8.6, the calculation is satisfactory. We have assumed the strain in the 12 mm diameter bars is large enough for them to develop their maximum stress. Strain in bars

$$= (0.0035/212.8) \times (450 - 25 - 6 - 212.8) = 0.003391$$

This strain is greater than 0.002, (see previously) so the maximum stress is realised. Then M_u is determined by taking moments about the line of action of N_c.

$$M_u = 2 \times 491 \times f_s(0.375 - 0.5 \times 0.9x)$$
$$+ 226 \times 460 \times (0.419 - 0.5 \times 0.9x)/1.15 = 244.6 \text{ kN m}$$

8.4.7 Top non-prestressed steel

Wires or handling reinforcement placed in the top of a beam are usually too inadequately anchored against buckling (see BS 8110) to withstand compression and be included in the ultimate resistance moment calculations. If compression steel is to be included in these calculations, it is included in the previous calculations in the same way as given in Chapter 3.

However as top steel, it may be more useful in some circumstances for taking tension at transfer, in which case it does not need to be restrained by stirrups to prevent buckling.

8.4.8 Ultimate limit moment due to flexure (unbonded tendons)

In this instance Sections 8.4.5–8.4.7 apply, except that BS 8110 reduces the force which can be developed in the tendons. The problem is that as loading is applied, instead of the force imposed in the tendon decreasing towards the support as with bonded tendons or reinforcement bars, the force in the tendon is always the same from end to end in an unbonded tendon. Towards failure the first crack occurs at the position of maximum bending moment. At this crack, instead of the tendon being highly stressed locally and anchored on either side of the crack so that its extension is limited (as would be the case if the tendons were bonded to the concrete), when the tendon is unbonded, this high stress extends along its whole length. The whole length thus extends *pro rata* and the extension is considerable, allowing the first crack to open excessively (few if any extra cracks form towards failure), precipitating earlier failure than occurs with a beam with a bonded tendon. In this paragraph the relatively small effect of friction between cables and ducts has been ignored (see Section 8.3 item 7).

The normal theories treat post-tensioning as if it were pretensioning and just modify the ultimate resistance moment for unbonded tendons empirically. However, the problem is basically different at pretensioning, working loads and ultimately. References 7 and 10 deal at length with this problem at tensioning and at working loads. They take account of pressures between tendons and their surrounding concrete, which can give high stress concentrations in the concrete.[7,11] (A failure has been reported where these pressures were considered to be too high and the concrete was under-strength.) Tendons cannot be deflected say vertically by beams without such forces existing and the theories of Refs. 7 and 10 calculate stresses and deflections for beams with tendons of various profiles.

8.4.9 Prestressed columns

It is rarely economical or necessary to prestress columns. One example of prestressing columns (designed by the author) is in the case of large span pitched-roofed portal frames; in this instance, however, the columns experience very small direct stresses relative to the bending stresses.

8.4.10 Prestressed ties

Prestressed ties are often extremely useful for space frames, arches, hyperbolic paraboloids, gable ties to barrel vault and folded plate roofs, suspenders to tied arch bridges and ties beneath prestressing beds. Extensions of ties are often desired to be as small as possible. This means a low strain is desirable in a tie, hence a steel tie or a prestressed concrete tie is designed, using a low stress. If the steel tie needs to be clad to resist fire or corrosion then the prestressed tie is often a more economical solution. One objection to unprotected steel ties to concrete structures is that their life and fire resistance is far less than that of the concrete members and if they fail a heavy structure collapses. Pretensioning was favoured for ties because the long slender members were considered to buckle as Euler's theory when post-tensioned Refs. 7, 11 and 12 show that the tendons restrain such buckling and a position of static equilibrium can be obtained when post-tensioning, so that if a certain unnoticeable curvature is allowed then the post-tensioned member can be designed accordingly and very economically. Post-tensioned ties have the economic advantage that they can easily be effected on site from existing scaffolding to shells, arched bridges, etc., when required. Pretensioned ties have to be delivered on time and threaded amongst the scaffolding and provided with special end attachments. Designs of pretensioned and steel ties are compared in Ref. 13 and post-tensioned ties are designed in Refs. 7, 11 and 12.

A reinforced concrete tie is the same in principle as a steel tie encased in concrete. The concrete cracks at the strain associated with the stress to which the steel is normally designed, and this cracking is undesirable, particularly if the ties are exposed, for example as hangers to bow-string girder bridges.

8.4.11 Shear resistance of prestressed concrete beams

At working loads for BS 8110 Class 1 and 2 structures, beams are considered as uncracked and hence principal stresses can be calculated in the usual manner by combining stresses due to prestressing, bending and shear. The concrete is usually well able to resist the principal compressive stresses, and can usually resist the principal tensile stresses; if it cannot, then the section or the amount of prestressing has to be altered, or shear reinforcement in the form of inclined tendons, or vertical or inclined stirrups, or vertical prestressing, has to be introduced. The principal stresses can be calculated from the well known expression

$$f = 0.5\{f_h + f_v \pm \sqrt{[(f_h - f_v)^2 + 4v^2]}\} \qquad (8.7)$$

where f_h and f_v are horizontal and vertical direct stresses (tensile positive) and v = shear stress.

In the early days of prestressed concrete it was only necessary to limit the principal tensile stress to zero or a small amount, say 0.5 N/mm² at working loads. This can still be done for a preliminary design. Research in shear generally shows that, with the kind of load factors used, if a beam is satisfactory with regard to its ultimate shear resistance then the diagonal cracks at working loads for reinforced concrete beams are adequately narrow and they are narrower still for prestressed concrete beams because of the prestressing forces tending to close such cracks. BS 8110 therefore regulates only the ultimate shear resistance and equation 8.7 is used for sections not cracked in bending on the basis that when the principal stresses become great enough to cause cracking this can be regarded as corresponding to ultimate failure in shear. BS 8110 treats sections experiencing cracks due to bending differently in shear. Research concerning ultimate shear strength,[14,15] as with non-prestressed concrete, is inconclusive and appears inconsistent; hence empirical formulae have to be agreed for codes and these have to err greatly on the side of safety in some instances, because of the erratic nature and sensitivity to many variables of shear failures.

BS 8110 recommends that the design ultimate shear resistance of the concrete alone should be considered at sections that are uncracked in flexure by using equation 8.7 as follows (for the usual case of no vertical prestress i.e. $f_v = 0$):

$$f_h = -0.8f_{cp}, \quad f = f_t$$

$$\therefore f_t = 0.5\{-0.8f_{cp} \pm \sqrt{[(0.8f_{cp})^2 + 4v^2]}\}$$

$$2f_t + 0.8f_{cp} = \pm \sqrt{[(0.8f_{cp})^2 + 4v^2]}$$

$$4f_t^2 + 3.2f_{cp}f_t + (0.8f_{cp})^2 = (0.8f_{cp})^2 + 4v^2$$

$$\therefore v^2 = f_t^2 + 0.8f_{cp}f_t$$

$$\therefore v = \sqrt{[f_t^2 + 0.8f_{cp}f_t]}$$

From classical theory, for a rectangular beam, this has a parabolic distribution so that the average shear stress for the depth of the section is $\tfrac{2}{3}v$ therefore

$$V_{co} = \tfrac{2}{3}vb_v h = 0.67 b_v h \sqrt{[f_t^2 + 0.8f_{cp}f_t]} \qquad (8.8)$$

and this equation is given in BS 8110.

BS 8110 recommends that in no circumstances should the maximum design shear stress exceed $0.8\sqrt{f_{cu}}$ or $5\,\text{N/mm}$ whichever is less (this includes an allowance of 1.25 for γ_m). It also recommends that V_{co} corresponds to a maximum design principle tensile stress of $f_t = 0.24\sqrt{f_{cu}}$.

In a pretensioned member the critical section should be taken at a distance from the edge of the support equal to the height of the centroid of the section above the soffit. Where this section occurs within the prestress development length, the compressive stress at the centroidal axis due to prestress to be used in equation 8.8 may be calculated from the following relationship:

$$f_{cpx} = \frac{x}{l_p}\left(2 - \frac{x}{l_p}\right)f_{cp}$$

where f_{cp} is the design stress at the end of the prestress development length l_p.

The prestress development length should be taken as either the transmission length or the overall depth of the member, whichever is the greater.

When sections are cracked in flexure BS 8110 recommends that V_{co} and V_{cr} be both calculated and the lesser value taken and the latter is given by the empirical formula

$$V_{cr} = \left(1 - 0.55\frac{f_{pe}}{f_{pu}}\right) V_c b_v d + M_o \frac{V}{M} \qquad (8.9)$$

The value of V_{cr} should be taken as not less than $0.1 b_v d \sqrt{f_{cu}}$.

Shear reinforcement can be omitted according to BS 8110 when V is less than $0.5 V_c$ or when V is less than V_c for members of minor importance.

When shear reinforcement is required and V does not exceed $V_c + 0.4 b_v d$, BS 8110 recommends that links should be provided to satisfy the empirical equation

$$\frac{A_{sv}}{s_v} = \frac{0.4 b_v}{0.87 f_{yv}} \qquad (8.10)$$

But when V exceeds the above value links should be provided to satisfy the empirical equation

$$\frac{A_{sv}}{s_v} = \frac{V - V_c}{0.87 f_{yv} d_t} \qquad (8.11)$$

Figure 8.7

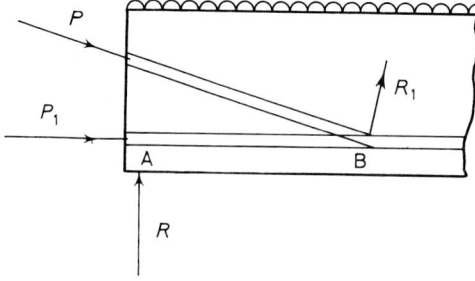

Limit state design of members

8.4.12 Inclined tendons

If a tendon is inclined upwards, at an angle α to the horizontal, towards and to the support, it is easy to imagine in the case of post-tensioning with a force P that reactions P and R_1 are imposed on the concrete (see Figure 8.7); thus for any section between A and B the shear force will be that due to the loading minus the vertical component of P, namely $P \sin \alpha$. Analyses giving shear forces and bending moments for beams with cables displaced upwards towards the supports with various profiles are given in Refs. 7 and 10.

8.4.13 Composite construction

An example of this is shown in Figure 8.8. The prestressed rectangular beam is propped until the *in-situ* reinforced concrete slab is mature. Figure 8.9(a) shows the final T-beam. Figure 8.9(b) gives the stress distribution in the prestressed concrete beam before the slab is cast. There are dowel bars between the beam and the slab so that, when the props are removed, the self-weight of the slab and future live loading are carried by the 'composite' T-beam. Figure 8.9(c) shows the stress distribution after the props have been removed, the beam having to carry the self-weight of the slab, and Figure 8.9(d) shows the stress distribution when the live loading is also being carried. The dowels required can be calculated by determining the horizontal shear stress at the junction between slab and rectangular beam (see Section 3.3). Composite construction is generally economic when a floor or bridge deck is desired to be *in situ* as opposed to precast, for robustness, and its total depth is required to be less than for *in-situ* reinforced concrete construction or when durability (absence of cracks at working loads) is required (for example bridge decks).

Figure 8.8

Figure 8.9

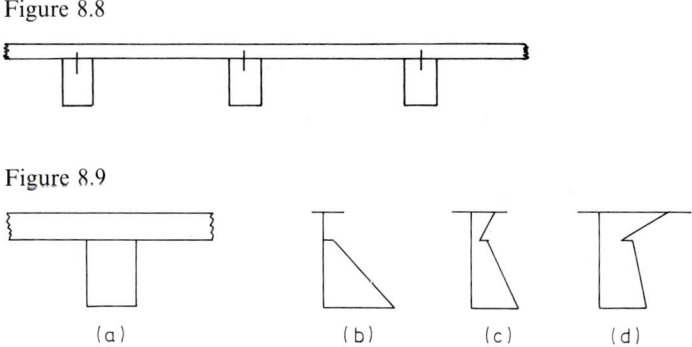

Sometimes prestressed precast beams as in Figure 8.8 are used without propping to support the shuttering to the *in-situ* slab (this can also be a hollow tile floor or roof, or a deck comprising precast units, where the portion over each prestressed beam is made *in situ* but the precast soffit is such that it can be supported by the prestressed beam). Holes to accommodate bolts (for example about 12 mm diameter) are cast through the prestressed beams to enable the shuttering to be supported. This method is useful when the headroom is high, in avoiding expensive scaffolding to support shuttering. In Figure 8.8 above it means that the composite T-beam supports the superimposed loading but not the self-weight of the slab.

Composite slab construction is often carried out (particularly in bridge work) without propping, using the prestressed precast beams as permanent formwork.

Example 8.9. A floor consists of precast pretensioned prestressed concrete units side by side with a 50 mm thick *in-situ* concrete topping. The precast units are each 1200 mm wide and of double-tee cross section. There are adequate steel dowel connectors between the precast units and the topping so that they can act as a composite section of the cross-section shown in Figure 8.10.

(A) Assuming that the precast units are propped whilst the *in-situ* concrete is placed and until it has matured, determine the longitudinal

Figure 8.10

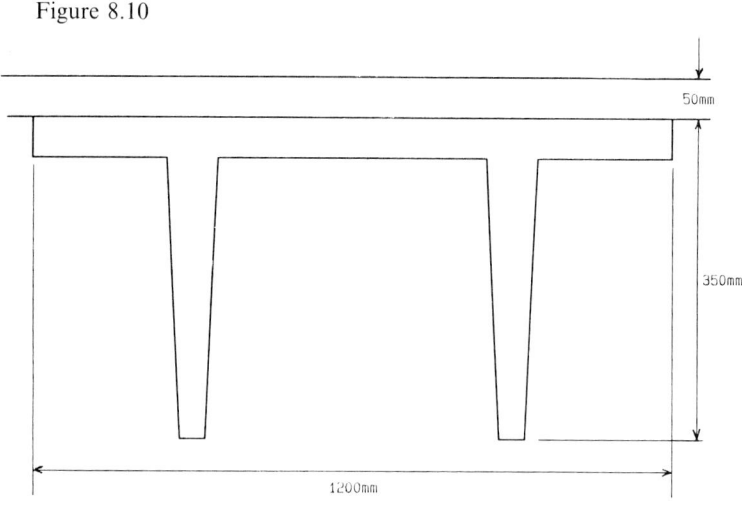

Limit state design of members 289

stresses at mid-span (making sketches approximately to scale showing these stress distributions and giving salient dimensions) for:

1. the prestressed concrete member carrying its self-weight only,
2. the composite section (i.e. precast unit plus topping) carrying its self-weight only,
3. the composite section carrying its self-weight and the live load.

(B) Assuming that the precast units are not propped as above, determine the longitudinal stresses at mid-span for:

(i) the prestressed concrete member carrying its self-weight and the weight of the *in-situ* topping,
(ii) the composite section carrying the live load.

The following data applies for a width of floor of 1.2 m:

> characteristic live load/unit length of span 3.0 kN/m^2
> span (simply supported) 8.0 m
> cross-sectional area of precast member 0.116 m^2
> weight density of concrete 24 kN/m^3
> height of centroid of precast member from its lowest extremity 0.254 m
> height of centroid of composite member from its lowest extremity* 0.298 m
> second moment of area about horizontal axis for precast member 0.001 19 m^4
> second moment of area about horizontal axis for composite member* 0.001 75 m^4
> each rib (stem) contains three Strands (i.e. six for each precast member) and some are deviated towards the supports
> prestressing force in each Strand after losses 50 kN
> eccentricity of centroid of tendons from neutral axis of precast member at mid-span 0.2 m

* These are equivalent figures allowing for the different value of Young's Modulus for the *in-situ* concrete, i.e. these numbers are for use with the Young's Modulus of the precast concrete.

(A)
(a) weight per unit length of precast member = $0.116 \times 24 = 2.784$ kN/m
mid-span bending moment due to this = $2.784 \times 8^2/8 = 22.27$ kN m

extreme fibre stresses due to this:

$$f_B = 2.27 \times 0.254/0.001\,19 = 4753 \text{ kN/m}^2, \text{ tensile}$$

$$f_T = 22.27 \times (0.35 - 0.254)/(0.001\,19 = 1797 \text{ kN/m}^2,$$
$$\text{compressive}$$

where f_B = bottom fibre stress and f_T = top fibre stress.
(b) extreme fibre stresses due to prestressing force:

$$f_B = 6 \times 50 \left\{ \frac{1}{0.116} + \frac{0.2 \times 0.254}{0.001\,19} \right\} = 15390 \text{ kN/m}^2$$
$$\text{compressive}$$

$$f_T = 6 \times 50 \left\{ \frac{1}{0.116} - \frac{0.2 \times (0.35 - 0.254)}{0.001\,19} \right\} = 2255 \text{ kN/m}^2$$
$$\text{tensile}$$

(c) For a uniformly distributed load of w/unit length (kN/m) on the combined section:
mid-span bending moment due to this $= w \times 8^2/8 = 8\,w$ kN m
extreme fibre stresses due to this:

$$f_B = 8w \times 0.298/0.001\,75 = 1362\,w \text{ kN/m}^2, \text{ tensile}$$

$$f'_T = 8w \times (0.4 - 0.298)/0.001\,75 = 466.3\,w \text{ kN/m}^2,$$
$$\text{compressive}$$

$$f_T = 8w \times (0.35 - 0.298)/0.001\,75 = 237.7\,w \text{ kN/m}^2,$$
$$\text{compressive}$$

where f'_T = stress at top fibre of topping.
(1) For prestressed concrete member carrying its self-weight only, add (a) and (b):

$$f_B = 15390 - 4753 = 10637 \text{ kN/m}^2, \text{ compressive}$$

$$f_T = 2255 - 1797 = 458 \text{ kN/m}^2, \text{ tensile}$$

these stresses are shown in Figure 8.11(a).
(2) For composite section carrying its self-weight only:
using (c), w = weight of topping = $0.050 \times 1.2 \times 24 = 1.44$ kN/m.

$$f_B = 1961 \text{ kN/m}^2, \text{ tensile. } f'_T = 671.5 \text{ kN/m}^2, \text{ compressive.}$$

$$f_T = 342.3 \text{ kN/m}^2, \text{ compressive.}$$

Limit state design of members

Figure 8.11(a)

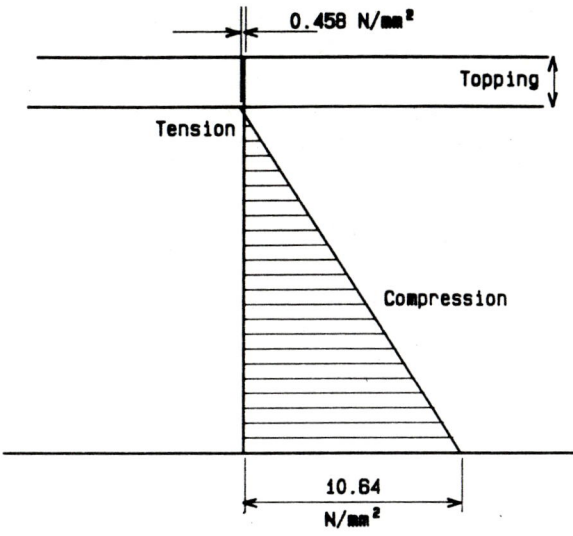

Figure 8.11(b)

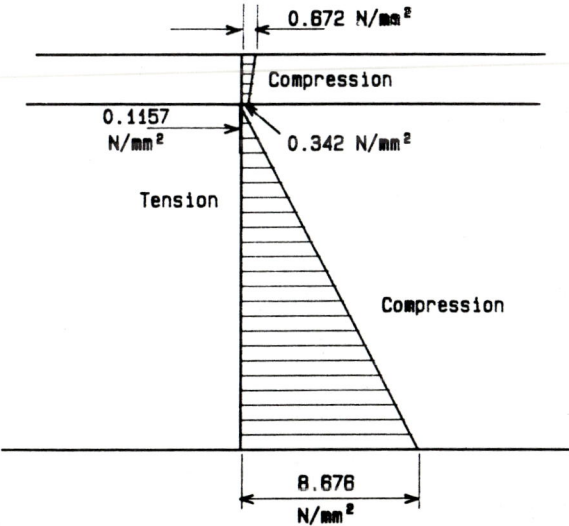

Figure 8.11(c)

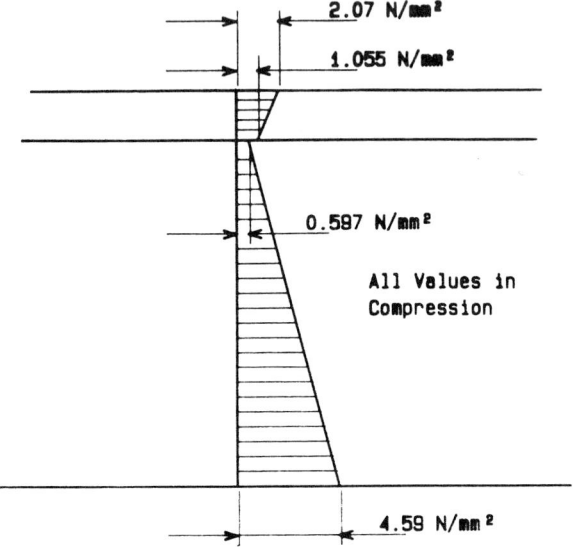

Figure 8.11(d)

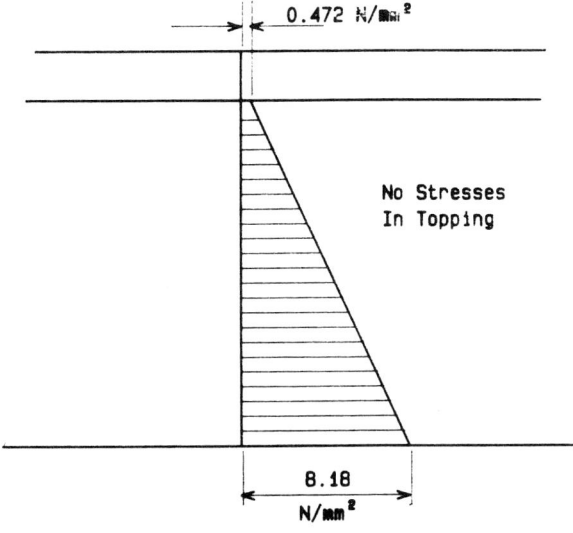

Limit state design of members

add these to (1),

$f_B = 10637 - 1961 = 8676$ kN/m², compressive.

$f'_T = 671.5$ kN/m, compressive.

$f_T = 342.3 - 458 = 115.7$ kN/m², tensile.

these stresses are shown in Figure 8.11(b).

(3) For composite section carrying its self-weight and live load: using (c), w = weight of topping + live load = 1.44 + 3.0 = 4.44 kN/m.

$f_B = 6047$ kN/m², tensile.

$f'_T = 2070$ kN/m², compressive.

$f_T = 1055$ kN/m², compressive.

add these to (1),

$f_B = 10637 - 6047 = 4590$ kN/m², compressive.

$f'_T = 2070$ kN/m², compressive.

$f_T = 1055 - 458 = 597$ kN/m², compressive.

these stresses are shown in Figure 8.11(c).

Figure 8.11(e)

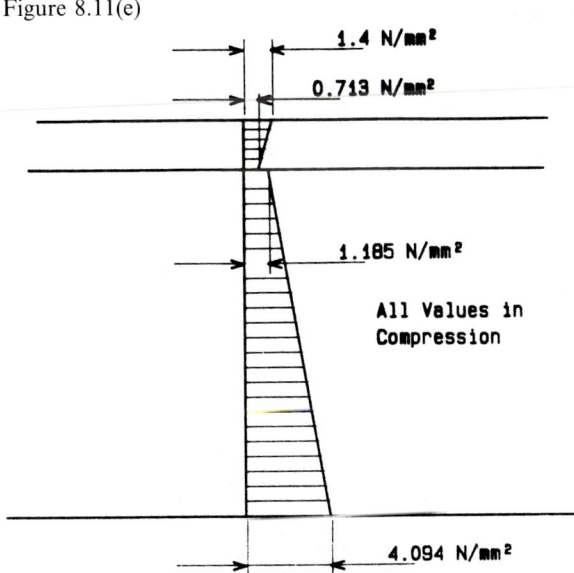

1.4 N/mm²
0.713 N/mm²
1.185 N/mm²
All Values in Compression
4.094 N/mm²

294   Prestressed concrete

(B)
(i) the *in-situ* topping weighs, see (A)(2), 1.44 kN/m.
extreme fibre stresses on precast member due to this, calculate *pro rata* to (A)(a),

$$f_B = 4753 \times 1.44/2.784 = 2458 \text{ kN/m}^2, \text{ tensile.}$$

$$f_T = 1797 \times 1.44/2.784 = 929.5 \text{ kN/m}^2, \text{ compressive.}$$

add these to (A)(1)

$$f_B = 10\,640 - 2460 = 8180 \text{ kN/m}^2, \text{ compressive.}$$

$$f_T = 930 - 458 = 472 \text{ kN/m}^2, \text{ compressive.}$$

these stresses are shown in Figure 8.11(d).
(ii) stresses on composite section due to live load (3 kN/m), using (c)

$$f_B = 4086 \text{ kN/m}^2, \text{ tensile.}$$

$$f'_T = 1400 \text{ kN/m}^2, \text{ compressive.}$$

$$f_T = 713.1 \text{ kN/m}^2, \text{ compressive.}$$

add these to (B)(i)

$$f_B = 8180 - 4086 = 4094 \text{ kN/m}^2, \text{ compressive.}$$

$$f'_T = 1400 \text{ kN/m}^2, \text{ compressive.}$$

$$f_T = 713.1 - 472 = 1185.1 \text{ kN/m}^2, \text{ compressive.}$$

these stresses are shown in Figure 8.11(e).

8.4.14 Continuity

This has problems in that for various combinations of live loads on different continuous spans the tendons ideally need to be in varying positions. Cables have to be waived over supports of continuous beams; this increases friction losses and can make grouting difficult. Many calculations of sequence of prestressing and different loading possibilities have to be made. The careful control of the sequence of prestressing makes this operation costly. A continuous beam shortens due to prestressing, so if the columns supporting it are *in situ*, ideally all but one need to be hinged at top and bottom so that some of the post-tensioning is not absorbed in bending the columns as opposed to post-tensioning the beam.

A continuous beam is very vulnerable to the slightest differential settlement of supports. They can be designed for some settlement of supports and this makes them less economic. This is done for cable-stayed prestressed concrete bridges, pioneered by Professor Leonhardt in Germany. Continuity is not favoured in mining subsidence areas; the jacking of supports requires too much attention and one could be caught out by sudden unpredictable settlement. (Continuous bridge beams with exposed tendons, which can be periodically inspected and eventually replaced if necessary, are analysed in Ref. 7.)

Relatively few prestressed concrete bridges have been designed with continuity in the U.K., yet many have been designed in the U.K. for overseas.

8.4.15 End splitting forces

Referring to Figure 2.13, the prestressing wire upon release increases its diameter at A, and thus splitting forces are created between, and normally to, a line between A and B. Designers have sometimes been unaware of this problem and have experienced splitting cracks in pretensioned members along the line between A and B. Other end splitting forces are caused by the end anchorages of tendons being, in effect, a system of irregularly distributed point loads on the end of a member. Each point load causes splitting forces normal to its line of action. Again failures have occurred.

This problem should be considered by the designer and BS 8110 gives simple empirical guidance.

8.4.16 Prestressed concrete tanks, pipes, domes, shells and piles

For circular tanks and pressure pipes, circular prestressing is provided to counteract the circumferential tension due to the loading. A residual circumferential compression can ensure no cracks developing, due to shrinkage and temperature change, and this increases the watertightness. The pipes also need longitudinal prestressing for handling purposes. The writer has been consulted concerning troubles with certain prestressed concrete pipes. From his considerable literature searches he would recommend Ref. 16 for determining soil pressures on pipes and Ref. 17 for guidance on the design of prestressed concrete pipes. Some research supervised by the author on this problem is given in Ref. 18.

With rectangular tanks the walls must be free at the base, otherwise the corners act rigidly as folded plates and prevent the post-tensioning

296   *Prestressed concrete*

imposing stresses along the walls (a very able designer overlooked this point).

Prestressing is useful for providing the ring tension to domes. The writer has rectified a dome, failing due to inadequate ring steel, by prestressing around the periphery.

Prestressed concrete piles are used for longer piles when handling stresses are a problem; end reinforcement details are important.

Prestressing is useful for normal and North Light barrel vault roofs longer than about 36 m and 27 m, respectively, assuming that the ratio of width to length is about 1:2.

8.4.17   Torsional resistance

The ultimate limit state for torsion is dealt with in the same way as for non-prestressed beams (see Chapter 3).

8.5   Load balancing

This method of design[19] helps suitable profiles for tendons of simply supported and continuous beams to be rapidly determined.

It is well known that for a simply supported beam carrying a uniformly distributed load the bending moment diagram is a parabola. If the centroid of the tendons traces a parabolic shape such that it has maximum eccentricity at mid-span and zero eccentricity at the supports, and, if the bending moment provided by its eccentricity at mid-span equals the maximum bending moment there due to the distributed loading, then there will be no resultant bending moment at mid-span, nor anywhere along the beam. No resultant bending moment means no deflection regardless of the modulus of elasticity or creep of the concrete (a great simplification for the designer).

If for each span of a continuous beam the tendons are placed in this way, then there are no resultant bending moments anywhere along the continuous beam. This also applies to continuous beams having spans of various different lengths.

This therefore gives an amazingly fast and simple method of design for continuous beams compared with the methods of design previously described in this Chapter.

In effect the parabolic shaped cables press upwards along a beam and equal the pressure downwards from the uniformly distributed loading. This must be the case as the two uniformly distributed loadings are to give equal, but opposite, bending moments at mid-span. Hence the term 'load-balancing' – the uniformly distributed loading is balanced by the uniformly distributed cable pressure. Designing for this balance of loading, means that there is no bending moment and no deflection. For extra loading to that already 'balanced' (e.g. for extra live loading) the total deflection is simply that computed for the extra loading only, acting on an elastic beam.

To summarise the methods of design so far described in this Chapter:

(1) An elastic analysis is made to give stresses and to limit the tensile stresses to zero or a small amount, considered to be able to be resisted by the concrete. The limitations depend upon which code of practice is used internationally. This method eliminates cracking at working loads and deflections are checked and limited if necessary. Ultimate bending and shear strengths are checked and are usually satisfactory.

(2) The ultimate strength is designed to be satisfactory. When the elastic stresses at working loads are assessed for this design it would usually be the case that the concrete would not be able to resist the tensile stresses. Persisting with this design, purely for economic reasons, would result in the method of 'partial prestressing' (pioneered by Dr Paul Abeles in the U.K. and used by him for several railway bridges) where the tensile stresses, as per elastic design (which does not truly apply as the concrete cracks in tension) are limited to quite large amounts, but these amounts are claimed to be such that the resultant design only experiences 'micro-cracking', according to Abeles, that is very small cracks considered to be allowable for the particular conditions of exposure. Partial prestressing, advocated by Abeles, has been included in CP 110 and BS 8110. At working loads, cracks and deflections are assessed and considered.

(3) The 'load-balancing' method makes designs to give zero deflection at whatever loading is decided upon. This could be for dead load or dead load plus some small (according to Lin) proportion of the live loading. In effect this means that a very significant part of the total loading is provided for, very rapidly and simply. In the case of a continuous beam with spans of various different lengths, this is a great asset to the designer. Then the design has to be checked for variants to this basic design caused by live load, if the load-balancing was for dead load only; or for extra and less applications of live load, if the load balancing was for dead and some proportion of the live load. This checking could be by elastic theory for deflections due to the non-balanced loading only, as the balanced loading causes no deflections. The checking for the bending moments, due to the non-balanced loading, should be for elastic stresses. Checking for ultimate

bending and shear strength and cracks at working loads would be for the whole system. It may be unnecessary to check for cracks at working loads if the extra loading to the 'balanced loading' keeps all longitudinal stresses as compressive.

The 'load-balancing' method is therefore very useful for rapidly designing continuous beams with varying spans, including the use of unequal prestress in the spans. The method is varied slightly from that described above, for continuous beams. Quoting from Lin 'the most economical cable location is one with the maximum sag so that the least amount of prestress will be required to balance the load.'

At a freely supported end of a continuous beam, the centroid of the tendon will coincide with the centroid of the beam cross section so as to have no bending moment at this section. At the continuous supports the tendons will be taken up as high as possible consistent with the specified top cover of concrete. Between supports the tendons will sag as parabolic shapes. Adjustments in the profile of the centroid of the tendons might have to be made, such as rounding off the peaks over the interior supports. The effect of such variations should be calculated or at least estimated.

Prestressed concrete rigid frames can be designed by the load-balancing method, very quickly and simply. Quoting from Lin 'The balanced-load approach quickly leads to a condition of uniform stress distribution for all members of a rigid frame. Since the relatively unfamiliar effect of prestressing has been eliminated by load-balancing, it is only necessary to analyse an ordinary elastic rigid frame subjected to the additional loading or to the effect of axial shortening. Since engineers are already familiar with such analysis, and since we are only concerned with the added loading or with the effect of axial shortening, without the effects of bending, the problem becomes easily controllable.'

Two-dimensional load-balancing can be applied to two way slabs, waffle floors and grillage systems.

Three-dimensional load-balancing can be applied to complex bridges and shell roofs.

Lin[19] gives worked examples of load-balancing applied to continuous beams, a two-way slab and a grillage and cites schemes he has designed using two-dimensional and three-dimensional load-balancing.

8.6 Post-compressing

Post-tensioning, tensions steel bars or cables causing the surrounding concrete to be compressed. The author would call the opposite of this with bars rather than cables 'post-compressing'.

Prestressing a beam is done to counteract the bending moment in it at working loads. But as the centroid of the tendons is normally within the depth of the beam, and is also a force and not a pure couple, it provides axial compression as well as a couple. This latter is fine for resisting the bending moment, but the axial compression can be regarded as uneconomic, as there is basically no axial tensile force to resist.

By post-compressing bars in the top and post-tensioning tendons in the bottom of a beam, a pure bending moment can be provided to equal the bending moment imposed upon the beam by the loading it is designed to carry. In this situation there are no flexural stresses on the beam.

Professor H. Reiffenstuhl[20, 21, 22, 23] of the Technical University of Vienna, Austria, showed the author details of a bridge he has designed using this principle, and which has been constructed in mainland Europe. The method allows the ratio of span to construction depth to be greater than for normal prestressed concrete construction. Certainly this is a most interesting and very novel idea of creditable originality. Of course, like everything else, advantages and disadvantages can be discussed.

References

1. Walz, K., 'Requirements Concerning the Grout for Prestressed Concrete Elements', *Bau und Bauindustrie* (Düsseldorf), **8**, 16, p. 468 (1957) (in German).
2. Leonhardt, F., On the injection of the grout into the ducts. *Proceedings of the Third Congress of the Fédération Internationale de la Précontrainte* (Berlin, 1958), Cement and Concrete Association, pp. 323–336 (1958) (in German).
3. Leonhardt, F., *Prestressed Concrete for the Practice*, Wilhelm Ernst und Sohn, 2nd edn, p. 45, Berlin (1966) (in German).
4. Benz, G. H., *Grout for the Ducts of Prestressed Concrete*, Sittler and Federman, Illertissen, Germany (1965) (in German).
5. Soroka, I. and Geddes, J. D., Cement grouts and the grouting of post-tensioned prestressed concrete. *Bulletin No. 26*, University of Newcastle-upon-Tyne, Dept. of Civil Engineering, June (1966).
6. Szilard, R., A survey of the art – corrosion and corrosion protection of tendons in prestressed concrete bridges. *Journal of the American Concrete Institute*, Jan. (1969).
7. Wilby, C. B., *Post-tensioned Prestressed Concrete*, Applied Science Publishers Ltd., London (1981).
8. Wilby, C. B., 'Precast Concrete Framed Roofs – Design of Joints and Use of Post-tensioning', *Indian Concrete Journal*, Feb. (1960).

9. Winter, G. and Nilson, A. H., *Design of Concrete Structures*, McGraw-Hill, U.S.A. (1973).
10. Wilby, C. B., *Prestressed Concrete Beams*, Elsevier Publishing, Amsterdam, London, New York; Applied Science Publishers, London (1969).
11. Wilby, C. B., Design of post-tensioned prestressed concrete ties for bridges and shell roofs. *Civil Engineering and Public Works Review*, Apr. (1972).
12. Wilby, C. B., *Elastic Stability of Post-tensioned Prestressed Concrete Members*, Edward Arnold (1964).
13. Evans, R. H. and Wilby, C. B., *Concrete – Plain, Reinforced, Prestressed and Shell*, Edward Arnold (1963).
14. Wilby, C. B. and Nazir, C. P., Shear strength of uniformly loaded prestressed concrete beams, *Civil Engineering and Public Works Review*, Apr. (1964).
15. Wilby, C. B. and Inman, P., The structural behaviour of jointed prestressed concrete beams. *Proceedings of the Institution of Civil Engineers*, Sept. (1972).
16. Clarke, N. W. B., *Buried Pipelines: a Manual of Structural Design and Installation*, Maclaren (1968).
17. Swanson, H. V., Design of prestressed concrete pressure pipe. *Journal Prestressed Concrete Institute*, Chicago, Illinois, **10**, Aug. (1965).
18. Johnson, P., 'Structural Design of Buried Pipes', M.Sc. Thesis, University of Bradford, U.K. (1973).
19. Lin, T. Y., *Design of Prestressed Concrete Structures*, John Wiley, New York (1963).
20. Reiffenstuhl, H., Druckspannbewehrung, ein neues Konstruktionselement zur Entscheidenden Steigerung der Tragfähigkeit von Stahl-Beton- und Spannbetonquerschnitten. Gehalten am 8 Juni 1972, Antrittsvorlesungen der Technischen Hochschule in Wein. *Verlag der Technischen Hochschule in Wein*, Vol. 31, Vienna, Austria (1973).
21. Reiffenstuhl, H., Die Fussgängerbrücke über die Rupert-Mayer-Strasse in Munchen Spannbetontragwerk mit Druckspannbewehrung. *Bauingenieur*, Vol. 64, Springer-Verlag (1989).
22. Reiffenstuhl, H., Das Vorsspannen von Bewehrungen auf Druck; Grundsätzliches und Anwendungsmöglichkeiten. *Beton- und Stahlbetonbau*, Vol. 77, pp. 69–73 (1982).
23. Reiffenstuhl, H., Eine Brücke mit Druckspannbewehrung – Konstruktion, Berechnung, Baudurchfuhrung, Messungen. *Beton- und Stahlbetonbau*, Vol. 77, pp. 273–278 (1982).

9

Shell roofs

9.1 Notation

There is not an internationally agreed nomenclature for the analysis and design of shell roofs. Nor is there a nationally agreed nomenclature suitable for this purpose. The nomenclature most commonly used and which is most suitable for shells is as given, for example, by Timoshenko[1]. This will be used in this chapter; it is listed below.

A	a constant (see Eqn. 9.38)
a	radius of spherical dome in Section 8, and dimension on Fig. 9.18
B	a constant (see Eqn 9.38)
b	dimension on Fig. 9.18, and dimension on Fig. 9.19
$C_1(\varphi)$	a function of φ
$C_2(\varphi)$	a function of φ
$C_3(\varphi)$	a function of φ
$C_4(\varphi)$	a function of φ
c	value of z when $x = a$ and $y = b$
E	Young's modulus of shell
g	$1 - H_0/H$
H	dimension on Fig. 9.38
H_0	dimension on Fig. 9.38
h	thickness of shell
k	a constant (see Eqn 9.31)
l	span
N_x	stress resultant at $x = x$ (see Fig. 9.10)
N'_x	$N_x + (\partial N_x/\partial x)\cdot dx$, i.e. value of N_x at $x = x + dx$
$N_{x\varphi}$	stress resultant at $x = x$ and $\varphi = \varphi$ (see Fig. 9.10)
$N'_{x\varphi}$	$N_{x\varphi} + (\partial N_{x\varphi}/\partial x)\, dx$, i.e. value of $N_{x\varphi}$ at $x = x + dx$

N_{xy}	stress resultant (in-planar shear) at $x = x$ and $y = y$ (see Fig. 9.18)
N_y	stress resultant (tension in direction $0y$) at $y = y$ (see Fig. 9.18)
N_θ	stress resultant at $\theta = \theta$ (see Fig. 9.14)
N'_θ	$N_\theta + (\partial N_\theta/\partial\theta)\,d\theta$, i.e. value of N_θ at $\theta = \theta + d\theta$
$N_{\theta\varphi}$	stress resultant at $\theta = \theta$ and $\varphi = \varphi$ (see Fig. 9.14)
$N'_{\theta\varphi}$	$N_{\theta\varphi} + (\partial N_{\theta\varphi}/\partial\theta)\,d\theta$, i.e. value of $N_{\theta\varphi}$ at $\varphi = \varphi$ and $\theta = \theta + d\theta$
N_φ	stress resultant at $\varphi = \varphi$ (see Figs 9.10, 9.14 and 9.17)
N'_φ	$N_\varphi + (\partial N_\varphi/\partial\varphi)\,d\varphi$, i.e. value of N_φ at $\varphi = \varphi + d\varphi$
$N_{\varphi x}$	stress resultant at $\varphi = \varphi$ (see Fig. 9.10)
$N'_{\varphi x}$	$N_{\varphi x} + (\partial N_{\varphi x}/\partial\varphi)\,d\varphi$, i.e. value of $N_{\varphi x}$ at $\varphi = \varphi + d\varphi$
$N_{\varphi\theta}$	stress resultant at $\theta = \theta$ and $\varphi = \varphi$ (see Fig. 9.14)
$N'_{\varphi\theta}$	$N_{\varphi\theta} + (\partial N_{\varphi\theta}/\partial\varphi)\,d\varphi$, i.e. value of $N_{\varphi\theta}$ at $\theta = \theta$ and $\varphi = \varphi + d\varphi$
p	total load on shell per unit of curved surface area
q	total load per unit area of curved surface
q_0	self-weight load per unit surface area
R	total loading on a symmetrically loaded shell of revolution about a vertical axis, above the horizontal section defined by an angle φ, under consideration (see Fig. 9.17)
r	radius of shell (see Fig. 9.10)
r_0	radius of horizontal circle (see Fig. 9.14)
r_1	radius of curvature of generating curve at point under consideration (see Fig. 9.14)
r_2	radius of principal curvature in plane perpendicular to other plane of principal curvature
u	displacement in longitudinal direction x at $x = x$ (see Fig. 9.12)
v	displacement in transverse tangential direction at $\varphi = \varphi$ (see Fig. 9.12)
w	displacement in radial direction at $x = x$ and $\varphi = \varphi$ (see Fig. 9.12)
X	component of total loading in plane of differential element (see Figs. 9.10 and 9.14) per unit area of element
Y	component of total loading in plane of differential element (see Figs. 9.10 and 9.14) per unit area of element
Z	component of total loading normal to plane of differential element (see Figs. 9.10 and 9.14) per unit area of element
$X(x)$	a function of x, independent of y
$Y(y)$	a function of y, independent of x
x, y, z	coordinates in three dimensions, at right-angles to one another
θ	horizontal angle locating point under consideration (see Fig.

Introduction 303

φ   9.14) but for Eqns 9.35 and 9.36, θ is angle of slope of differential element with respect to axes Oy and Oz (see [1])

φ   angle between point under consideration and vertical (see Figs. 9.10, 9.14 and 9.17), but for Eqns 9.35 and 9.36 φ is angle of slope of differential element with respect to axes Ox and Oz (see [1])

ω   angle between Ox and Oy axes (see Fig. 9.18)

9.2 Introduction

In this section, a brief description of the most popular types of shells used, the reasons for their use (excluding economics, which are dealt with in another section), their designation and application will be given. A photograph will show an example of a certain type of shell used for a certain purpose. Other purposes for which this type of shell is often used will be mentioned. Because of the limitations upon space in this Chapter, references will be given to photographs in other publications, for the benefit of the reader interested in a particular type of shell, perhaps with a usual or unusual application. These references will be spread, to some extent, internationally to help readers internationally. For example, for a particular type there may be three references quoted; one British, one American and one by a non-anglo-american author. Sometimes, alternative references may be given for the same photograph or scheme as the reader will usually be able to locate one easier than one of the others.

9.2.1 *Types of shells*

Various types of shells in popular use will be described in this sub-section.

Consider Figure 9.1; diagram (a) shows the cross section of an *in-situ* reinforced concrete roof construction, the slab spanning between the beams shown, which are of 10 m span and simply supported. Diagram (b) shows how the latter mentioned slab can be folded so that it is stronger in that, normal to its plane, it carries only a component of the vertical roof loading and not its full amount. It can therefore be thinner. Although the slab is thinner, the overall depth of construction is much greater than that for the beam–slab construction of diagram (a) and it can therefore span further. This is a 'folded plate', 'hipped plate' or 'prismatic' construction, and it spans 20 m (simply supported). Diagram (c) now curves the slab of diagram (a) above to a cylindrical shape (the soffit is cylindrical). Because the slab is arching, it can now be thinner still, much of the roof loading

304 Shell roofs

being carried by compressions in the arch, reducing the bending moments. As for diagram (b), its substantial overall depth allows it to span further than the construction of diagram (a). This (c) is a 'cylindrical shell roof' or 'barrel vault roof' and it spans 20 m (simply supported).

In comparing the roofs of Figure 9.1, it can be seen that the constructions (a), (b) and (c) are mainly 125, 100 and 65 mm thick, respectively. Thus the shell is of a lighter-weight construction capable of spanning further and using less concrete. With roofs, the self-weight of the concrete is the main proportion of the total loading; in cases (a), (b) and (c) the self-weight would be about 83, 80 and 70% of the total loading

Figure 9.1 Cross-sections of *in-situ* reinforced concrete roof constructions.

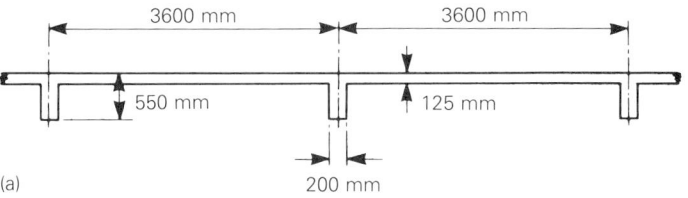

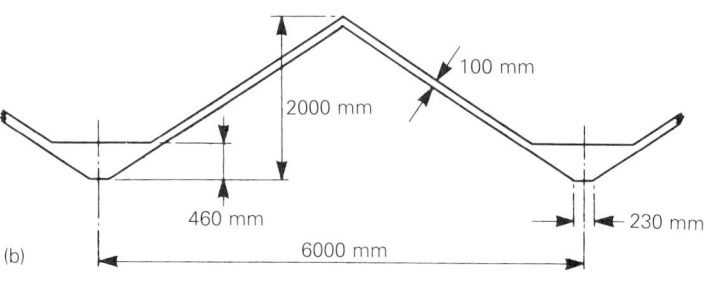

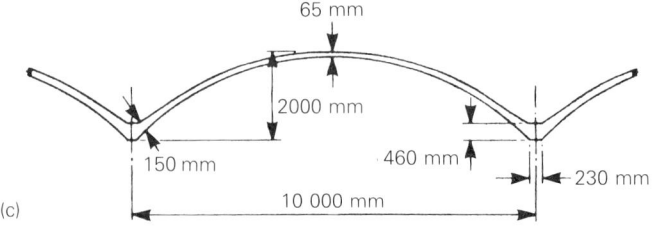

Introduction

respectively for the concrete thicknesses of 125, 100 and 65 mm, respectively. Also concrete is a heavy structural material. Therefore, the amount of concrete is very important when it is required to use it to span larger distances.

The dimensions shown in Figure 9.1 were assessed approximately using, respectively, Table 7.1, Art. 9.1.2 Proportioning of cylindrical shells and Art. 9.2.4 Design tables for folded plates in Ref. 2.

Figure 9.2 shows part of a scheme of cylindrical shell roofs designed by the author. This scheme at Blackburn College of Technology (England) includes many different spans and proportions of 'ordinary cylindrical shell roofs', called 'ordinary' to distinguish them from 'N-light shell roofs'. Refs. 3 and 4 show photographs of an ordinary cylindrical shell over machinery for the manufacture of leathercloth in the U.K. Each of these shells has a special feature in that it has steel sleeves through the 'valley beams' (see Figure 9.9 for description) and bolts projecting from the soffit of the shell to enable light-weight pipes, electrical conduits, pieces of machinery, etc., to be supported from the roof in any location. In this type of factory, the optimum positions of many of these fitments, etc., are decided from time to time after occupation of the factory. The structural steel roof in competition with this shell-roof scheme naturally gave this facility, which was demanded by the mechanical engineers of the

Figure 9.2 Blackburn College of Technology, England.

concern. Ref. 5 shows a photograph of ordinary cylindrical shells used in an unusual way; essentially this roof at St Louis Airport, U.S.A. is the conjunction only of four shells, making an interesting structure architecturally. Ref. 5 also shows a scheme of ordinary cylindrical shell roofs, at Barcelona Airport, Spain, without 'end diaphragms' or 'end stiffeners' (see Figure 9.9 for description), an uncommon feature, just with the valleys (see, Figure 9.9 for description) tied together at the supports instead of diaphragms. Refs. 6 and 7 show the same photograph of a scheme of prestressed concrete ordinary shell roofs in the U.K.

Ordinary shell roofs of the type shown in Figure 9.2 can span up to about 36 m simply supported, without the need for prestressing.

Overhead cranes have sometimes been supported from 'edge beams' (see Figure 9.9 for definition) by making a boot lintel feature of the bottom of each edge beam. Otherwise (Denton Gas Works, U.K. designed by the author for his former employer), the crane can be supported from corbels from the columns carrying an ordinary cylindrical shell at

Figure 9.3 North-light scheme for a printing works in Bradford, England.

Introduction

diaphragm supports and from intermediate columns, for this purpose, between the diaphragm supports and supporting the shell edges.

Figure 9.3 and Ref. 3 show a scheme of N-light (north-light) shell roofs designed by the author. This scheme, a printing works for Sharpe's at Bradford (England), was required to maintain an attractive and clean internal environment, the former to impress important visitors and the latter to protect the expensive machinery for high-quality printing which is very sensitive to dirt. This particular scheme had half the normal number of columns in the north-to-south direction. This necessitated top truss members which were kept low down so that normally they could not, for aesthetic reasons, be seen on the top of the shell roofs. In practice, N-light schemes, for reasons of aesthetics and natural lighting, limit substantially the normal span between valleys. That is why the scheme just described halved the number of supports in the north-to-south direction. In the longitudinal direction, the shells were continuous over three spans.

Very large spans have been constructed in reinforced concrete by using large arches, the roof between comprising cylindrical shells for lightness in weight as opposed to slabs. (An arch is a very efficient reinforced concrete member for large spans as it tends to carry loading by axial compression rather than bending, often eliminating resultant tension. Concrete is, of course, very economic for carrying compression but not tension, as this has to be resisted by reinforcement, the concrete having cracked and thus become useless for resisting tension.) To keep the total height of the construction down for aesthetic and economic reasons, the 'stiffener beams' (see Figure 9.9 for description) are integral with the large arches. Thus, the arches have considerable depths and can span remarkably great spans for reinforced concrete roof construction. One such scheme, an aircraft hangar in S. Dakota (U.S.A.) comprises arches of 102 m span and at 7.5 m centres. Between these arches, there is a 127 mm thick shell. Refs. 6 and 7 show the same photograph of a scheme at the Bank of England,

Figure 9.4 In-planar forces supporting loads.

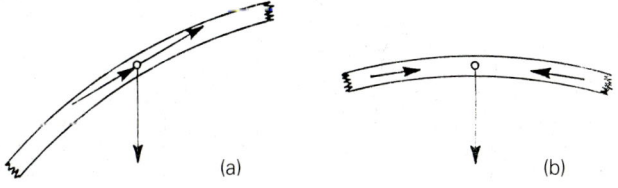

308 *Shell roofs*

in which large arches carry N-light shell roofs, the stiffener beams of the N-light shells being integral with the arches.

In the case of the cylindrical shell, in the curved direction each element of loading is supported by forces, as shown in Figure 9.4. These are called 'in-planar' forces, that is forces in the plane of a shell. A thin shell is strong at resisting these in-planar forces but not strong in bending. The curvature is thus a great asset in providing in-planar resistance to supplement very substantially the weak bending resistance, to carry the loading.

Now, a doubly curved shell has this benefit of curvature in three dimensions. Hence, each element of loading is supported very effectively by 'in-planar' forces in three dimensions. This means that doubly-curved shells are more rigid, can be thinner and thus lighter in weight than cylindrical shells, already described.

Figure 9.5 shows a dome (a doubly-curved shell) over a chemical-waste disposal tank in the U.K. designed to resist explosive gases under pressure. An external post-tensioning system, suitably protected against corrosion, was designed and the construction then supervised by the author for the consultants because a mistake had been made in providing inadequate reinforcement in the lower periphery, i.e. to resist 'ring tension', and the dome was spreading out minutely but sufficiently for cracks to develop in the shell of the dome, enabling explosive gas to leak. Ref. 5 includes a photograph of a spherical roof covering a building of rectangular shape in plan to give an attractive auditorium in the U.S.A.

Figure 9.5 Chemical-waste disposal tank with dome cover, England.

Introduction

Figure 9.6 shows a 'hyperbolic paraboloidal' or 'hypar' shell roof (a doubly-curved shell) over a market in Huddersfield. A different aspect of the same roof scheme is illustrated in Ref. 3. Ref. 3 also shows a photograph of a model of this inverted umbrella type of hypar shell under the supervision of the author, Ref. 8.

Refs. 6 and 9 show photographs of a scheme including four hypars with their lower extremities tied together with a prestressed concrete tie and also

Figure 9.6 Hypar shell roof over a market in Huddersfield, England.

Figure 9.7 Scheme of conoidal shell roofs (From: C. B. Willby, *Concrete for Structural Engineers*, by permission of Newnes-Butterworth, London).

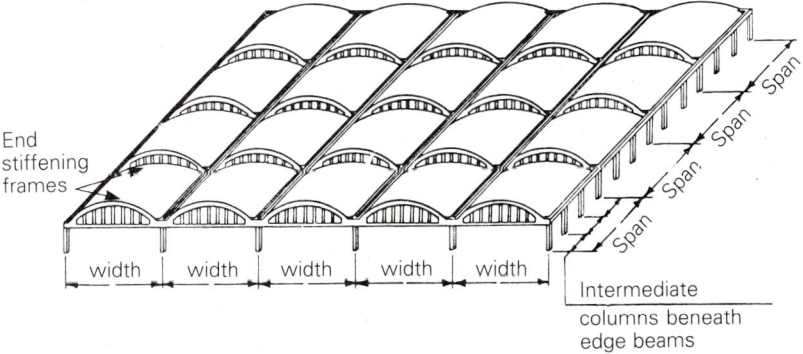

a scheme of hypars, where each group of four looks like two folded plates; both schemes are in the U.K. In Refs. 5 and 10 there are photographs of various interesting types of hypar used in the U.S.A. and India, respectively.

Figure 9.7 indicates a scheme of conoidal shell roofs (conoids). A photograph of a scheme of conoidal shell roofs over a motor-car factory in India can be found in Ref. 10, which also shows an electrical equipment factory in India whose roof scheme looks as though it is covered with conoidal shells but is actually composed of tilted, ordinary cylindrical shells.

9.2.2 Designation of shells

A classification of shell roofs was proposed by Bharucha[11] but is too long and comprehensive to include in this chapter. Shells may be classified from various viewpoints:[11] (a) shells of single or double curvature, (b) shells of translation or rotation, (c) shells that can or cannot be ruled, and (d) single form or combination shells. Of these, (a) is regarded as the most important[11] classification and Billington[12] states that the most general classification is by Gaussian curvature. Most engineers are familiar with curvature being the reciprocal of the radius of curvature for two-dimensional structures. For three-dimensional shells, the Gaussian curvature at any point is the product of the principal curvatures[13].

Shells of positive Gaussian curvature, Refs. 11 and 12, 'synclastic shells' are such that the surface curves away from a tangent plane at any point on the surface and lies completely on one side of this plane, e.g. spherical domes and elliptic paraboloids.

Shells of negative Gaussian curvature, Refs. 11 and 12, 'anticlastic shells', are formed by two families of curves, each curved in opposite directions; e.g. hypars, conoidal shells or conoids[13] and hyperbolas of revolution.

Shells of zero Gaussian curvature, Refs. 11 and 12, or singly-curved shells, lie between shells of positive and negative Gaussian curvature, e.g. cylinders and cones.

Classification (b) is fairly popular. Shells of rotation[11,12] are formed by the rotation of a curve about any straight line. The curve and axis of rotation are normally in the same plane. Examples are domes and tanks. Shells of translation, Refs. 11 and 12, are formed by the translation (i.e. movement) of a curve in one plane such that its ends move along two curves which may be the same or different; the plane of the curve being

Introduction

translated, called the generating curve, always remains parallel to its original position. Examples are cylindrical shells, elliptic paraboloids and hypars.

9.2.3 *Applications*

Generally, a prime reason for using shells seems to be aesthetic. The photographs in Section 9.2.1 and its references indicate a range of aesthetic possibilities. With good architectural treatment, most interesting and attractive-looking roofs can be produced – a delightful change from the box type of buildings which are a bore to our present environment. Shells can easily be made to look attractive. The only type of shell scheme which the author regards as being ugly is that which gives the impression of giant oil drums laid on their side on the landscape. This type of scheme involves ordinary cylindrical shells covered with black or very dark green roofing felt. The author prefers the avoidance of this type of colour scheme and the use of white, light green or even pink mineral-finished roofing felts. Light-coloured felts also resist transmission of radiant heat better than black or dark coloured felts.

Probably the second most important reason for using shells is that a good robust roof is produced, which can span comparably large spans to those possible with lighter constructions using strong materials, e.g. structural steelwork supporting light-weight roof sheeting. In the U.K., a shell will often be 65 mm thick. It will contain four layers of reinforcement in parts; its minimum reinforcement may be two layers of fabric, one near each surface, and a layer of reinforcement between these fabrics. It will be covered with either 25 mm thick corkboard (or 50 mm thick vermiculite) and three-layer built-up roof roofing felt, the top layer being mineral finished. The soffit may be plastered, or not, and painted with emulsion or distemper. This quality of roof is thus superb compared to any roof sheeting fastened to roof purlins. An alternative reinforced concrete roof of large span not using shell construction would have to be constructed by using prestressed concrete beams, close together and supporting a one-way continuously spanning *in-situ* reinforced concrete (or hollow tile or precast concrete) slab. The overall depth of this construction is much less than that of a shell and its weight is greater, hence it is much less of a practical proposition for spanning large spans because of both strength and deflection limitations. The author once checked the design of a roof comprising *in-situ* post-tensioned rectangular beams cast between hollow concrete caissons. It surprisingly achieved a simply supported span of

17.7 m with an overall depth of only 381 mm but its deflection was considerable and because of this it would not have been acceptable in the U.K. but it was satisfactory in the country concerned.

Ordinary cylindrical shells have been used where large (for reinforced concrete) spans were required for factories, bus stations, etc. They have also been used where shorter spans would have been satisfactory, for offices, schools, colleges, etc.

North-light cylindrical shell roofs have more limited spans than ordinary cylindrical shell roofs because they are not as strong or rigid in the north-to-south direction and also because of the aesthetic layout required because of the north lighting in the east-to-west direction; for example, if the N-lights are long and deep, one either has undesirably deep glazing areas which become relatively expensive per unit area of glazing or obtains less lighting per unit area covered. They have been built over colleges, schools, waterworks, factories, etc.

Conoids give a good distribution of light over the area covered and have been used as alternatives to N-light cylindrical shells. They are often more attractive, use less materials and are thus lighter in weight but are more expensive for shuttering and steel fixing, and very slightly more expensive for concreting. They have been used, for example, over factories[10] and the testing laboratories at Delft.

Domes have been used greatly over circular tanks for sewage, water and chemical waste (Figure 9.5). Although bad for acoustics, it was fashionable to use them over certain prestigious libraries constructed in the U.K. before the Second World War. Domes, not necessary spherical, have been used over many prestigious buildings, such as exhibition halls, temples, mosques and synagogues.

Cylindrical shells tended to be favoured because the shutter was straight in one direction and conformed to the simplest curve – a circle – in the other direction.

Hypars found favour in that they were and had the advantages of doubly-curved shells (see Section 9.2.1) and yet could be formed with straight boards or planks. Most are formed with straight boundaries. In time, the fact that the straight boards needed to be tapered or tapered occasionally became rather a disenchantment. However, they are possibly the easiest of the doubly-curved shells to construct. They have been used for many aesthetically attractive structures over factories, garages, markets, churches, exhibition halls, etc.

9.3 Economics

There are many different types of construction which can be used for roofs. These may sometimes be compared only on first cost if this consideration is so desperately overriding. In this instance, the type of shell satisfactory in the U.K. would never be less expensive than competitive forms of construction. For the general roof area covered, Big 6 corrugated asbestos sheeting with a fibre-board lining for insulation, or equivalent, would be much cheaper than the 65 mm thick shell described in the second paragraph of Section 9.2.3. It would also be lighter in weight, helping the economy of the supporting structure.

First cost is not the only practical consideration and often not the most important. The quality of the roof matters with regard to its aesthetics and its resistance to watertightness, heat transmission, fire, small knocks and forces, for example, from maintenance men, craftsmen engaged in internal alterations, etc. The life of the roof and its maintenance cost are also important.

If one wanted a column layout of say 9 m by 18 m and to obtain comparable (even though not as good) qualities to the shell by using wood wool slabs supported by precast concrete purlins and frames and assuming there were say 20 of these bays (9 m by 18 m), then ordinary cylindrical shells would most probably be the cheaper solution of these two alternatives. In addition the shell would be safer for workmen to walk and work upon than the wood wool slabs. It is likely that the shell could be beaten on initial cost in the U.K. if structural steel work were used with steel purlins supporting say a patent metal sheeting. However, this latter construction would require periodic painting, its fire resistance would be relatively poor and its life would probably be less.

The author designed, estimated and constructed some barrel shells like the ones just mentioned, in the U.K. The alternative steelwork solution was slightly less in cost even if one allowed for the cost of painting it for fifteen years. However painting could only be effected in a two week period each year, the works holiday. The people responsible for maintenance found the task of engaging a tremendous number of structural steel painters for this short period in the particular locality, at a time of full/over employment, exceedingly difficult, so the shells were used. All the windows had reinforced concrete frames. The shells were emulsion-painted after construction, and nineteen years later had not been redecorated; their condition was regarded as satisfactory, and during this period nothing was spent on maintenance. This is an accurate case study;

a normal profit was made on the shell construction which was well done, and the steelwork price was as expected and accurately assessed.

In the early days of shell-roof popularity in the U.K., there was – in the author's experience – a particularly wide variation in contractors' prices for constructing shells. All contractors thought that curved shuttering was a formidable and expensive proposition. At one extreme, some would price very high, either over-reacting to this problem of curved shuttering or being realistic about their firms' lack of ability to be efficient with a type of structure which none of their personnel had tackled before. However, at the opposite extreme, some would believe that this type of construction was a coming thing and that they should gain experience with it, even if they had to pay to some extent for this experience. Hence, they would price very optimistically and not worry too much if there was a risk of making very little profit.

The latter type of contractor would win the contract. Often, he would set about the construction in most unskilled, uneconomic ways and fight hard not to lose money or, at worst, to contain his losses. Then one would find that the next time this contractor quoted for a shell, he would be sky high because he was basing his costing on a most inefficient operation; his first attempt.

It is thus difficult to generalize about costing from case studies unless one knows a great amount of detail in each case, e.g. whether a profit or a loss was made.

Shells became easier to design mainly for the few who knew how to design them and had experience of this work. This knowledge did not spread easily because of the complexity of shell analysis compared to the mathematical limitations of the average designer or engineer graduate. Shells also became easier and cheaper to construct. The curved shuttering became less expensive because various firms specialized in this work.

In the last shell scheme which the author designed, estimated and constructed, the final cost per unit area of the sub-let curved shuttering was the same as the cost per unit area of the flat shuttering. Admittedly, the latter price was not the most competitive, but even so this demonstrates that the cost of curved-shuttering is not excessive. In estimating shells, the author tries to see how the cost will be distributed and where the major expenditure will be incurred. Generally, it is difficult to point to any particular item of extravagance. However, in the author's experience the cost of a valley beam (Figure 9.9) for a scheme of cylindrical shells tends to be very high indeed. The valley beam plus local thickening of the shell, special reinforcement at this location, etc., can account for about one third of the cost of the whole shell plus the valley beam (excluding end stiffeners and columns). So if it is possible to manage without valley beams, it is better to do so.

The difficulty of generalizing about U.K. costs becomes worse when attempting to generalize internationally. It is complicated by the fact that there is a different family of less expensive shells which can be used in some countries. These are not in total construction as watertight (maintenance is effected when necessary) and as resistant to heat transmission, fire and localized forces and minor knocks from, for example, maintenance men, etc. These shells are often much thinner than the 63.5 mm (2.5 in) minimum thickness generally used in the U.K. Practice in the U.S.A. is similar to the U.K. In some countries, hypars have been constructed only 32 mm thick and without finishes such as roofing felt, thermal insulation or plaster. In some of these, it is very hot and seldom rains. When it does, the rainfall is often very powerful for a short duration. Then, the sun comes out and rapidly makes the concrete very hot and dry. If the storm exposes any leaks, they are grouted up on the top of the roof. A restaurant, for example, may be inexpensively but attractively decorated before the beginning of each tourist season by distempering or emulsion painting the concrete ceiling; the odd leak causes little difference to the general effect. The floor is tiled, so that any leakage can be quickly mopped up and no damage is done.

The U.K. and similarly cold and wet areas of the world pose a different problem. In the U.K., building surfaces, such as concrete and brickwork, are rarely very hot and dry. They are usually damp and there is so much rain-fall that it takes a long (by U.K. standards) freak heat-wave to dry out the ground to more than about 150 mm depth. During this process, water vapour is evaporated continuously so that the humidity is high in the best hot weather as well as when it is raining or drizzling – which it does frequently day and night. If a leak occurred in a shell roof in the U.K. then, usually, expensive finishes under the roof would be damaged. In a restaurant, for example, fitted carpets and underfelts would be damaged. Not only would the architect, structural designer and contractor have to make the roof watertight for the next six years, but they would have to make good the damaged decorations and furniture.

So, in some countries, the standards reasonable for the climatic conditions, way of life, etc., allow shell roofs to be cheaper than in the U.K. Also, in some of these countries, the ratio of the cost of structural steel-work to that of reinforced concrete is greater than that in the U.K. Furthermore, the cost of labour relative to materials is lower than that in the U.K. so doubly-curved shells, for example, which are very efficient at using the minimum reinforced concrete materials for any large span, can be more economic than in the U.K.

For work in the U.S.A., Ketchum[14] gives a (two-part) figure – prepared from a report made by his firm, Ketchum and Konkel – relating cost of structure to span, for spans between 9 and 30 m, for bays of 6 and 12 m width, respectively. The comparison is made between ordinary cylindrical shells, hypars and a structure with steel frame, purlins and deck (or covering or sheeting). The cylindrical shells are about 20 and 8% more expensive than the steelwork for the 6 and 12 m bay widths, respectively. However, Ketchum[14] reports that an insurance charge is required in the U.S.A. for the steel structure. This then makes the costs of the cylindrical shell and steelwork structures the same for the 6 m bay width; for the 12 m bay width, the cylindrical shells are about 8% less in cost than the steelwork roofs. Ketchum gives a table of quantities for various folded plates, barrel (ordinary cylindrical) shells, hyperbolic paraboloid inverted umbrella shells and hyperbolic paraboloid dome shells. With this table, a reader anywhere in the world can use the various costs relevant to where his structure is to be built and obtain an estimate of the total final cost of each of these types of construction. The roofs were designed for a firm in Denver, Colorado, 'for a uniform live load of 1.436 kN/m^2 (30 psf) so for the smaller live loading used in the south or on the west coast [of the U.S.A.] the reinforcing steel quantities may be too high and must be reduced' in the table. The hypars are square in plan and are only considered for spans between 9 and 18 m. The hypars are square in plan and are only considered for spans between 9 and 18 m. Generally, they are about 8% cheaper than the steelwork structures which have a bay width of 12 m.

Ketchum makes remarks for the U.S.A. similar to those made earlier in this sub-section regarding the high overall quality of concrete shell roofs relative to alternative roofs in steelwork. He also makes observations for the U.S.A. similar to those made earlier in this sub-section on the great variations in prices quoted for shell schemes by contractors.

Whitney[15] agrees with Ketchum and the present author regarding the high quality of shells and cites the further advantages of 'superior wind and blast resistance and low amortization costs'.

Ketchum[14] considers forming (shuttering) costs and concludes that, in the U.S.A.,

> Unless forming systems can be devised that will permit a pour of reasonable size with five to six re-uses, then it is better to stick to a single-use form.... The trade secret in reducing the cost of single-use forms is to persuade the contractor that he should not write off all the cost of form material against this single job and to design the forms so that there is a minimum of plywood forming material so most of it may be salvaged. A large percentage of the shell structures designed by the writer's [Ketchum's] firm have been built with single-use forms often at the option of the contractor and in preference to movable forms.

This is opposite to the experience of the author both with firms of designers and designer-contractors in the U.K. This is probably because timber is very expensive in the U.K. and also because the labour force and capital investment in shuttering, scaffolding and plant would be less in the U.K., the construction time longer and more use and value obtained from the aforementioned capital investment. Based on his experience in the U.K., the author recommends that good shutters be made and re-used as many times as possible. Hired specialist shuttering is often economic, even though it has often to be transported up to about 200 miles. This is because the same shutters are used many times. In the U.K., the salvage value of timber shuttering is nil: transportation and cleaning of the timber are involved and joiners are very reluctant to make shutters with second-hand timber – it takes them much longer and the shutters thus cost more to fabricate.

However, Whitney[15] cites U.S.A. experience favouring as many uses as possible being obtained from the same shutters.

Ketchum[14] seems to be particularly impressed with the low cost of hypars in the U.S.A. Considering a column grid 12×12 m and taking the cost of a hypar umbrella shell as 100 units per unit area covered, then the pro-rata cost for various structural systems would be: timber frame 91, steel frame with wood joists 78, steel frame with wood purlins 77, steel frame with steel purlins and deck 104, steel frame with open web joists 90, and prestressed concrete 101. This comparison favours hypars when the maintenance cost and relative qualities are taken into account.

When the author worked for a firm of designer-contractors he designed and estimated certain schemes for precasting shell roofs, and is convinced that, in the U.K., any precast shell would be much more expensive than an *in-situ* shell. This is because of the costs involved in precast work which do not occur in *in-situ* work, namely, factory overheads, handling, stacking at factory, loading, transportation, unloading, stacking at site, and erection. Ketchum's paper[14] seems to indicate less difference in price between *in-situ* and precast construction for the U.S.A. Generally, *in-situ* appears to be slightly cheaper than precast in the U.S.A. but Ketchum has the opinion that barrel shells of spans up to 15–18 m might be economic to precast.

Earlier in this sub-section, it was mentioned that not many designers, because of their mathematical limitations, became able to design shells. Ketchum comments similarly for the U.S.A. and goes on to say 'shells are not used in many situations where they could be the best solution'. The author's UK experience fully agrees with this statement.

Whitney[15,16] gives examples of the costs of shells in the U.S.A. Generally, these favour shells compared to alternative constructions for long spans, particularly those requiring good quality roofs. Whitney[15] concludes with a good piece of advice:

> The conception of an economical concrete shell structure must be based primarily on construction considerations. After it has been determined how it can be built most cheaply, the structural design follows logically. The cost lies not so much in the quantity of concrete and reinforcement steel needed, as in the cost of the process of getting them into place.

The author has experienced an economic use of shells where large concrete spans were required over a service reservoir. Large spans were used because of bad ground conditions, i.e. the fewer columns there were, each of which required an expensive foundation, the less was the total cost of the scheme.

In many underdeveloped countries, labour is less expensive relative to materials than in the more advanced countries. This point is made by Chatterjee[17] with regard to India. He gives details of factors affecting the costs of a considerable range of shells constructed in India and and these should be of use to those designing and/or constructing shells in underdeveloped countries. The shells quoted range in thickness from 50 to 100 mm.

9.4 Design

Designing shells comprises assessing their dimensions and then analysing them elastically for internal forces and moments.

Recent British and U.S. codes of practice have superseded the designing of sections by elastic theory only, but these and other codes do not deal with shells. It can be argued that there is insufficient research and practical experience to make the change from the long established practice of designing shell sections elastically, to resist forces and bending moments obtained by elastic analysis, to a limit state analysis like the present British and U.S. codes of practice recommendations for reinforced concrete slabs, beams and frames.

The established practice of shell design, Refs. 2, 3, 10, 12, 13, 18 and 19, has been found to be satisfactory and reliable over a period of many years. Where cracks have occurred, this has led to local thickenings[14] being used. The elastic analysis endeavours to control cracks by limiting steel and concrete stresses. Deflections can be calculated from the elastic analysis and these should always be considered when, for example, walls and windows may be damaged by shell deflections. In the case of windows or fragile walls, etc., built up to the shells, sliding arrangements need to be devised so that shells can deflect freely without damaging them. The established practice has a good record with regard to safeguarding against failure. The author does not know of any failures where proper elastic design has been used; indeed, all the indications are that shells designed elastically are particularly robust, and even though we probably would not design them thinner than 64 mm in the U.K., they often seem to be conservatively designed at this thickness with regard to ultimate strength.

If, for a shell of the dimensions assessed, the reinforcement is too heavy to fit into the sections, or if the concrete stresses or deflections are too great, then the dimensions of the shell have to be altered and the analysis repeated until a satisfactory solution is found.

The dimensions can be initially assessed from experience and/or looking at similar shell solutions reported in architectural and structural engineering publications, and/or using approximate designs such as the beam analogy,[20] or the beam–arch approximation, for cylindrical shells.

For ordinary and N-light cylindrical shells, Refs. 2 and 19, give proportions for shells, in British Imperial (U.S.A.) and S.I. (metric) units, respectively, where the width-to-length ratio is 1 to 2. This ratio is regarded as an optimum, bearing in mind both economy and appearance. Ref. 19 then gives an approximate but fairly accurate method of estimating the reinforcement in these shells. The method yields the kind of accuracy required by firms which design and supply reinforcement. Contractors and consultants do not usually require this accuracy, as the item in the bill of quantities will be subject to final measurement, whereas the designer-suppliers usully quote a lump sum price for the reinforcement, so Ref. 2 gives a simple-to-use group of graphs, for estimating the reinforcement. This is reproduced in Figure 9.8. Curves SH and AH are

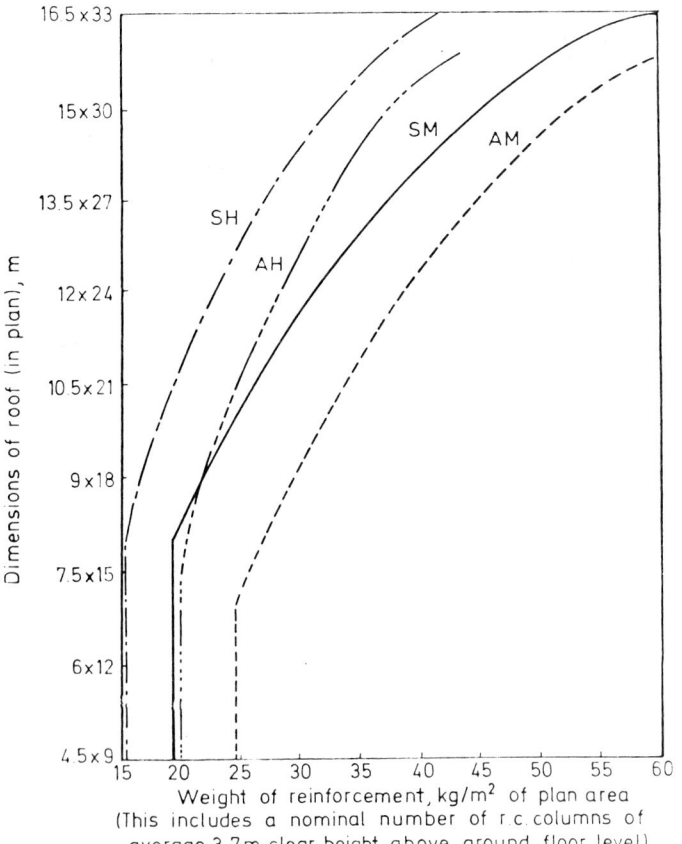

Figure 9.8 Graphs for estimating the reinforcement in cylindrical shells. Curves SH and AH are respectively for normal and north-light shells using high-yield reinforcement apart from mild steel for L-bars, stirrups and column bars. Curves SM and AM are respectively for normal and north-light shells using mild steel throughout except for shell fabrics and their spacer bars (From: C. B. Wilby, *Concrete for Structural Engineers*, by permission of Newnes-Butterworth, London).

for normal and N-light roofs respectively, where high-yield reinforcement is used apart from mild steel for L-bars,[5] stirrups and column bars. Curves SM and AM are for normal and N-light shells respectively, where mild steel is used throughout except for shell fabrics and their spacer bars.[5]

Ref. 14 includes a table giving quantities of steel and concrete for typical interior bays of shell roof structures: (a) barrel shells 15–30 m span, width 9 m and thickness 89 mm; (b) hyperbolic paraboloid inverted umbrella shells, 9–18 m square, and of thickness 57 mm; (c) hyperbolic paraboloid dome shells, 12–30 m square, and of thickness 57 mm.

9.5 Analysis

A common way of analysing a shell is to consider an infinitesimal element of it with all the forces (longitudinal and shear), moments and torsions in three dimensions which could conceivably occur on the element; in other words, a differential element. These forces and moments are equated to one another by general equations of static equilibrium, and are related to strains, stresses and displacements, e.g. see Ref. 12, or for cylindrical shells only Ref. 2.

When the differential equation finally obtained is applied to the case of a uniformly distributed loading, the result is very similar to that obtained from a much simpler statically determinate membrane analysis. The latter is an analysis which assumes the shell has negligible bending rigidity and that all the loading is carried by in-planar forces only, i.e. forces and shears in the plane of the shell. For example, Billington[12] (pp. 171–2) shows for a cylindrical shell the negligible error involved in using the membrane theory. This is a mathematical explanation. A physical explanation would be to imagine, say, a cylindrical shell as analogous to a large number of folded plates[12]. Now, the loading is carried transversely between folds and the consequent reactions at each fold are resisted by the adjacent two plates acting as beams. If the folds are close together, analogously to a cylindrical shell, then the transverse bending moments are negligible for the negligible spans between folds and the loads are carried entirely by plate action, i.e., in-planar forces.

The loadings considered for shells are often approximated to uniformly distributed loadings. In the U.K., it is common to consider the snow load of 0.75 kN/m² (16 psf) of horizontal area as a load per unit of curved surface area, so that it can be dealt with in the same way as the self-weight loading. Wind loading is often suction and not as great as the self-weight of a reinforced concrete (not a laminated timber) shell so that it can normally be ignored in the U.K.; similarly for the U.S.A. Refer to Section 9.3 for points regarding the use of thinner shells in other countries and the loadings used in various parts of the U.S.A.

Thus, for practical purposes, shells are often just analysed for a uniformly distributed loading per unit of curved surface area, comprising self-weight and snow load, and to begin with this loading is assumed to be carried by the membrane analysis (i.e. by in-planar forces only). This analysis would only be adequate if, at the boundaries where a shell became discontinuous, the supports provided forces equal and opposite to the membrane forces in the shell at these locations and if the support displacements were the same as those determined at these locations by the membrane theory applied to the shell. Not many practical shells adhere adequately to these conditions. For example, ordinary and N-light cylindrical shell roofs with edge beams (see Figure 9.9 for description) free to deflect vertically and horizontally do not, but a fairly impractical shell, a hemisphere, supported vertically only by, say, a tank wall all the way round, could be analysed by membrane theory only.

For the majority of shell roofs which cannot be analysed by membrane theory only, therefore, after the membrane theory has been used for determining the in-planar forces, the edge displacements, displacements in three dimensions, are determined according to this membrane theory. The support to the edges of the shell, be it an edge beam, ring beam, etc., has to resist the membrane forces at the edge of the shell in addition to its self-weight. In doing this, it will, almost without exception, realise different

Figure 9.9 System of construction joints (broken lines) for a scheme of barrel shells (Adapted from: C. B. Wilby, *Design Graphs for Concrete Shell Roofs*, Applied Science Publishers Ltd. London, with permission).

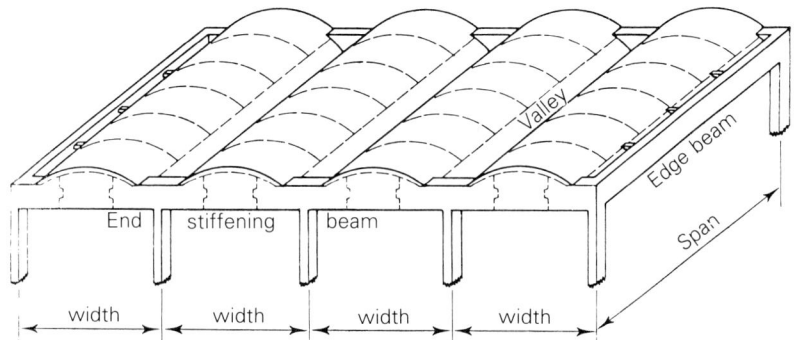

Analysis

displacements to those of the shell edge. As the edge or ring beam, etc., and the shell do not part company, unknown equal and opposite forces must be considered as acting between them by the analysis. For the shell, these are called 'edge loads' or 'line loads'. They are line loads occurring at the shell edges or boundaries. Now, whereas we said previously that uniformly distributed loads over the whole shell surface could be resisted with negligible error by in-planar or membrane forces, these edge loads cannot be considered as being resisted by in-planar forces without significant error. They cause bending moments, torsions, longitudinal and shear forces in three dimensions and the analysis is indeterminate for edge loads, not determinate as it was for membrane forces. To simplify the analysis, unit line or edge loads are considered one at a time and in each case the forces and moments at various locations in the shell and edge displacements are determined.

Also, unit line loads equal and opposite to the unit edge loads just mentioned are applied to the edge or ring beam, etc., and the displacements obtained at the junction with the shell in three dimensions. Bending moments, etc., for the design of the edge member when the unknown line forces are determined later can be determined at this stage.

Then, equations are constructed to obtain the unknown line loads to match up the displacements of the shell edge and edge or ring beam, etc., in three dimensions.

Once these equations are solved, the forces and moments due to the edge loads on the shell are, to obtain resultant forces and moments for our design, added to the membrane forces due to the uniformly distributed shell loading. The edge members are designed for these line loads as well as to carry the forces to resist the edge shell membrane forces and the self-weight of the edge members.

The designer without prior knowledge of shell analysis/design will appreciate that the above is arduous and complicated out of all relation to the simplified design of slabs and frames, e.g. Ref. 2. Thus, if need be a shell has to be designed in a tightly limited time (i.e. not allowing time for considerable study), there must be recourse to specialists and/or design tables and/or computer programs (see later).

If a designer were to ignore the difference in displacements of a shell and its supporting edge member due to the shell carrying its total uniformly distributed loading by membrane forces, and the edge member carrying its self-weight and providing reactions to the membrane forces at the edges of the shell, he would obtain a structure where the equilibrium of the forces is all right but the ignoring of the displacements would require considerable redistribution of the forces, involving much cracking. Hopefully, the redistribution would not be so severe as to require complete disintegration of one portion to allow the necessary forces to be carried by another portion. It would be wrong, according to the author's experience of tests, to assume that such considerable redistribution can easily or normally take place. The author has experienced the sensitivity of various types of shells to different modes of failure; for example, a shell can fail in one direction without being anywhere near failure in another direction.

However, many doubly-curved shells, particularly the hypar, have been designed in this way, often without even assessing any displacements. The amount of cracking which occurs depends upon how capable the designer is at placing nominal reinforcement, extra to the membrane design requirements, in the places where the shell and/or edge member is most likely to crack. Then, there is the consideration of whether the cracking really matters if the structure is a piece of sculpture (hypars have been used in this way) or a roof over a building in a very hot country (see Section 9.3 regarding acceptable occasionally leaking roofs).

Although the analysis of a shell is commonly split into a membrane and edge (or line) load problem, the author found it necessary to keep the problem as one combined analysis when using the electronic analogue computer for analysing cylindrical shells, see Refs. 2, 4, 21 and 22.

9.6 Construction

The contractor needs to know where he may make joints in the structure. There are two types of joint, namely, expansion and construction joints.

Expansion joints are made so that the structure is not too long in any horizontal direction as to be troubled by temperature movement. The coefficient of thermal expansion of both concrete and steel is about 0.000 009 per °C. A reinforced concrete member of length, say, 20 m, in a U.K. location where the maximum annual range of temperature is 30 °C, would experience a change in length of 5.4 mm, or nearly 3 mm at each end. If, for example, a cylindrical shell of this length were constructed between two existing buildings, then sheets of a special compressible material can be fastened to the existing walls in suitable locations and the shells cast against these sheets.

For a scheme of ordinary cylindrical shells, if the transverse end or stiffener beams (see Figure 9.9, for descriptions) have flat soffits, then every, say, 36 m, a sheet of compressible board about 25 mm thick will be used to make a discontinuity in the end stiffener and this will be continued to split the valley or valley beam for about 0.1 of the span from each end. Where each end stiffener beam is jointed, the column will be doubled. These can be constructed by casting one column, fastening 25 mm thick compressible board to the relevant face and then casting the second column against this board. The crowns of the central parts of the very flexible shells just rise and fall in accordance with temperature variations. This was not possible for the rigidly straight soffited end stiffener beams. Had the end stiffeners been slender arches and the crowns of these had been able (i.e. without damage to walls) to rise and fall with temperature movement, then they may have continued for a distance of, say, 60 m without an expansion joint. This joint would then be taken completely through the structure (i.e. splitting the whole valley or valley beam).

Thus, where curvature allows the temperature movement of a shell and/or arch to be accommodated by rise and fall at its crown rather than by creating a tremendous horizontal force, then expansion joints can be placed further apart then would happen with straight work.

The distance between expansion joints for reinforced concrete buildings varies somewhat between different designers and can depend upon the layout of the building in plan. The author's experience has generally been to place expansion joints at about every 36 m for buildings of rectangular shape in plan. Thus, for cylindrical shells, expansion joints should be made every 36 m or less in the straight direction, and the same in the direction of the stiffener beams if they have flat soffits but, say, every 60 m for flexible construction as just described.

Similar ideas can be applied to doubly-curved shells. An inverted umbrella type of hypar has straight horizontal edges, therefore joints are required every 36 m or less, whereas a spherical dome is flexible, and so the joint can be more widely spaced. In these cases 'joints' are shell boundaries.

Construction joints serve two purposes. First, they enable the structure to be cast in practical amounts; secondly, they relieve shrinkage stresses.[2] Designers in the U.K. specify and detail expansion joints but commonly ignore construction joints or deal with them by using rather loose general statements. The designer usually cannot know where the contractor eventually appointed will consider the construction joints should be made for optimum efficiency. The contractor may often agree the position of construction joints with the designer or ask the designer where he should make them.

Although expansion joints for temperature movement automatically relieve shrinkage locally to some extent, they do not relieve shrinkage generally. But shrinkage[2,19] is different from temperature movement in that the concrete moves but the steel does not, except for a small amount of compression strain induced in it, away from shrinkage cracks, by the concrete shrinking. Taking a shrinkage coefficient of 0.0005, from and with the reservations of Ref. 2, then for a reinforced concrete member of length 20 m, the concrete will want to move 10 mm, or 5 mm at each end (cf. a temperature movement of 3 mm cited at the beginning of this section) to avoid developing internal tensile stresses, and ideally the concrete should freely slide along the reinforcement. In practice, this cannot happen. The shrinkage coefficient of 0.0005 multiplied by a value of Young's modulus for concrete of,[2] say, $28\,000$ N/mm^2 (4.06×10^6 psi) would mean a tensile stress of 14 N/mm^2 (2030 psi) developing in the concrete if the reinforcement completely restricted its shrinkage. Now, the ultimate tensile stress of this concrete would be,[2] say 2.8 N/mm^2 (406 psi) so it would certainly crack. In practice, cracks occur periodically at right-angles to a reinforcement bar, the bond[2] stress between bar and concrete being nil at a crack (slip having occurred locally) and having a maximum value half-way between two cracks. Bond resistance depends upon various factors[2] and is greater for bars which mechanically bond to the concrete. (Such bars are now widely used in the U.K.) Then there is the problem of differential shrinkage. When a thin shell dries out in a drying wind and sunshine, the surface concrete shrinks very much faster than the concrete in contact with, say, a steel shutter, where the moisture will be well retained and thus this concrete may not shrink at all. This action can cause deep surface cracking, some of which may at some later date, after stripping the soffit shutter, penetrate the complete thickness of the shell.

The author recalls studying various mathematical works about 1948 concerning a differential element and partial differential equations for trying to predict shrinkage stresses, etc. These analyses have never given practically useful results nor have they agreed adequately with experimental behaviour. This is because of the already mentioned practical difficulties. For practical design purposes, therefore, the elastic shell analysis has been used ignoring temperature and shrinkage effects. Temperature and shrinkage effects have been dealt with adequately enough to obtain satisfactory shells in practice by detailing expansion and construction joints, respectively. It should be remembered that shell construction led the analytical and experimental work which the more academic would have deemed adequate to justify construction. This is certainly not an unusual way of making progress, judging from history and present practice in many technologies. Analyses can, of course, be concocted for both temperature and shrinkage effects but have not been necessary to produce satisfactory shell structures.

With all reinforced concrete structures, it is to some extent debatable where to place construction joints to do the least harm to the structure. Ideally, there should be no joints at all, yet they can help in reducing shrinkage stresses to some extent. What are the principles? Of first importance, a construction joint should be made where the shear stresses are the least and then, of second importance, where the bending moment is a minimum. In a beam, there are horizontal as well as vertical shear stresses. For economy of construction, a T-beam, where the top flange is an *in-situ* slab, is often jointed just below the slab, where the horizontal shear stresses are a maximum, hence suitable stirrups have to be present to take this horizontal shear which the joint cannot be relied upon to resist. Similarly, joints have to be made in shells at undesirable locations and suitable dowel bars used to make up for the weakness of the joint.

Sometimes, a joint may be not straight but provided with, say, a tongue to help key and resist shear.

When a joint is at a section of high bending moment, there is no difficulty as regards bending strength. In flexural compression, the joint causes no weakness, the surfaces just push together and take compression. In tension, the reinforcement takes the tension but the crack size will be increased at the joint. So, joints can be made where there are bending moments but ideally, if the bending moments are less, the crack size will be more favourable.

Shells tend to crack at changes of shape, e.g. at the junctions between shells and edge beams. Hence, shells are usually thickened[2,4] towards these junctions and construction joints made where the shells begin to thicken, well away from the junctions. Thus, an edge member is cast integral with the thickening, and starter bars protrude from the end of the cast to lap with the bars of the shell when it is cast and to act as dowel bars to strengthen the construction joint.

The shell itself will often be cast in portions to reduce shrinkage stresses and to enable construction. A thin shell can dry out very rapidly if a drying wind and/or strong sunshine occur just after casting. This results

328   Shell roofs

in a very high shrinkage in the liquid-cum-solid state and can cause very wide and deep cracks say 10 minutes after placing, i.e., well before the concrete is hard enough to cover with a polythene sheet. The author has experienced this kind of severe cracking in the U.K., just because the sun came out whilst trowelling.

Construction joints are placed along the length of edge beams, valley beams, end arches, etc., in the same way as for normal reinforced concrete construction, as described earlier in this section.

Figure 9.9 shows a suitable system of construction joints (broken lines) for a scheme of barrel shells.

The shuttering of shells is the most important constructional problem. Some pertinent points have already been made in Section 9.3 regarding the re-use of formwork. However, for accurate work (say to ± 3 mm), shell formwork needs to be strong and well made to withstand steel fixing and concreting operations without suffering distortion. This means that it is expensive and capable of lasting for many jobs (hired formwork is an extreme example of multiple re-use). Therefore, generally the objective is to obtain as many uses as possible out of each shutter. Again, as described in Section 9.3 the author has found the use of specialist shuttering sub-contractors economic.

One such system comprises T-shaped steelwork sections which, for a barrel shell, are bent to the required radius at the works and then sent to the site together with the necessary steel plates. These plates are laid on the T-sections so that adjacent plates abut on the top flange of each T-section and at their unsupported sides. After casting, a small whisker of mortar between each plate pulls off the soffit; these locations can be improved by minimum rubbing with a carborundum stone. This soffit is good enough for emulsion or distemper painting and the slight pattern of the plates is aesthetically pleasing. Another system of hired shuttering uses plywood instead of steel plates.

With a scheme of barrel shells (Figure 9.9), it is possible to cast the first shell and the first valley, terminating the concreting at the first longitudinal construction joint in the second shell just after the first valley. The framework of the shuttering needs to be in place for the second shell but it only requires a portion of the soffit sheeting. Then, when the concrete is sufficiently hard, the shuttering from the first shell can be moved to the third shell and the soffit sheeting removed from it to complete the shuttering of the second shell and to shutter a portion of the third shell. Then, the second cast completes the second shell and casts a portion of the

Construction

third shell. Whilst the operation is continued, it is necessary always to keep the last cast valley propped. This is a way to maximize the use of the shuttering for multi-barrel shells. With this scheme, when the end stiffeners have horizontal soffits level with the valley soffits, the deep inner shutters to the sides of the end stiffeners can be lifted out of the large, long strips voids required for metal-framed lantern-lights at the crown. But if the lights are individual – see, for example, frontispiece photograph in Ref. 4 or Plate 3 in Ref. 3 – and say about 1.2 m square or circular, then this is not possible and the shutter has to be made in three or four pieces which will unfasten easily so that they can be man-handled beneath the cast shell into their next position.

For short wide shells continuous in the span direction, moving shutters are sometimes used. This was done over a turbine house in the U.K. where movable steel shuttering used crane rails which were supported by reinforced concrete beams supported by reinforced concrete brackets from the columns. These crane rails were later to support an overhead crane. As the headroom was considerable, this arrangement saved an enormous amount of scaffolding. At each support, there was an arch, therefore the shell shutter had to be lowered below this, moved forward and then jacked up again.

Sometimes, shells continuous in the longitudinal direction are favoured with upstand diaphragms at the supports, so that a movable shutter can very easily slide forward with the minimum difficulty.

In the early days of shells in the U.K., cylindrical shells were favoured because the shuttering was straight in one direction and to the simplest curve for setting out – namely a circle – in the other direction. They were also thought to be much simpler than doubly-curved shells to construct and design.

In Mexico, Felix Candella created many architecturally very attractive structures with hypars. He appeared to demonstrate that many of these were doubly-curved shells formed by straight planks supported on straight boundaries, had the strength advantage of doubly-curved shells over singly-curved cylindrical shells and were adequately designed by very simple analyses, much simpler than had been developed for cylindrical shells.

Although the shuttering comprises straight planks, these need to be tapered. Ripping down the whole length of each plank is very expensive. This cost can be reduced by tapering more severely only one in every group of four or five planks. However, after this kind of procedure had been going on for some years in the U.K., the author came across one contractor who preferred to shutter hypars the curved way for economic reasons, i.e., obliterating the principal economic reason for using hypars in the U.K.

If the valley of a cylindrical shell were cast strictly horizontally, for optical reasons it would appear to sag. There would be, of course, a slight deflection due to self-weight, but it would be exaggerated visually. Aesthetically this is undesirable, so it is best to calculate the total deflection under live and dead loading and to give the valley shutters slightly more camber than this quantity. Often a rule-of-thumb guide can be used, such as that given in Table 9.1, where OCS and NCS are abbreviations for ordinary and north-light cylindrical shells, respectively.

Casting concrete shells is more difficult and consequently less economic as the surfaces are steeper. In the U.K., a 1:1.15:3 mix by dry volumes of cement:sand:gravel (10 mm down, i.e. maximum size of particles is 10 mm) has been found to be satisfactory for shells. The concrete is built up from the lower parts of a shell, full thickness as one goes, until the crown is reached. The concrete is as dry as possible consistent with full compaction. Poker and sausage vibrators may be put in the concrete to help the placing around reinforcement and in difficult corners, etc., but these will only be able to compact by vibration a very small amount of the concrete. So, essentially the concrete will have to compact itself easily under the weight of patting with a shovel and trowelling only. To compact easily, it must not be too dry. As the concrete is built up the slopes, wetness will rise to the top and it may be possible to mix the concrete drier because it will receive wetness from the already laid concrete when worked into position. Should it unexpectedly rain during concreting, the concrete added toward completion may need to be of an exceedingly dry mix because, by the time the rain has mixed with it during placing and wetness has risen from the previous concrete, the freshly laid concrete is adequate in water content. The shell may contain two layers of fabric reinforcement, one near each face. To prevent the concrete flowing easily down the shell

Table 9.1.

Span, m	15	18	24
OCS camber, mm	25	35	50
NCS camber, mm	35	50	75

slopes, extra 'spacer bars'[19] are used which not only help the casting but also help the fixing of the fabrics, strengthen the end laps of the fabrics and act as bars to help distribute cracks due to shrinkage and temperature, i.e. encourage a large number of smaller cracks. For maximum bulk for cost, the author found that the square twisted bar was the most economic and had a mechanical bonding action. A guide to the choice of spacer bars[19] for cylindrical shells is as follows: for shells up to a span of 18 m, use 8 mm square twisted (or 10 mm diameter) bars at 380 mm centres; for shells of longer spans, use 10 mm square twisted (or 12 mm diameter) bars at 380 mm centres. Spacer bars are briefly mentioned in Refs. 2 and 3. The edge and stiffener beams and columns are cast with the same type of specification as non-shell work.

Reinforcement fixing tends to be thought of as difficult because of the sloping curved shell surfaces. However, the cost of steel fixing is very small compared to the total cost of the contract and the difference in cost between fixing reinforcement on shell surfaces and to, say, normal slabs is not very great. Consequently, the overall cost effect is fairly insignificant in the U.K. with skilled fixers.

Thermal insulation on top of the shell is, in the U.K., usually either 50 mm thick vermiculite concrete or 25 mm thick corkboard. A good practical tip with the former is to waterproof the surface with about 3 mm thick layer of cement–sand. This is because, if it rains before the roofing felt is stuck to its surface, and this is not waterproofed, the vermiculite concrete will be saturated. Then, in hot weather, the water evaporates, causing the roofing felt to bubble and become unstuck. In any case, every precaution is taken to lay the roofing felt during a dry period – but this can be difficult in unpredictable climates, such as the U.K.'s.

Corkboard does not have this trouble. It is stuck down with bitumastic by those also laying the roofing felt. The felting follows immediately. All this is effected rapidly when the concrete shell is dry.

The author has found that laying corkboard is to the benefit of the subcontractor, and that laying vermiculite rather than corkboard is more economic for the concrete contractor because it dispenses with the subcontractor's services. Even to a consultant specifying the respective duties of the contractors, the corkboard tends to be more expensive overall, but this should be checked as prices vary considerably from time to time.

9.7 Generalized membrane analysis

The reasons for desiring a membrane analysis are described in Section 9.5. It is usual – see for example Refs. 5, 23, 24 and 25 – to consider three membrane analyses: (a) cylindrical shells, (b) shells of revolution (e.g. those used for spherical domes), and (c) shells of arbitrary shape (e.g. those used for hypars and conoids).

9.7.1 Cylindrical shells

The membrane analysis is only part of the total analysis for cylindrical shells (see Section 9.5) and can very rarely be used on its own. This analysis assumes that Poisson's ratio is zero.

Figure 9.10 shows a differential element of the surface of a cylindrical shell roof. Consider the longitudinal force per unit length of shell, or 'stress resultant', N_x at $x = x$; at $x = x + dx$, this increases to

$$N'_x = N_x + \frac{\partial N_x}{\partial x} dx$$

Similarly, $N_{\varphi x}$ at $\varphi = \varphi$ increases to

$$N'_{\varphi x} = N_{\varphi x} + \frac{\partial N_{\varphi x}}{\partial \varphi} d\varphi$$

at $\varphi = \varphi + d\varphi$. Similarly, for the other stress resultants shown in Figure 9.10.

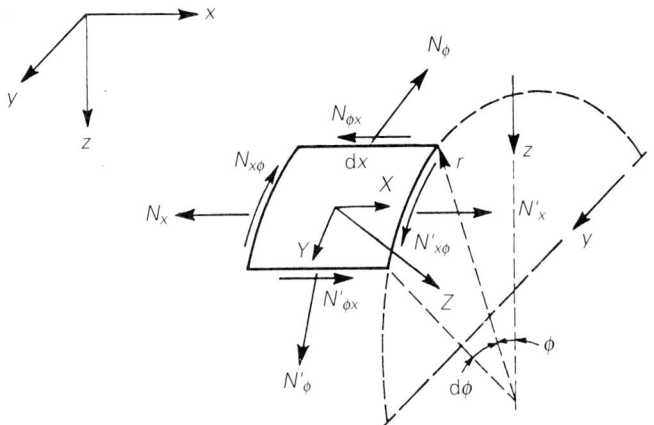

Figure 9.10 Differential element of the surface of a cylindrical shell roof.

Generalized membrane analysis

The total loading on the element has components X, Y and Z, as shown in Figure 9.10, all taken as forces per unit area. (X and Y are in, and Z normal to, the plane of the element.) Resolving in the x-direction,

$$\left(N_x + \frac{\partial N_x}{\partial x} dx\right) r\, d\varphi - N_x r\, d\varphi + \left(N_{\varphi x} + \frac{\partial N_{\varphi x}}{\partial \varphi} d\varphi\right) dx$$

$$- N_{\varphi x} dx + Xr\, d\varphi\, dx = 0 \quad (9.1)$$

where, as N_x is a force per unit length and the length of the side of the element upon which it acts is $r\, d\varphi$, it gives a force on this side of the element of $N_x r\, d\varphi$, and as $N_{\varphi x}$ is a force per unit length and it acts on the side of the element of length dx, then it gives a force on this side of $N_{\varphi x} dx$; and similarly for the other stress resultants; again X is force per unit area and the area of the element is $r\, d\varphi\, dx$, so this gives a total force of $Xr\, d\varphi\, dx$.

Cancelling like terms and dividing throughout by $r\, d\varphi\, dx$ gives

$$\frac{\partial N_x}{\partial x} + \frac{1}{r}\frac{\partial N_{\varphi x}}{\partial \varphi} + X = 0 \quad (9.2)$$

Figure 9.11 Forces to be considered when resolving perpendicularly to the plane of an element.

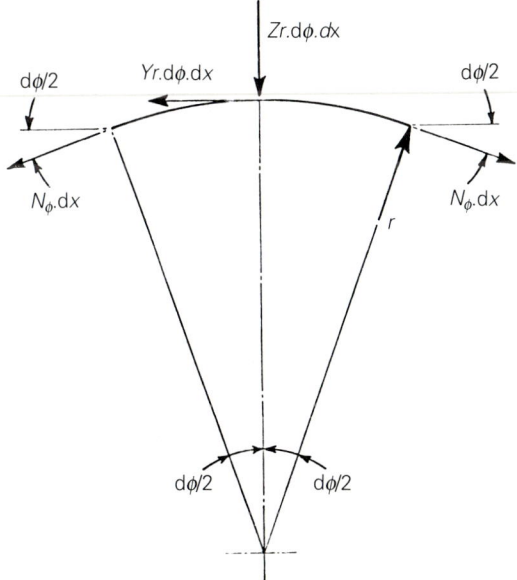

Resolving in the y-direction gives a similar equation to equation 9.1 which, with cancelling and dividing throughout by $r\,d\varphi\,dx$, gives

$$\frac{\partial N_{x\varphi}}{\partial x}+\frac{1}{r}\frac{\partial N_{\varphi}}{\partial \varphi}+Y=0 \tag{9.3}$$

Figure 9.11 shows the forces to be considered when resolving perpendicularly to the plane of the element, namely,

$$\left(N_{\varphi}+\frac{\partial N_{\varphi}}{\partial \varphi}d\varphi\right)dx\sin(\tfrac{1}{2}d\varphi)+N_{\varphi}\,dx\sin(\tfrac{1}{2}d\varphi)+Zr\,d\varphi\,dx=0$$

In the limit, as $d\varphi \to 0$, $\sin(\tfrac{1}{2}d\varphi) \to \tfrac{1}{2}d\varphi$. Therefore

$$N_{\varphi}\,dx\,d\varphi+\frac{1}{2}\frac{\partial N_{\varphi}}{\partial \varphi}dx(d\varphi)^{2}+Zr\,d\varphi\,dx=0$$

Therefore

$$N_{\varphi}+\frac{1}{2}\frac{\partial N_{\varphi}}{\partial \varphi}d\varphi+Zr=0$$

In the limit $\partial N_{\varphi}/\partial \varphi \to 0/0$, which is neither 0 nor ∞ but a normal quantity, and $d\varphi \to 0$, therefore

$$N_{\varphi}+Zr=0 \tag{9.4}$$

Then, taking moments about the normal to the centroid of the element,

$$(N_{x\varphi}r\,d\varphi)\tfrac{1}{2}dx+(N'_{x\varphi}r\,d\varphi)\tfrac{1}{2}dx=(N_{\varphi x}\,dx)\tfrac{1}{2}r\,d\varphi+(N'_{\varphi x}\,dx)\tfrac{1}{2}r\,d\varphi$$

Dividing through by $\tfrac{1}{2}r\,d\varphi\,dx$ gives

$$N_{x\varphi}+\left(N_{x\varphi}+\frac{\partial N_{x\varphi}}{\partial x}dx\right)=N_{\varphi x}+\left(N_{\varphi x}+\frac{\partial N_{\varphi x}}{\partial \varphi}d\varphi\right)$$

In the limit, $\partial N_{x\varphi}/\partial x$ and $\partial N_{\varphi x}/\partial \varphi \to 0/0$ and dx and $d\varphi \to 0$, therefore

$$N_{x\varphi}=N_{\varphi x} \tag{9.5}$$

which is similar to the well-known equality of complementary shear stresses for two dimensional work, e.g. beams. The four equations 9.2 to 9.5 can be solved for the four stress resultants N_x, N_{φ}, $N_{x\varphi}$ and $N_{\varphi x}$.

With regard to longitudinal strain, the element of length dx (Figure 9.12) has a longitudinal displacement of u at $x=x$ and of $u+(\partial u/\partial x)\,dx$ at $x=x+dx$. Hence, the change in length is $(\partial u/\partial x)\,dx$ and the strain

$\partial u/\partial x$. If the shell thickness is h, the corresponding stress is N_x/h. Therefore, using Hooke's law and assuming Poisson's ratio is zero,

$$\partial u/\partial x = N_x/(Eh) \tag{9.6}$$

where E is Young's modulus for the concrete.

Figure 9.12 shows the element distorted by shear stress. From this figure, the shear strain is

$$\frac{1}{r}\frac{\partial u}{\partial \varphi} + \frac{\partial v}{\partial x}$$

As the elastic shear modulus is equal to $E/2$ and the shear stress is $N_{\varphi x}/h$ or $N_{x\varphi}/h$ then, applying Hooke's law and assuming Poisson's ratio is zero,

$$\frac{1}{r}\frac{\partial u}{\partial \varphi} + \frac{\partial v}{\partial x} = 2\frac{N_{\varphi x}}{Eh} \tag{9.7}$$

The tangential strain can be considered by reference to Figure 9.13. The increase in length of the element due to the tangential displacements v and $v+(\partial v/\partial \varphi)\,d\varphi$ of its ends is $(\partial v/\partial \varphi)\,d\varphi$. Because of the radial displacements, w, of these ends (neglecting the very small difference between them), the length of the element decreases by an amount $r\,d\varphi - (r-w)\,d\varphi = w\,d\varphi$.

Figure 9.12 Longitudinal strain.

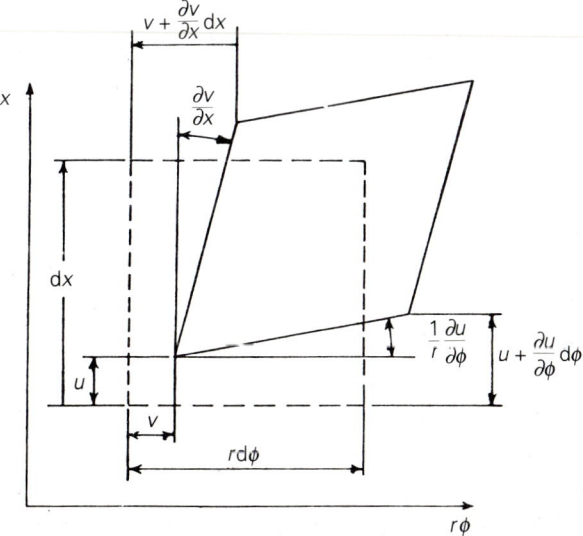

Thus, the total change in length of the element in the tangential direction is

$(\partial v/\partial \varphi)\, d\varphi - w\, d\varphi$

Dividing this by the initial length $r\, d\varphi$, the strain is

$$\frac{1}{r}\frac{\partial v}{\partial \varphi} - \frac{w}{r}$$

The tangential stress is N_φ/h, hence, assuming Poisson's ratio is zero and applying Hooke's law,

$$\frac{1}{r}\frac{\partial v}{\partial \varphi} - \frac{w}{r} = \frac{N_\varphi}{Eh} \qquad (9.8)$$

Now we have to explain the importance of using load harmonics (i.e. expressing loads and forces in terms of Fourier's series). It is difficult to explain to most engineers why this has to be done when a uniformly distributed load seems so simple to deal with. Indeed, beginners designing shells often just accept the use of load harmonics without really understanding why this procedure is necessary.

According to Evans and Wilby[19], under the heading Load harmonics: 'To solve the various mathematical equations employed in the design of shells it is necessary to express the loading on the shell in terms of a Fourier's series'.

And Chronowicz[26] states: 'The analysis of the deformations of the

Figure 9.13 Tangential strain.

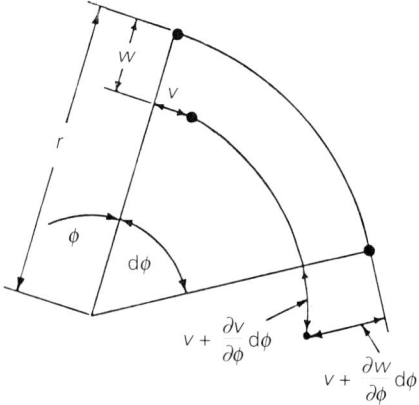

membrane is considerably simplified if the uniformly distributed loading is expressed by Fourier's series'.

Gould[5] maintains: 'Since we have a partial differential equation to solve, we try the standard method of separation of variables. Specifically, we apply the Fourier series technique, whereby all loadings and dependent variables are taken in the form of a Fourier's series'.

According to the ASCE's manual on the design of concrete shell roofs[18]:

> Although it is possible to obtain the internal forces produced by any surface loads, because of mathematical difficulties, the corrective line loads [see Section 9.5] applied along the longitudinal edges are expressible only as a Fourier's sine or cosine series.... To avoid confusion in the application of the line loads [see Section 9.5]... it is expedient to regard the surface load on the shell as the sum of partial loads (i.e. terms of a Fourier series)... comparing the values obtained by the partial loads with those obtained by the uniform load, it is seen that good agreement is obtained for the longitudinal force N_x. The shear and transverse force components obtained by the sum of the first two sinusoidal loads are approximately 10% and 15% too small. In designing shells by partial loads, this discrepancy should be borne in mind and adjustments should be made to the final values.

Tottenham[27] states:

> The basic differential equation for the edge load problem [see Section 9.5] is only soluble by considering the edge load in the form of a Fourier series.... Fortunately the effect of the first term is predominant and this is the only one considered here.... The loads we have applied have been the first term in the Fourier series for unity.

According to Gibson and Cooper[28]:

> To be of direct operational use the uniformly distributed loading must be put in the form of a Fourier half range series... only the first term of the Fourier series will be utilised as this dominant.... This (latter) tends to slightly overestimate the magnitude of the moments and consequently errs on the side of safety.

Essentially, doubly integrating a cosine gives the same cosine. This does not happen with, for example, an algebraic polynomial; the double integral becomes more complex and unlike the original expression. This device of using Fourier's series has been used for solving many elastic analyses, e.g. folded plates, grillages, columns with lateral loading, soil foundation problems, problems in continuum mechanics, etc.

Continuing the quotation from Ref. 19:

> In practice (in the U.K.) shells are often designed to carry their self weight and a snow load of 0.75 kN/m² (16 psf) of curved surface area. This is slightly more conservative than taking the snow as 0.75 kN/m² of plan area, but simplifies the design, as all the loading can then be considered as per unit of curved surface area. Thus, assuming the origin of x to be at the centre of the span l the loading can be expressed as:
>
> $$p\frac{4}{\pi}[\cos(\pi x/l) - \tfrac{1}{3}\cos(3\pi x/l) + \tfrac{1}{5}\cos(5\pi x/l)\ldots] \quad (9.9)$$
>
> where p is the total load on the shell per unit of curved surface area.
>
> Experience shows that generally it is satisfactory to consider only the first term of this series; hence the loading is considered as $(4/\pi)p\cos(\pi x/l)$ per unit area of the surface.

For the differential element, the vertical loading on a unit of area is thus $(4/\pi)p\cos(\pi x/l)$. On the same unit of area, the component perpendicular to the element (Figure 9.10) is

$$Z = (4/\pi)p\cos(\pi x/l)\cos\varphi \quad (9.10)$$

and on the same unit of area the component tangential to the element (Figure 9.10) is

$$Y = (4/\pi)p\cos(\pi x/l)\sin\varphi \quad (9.11)$$

Substituting these values in equation 9.4 gives

$$N_\varphi = -r(4/\pi)p\cos(\pi x/l)\cos\varphi \quad (9.12)$$

Inserting this value of N_φ in equation 9.3 gives

$$(\partial N_{\varphi x}/\partial x) + (4/\pi)p\cos(\pi x/l)\sin\varphi + \frac{4}{\pi}p\cos(\pi x/l)\sin\varphi = 0$$

Therefore,

$$(\partial N_{\varphi x}/\partial x) = -(8/\pi)p\cos(\pi x/l)\sin\varphi \quad (9.13)$$

Generalized membrane analysis

Many designers, Refs. 19 and 27, use $N_{\varphi x}$ in this form, i.e. as a stress, rather than as $N_{\varphi x}$. However, integrating equation 13 gives

$$N_{\varphi x} = -2\frac{1}{\pi}\frac{4}{\pi}p\sin(\pi x/l)\sin\varphi + C_1(\varphi) \tag{9.14}$$

where $C_1(\varphi)$ is a function of φ, but an arbitrary constant as far as x is concerned. Then, equation 9.2, at $X = 0$, gives

$$\frac{\partial N_x}{\partial x} - \frac{2}{r}\frac{1}{\pi}\frac{4}{\pi}p\sin(\pi x/l)\cos\varphi + \frac{1}{r}\frac{\partial}{\partial\varphi}C_1(\varphi) = 0$$

Therefore,

$$N_x = -\frac{2}{r}\frac{l^2}{\pi^2}\frac{4}{\pi}p\cos\frac{\pi x}{l}\cos\varphi - \frac{x}{r}\frac{\partial}{\partial\varphi}C_1(\varphi) + C_2(\varphi) \tag{9.15}$$

where $C_2(\varphi)$ is a function of φ, but an arbitrary constant as far as x is concerned.

In the case of a symmetric shell, when $\varphi = 0$, i.e. at the crown, $N_{\varphi x}$ must be zero as it is an asymmetric (or antisymmetric) force. Therefore, from Eqn 9.14,

$$C_1(\varphi) = 0 \tag{9.16}$$

In the case of a simply supported shell, the longitudinal forces, N_x, must be zero at the ends, i.e. where $x = \pm l/2$, therefore equation 9.15 gives

$$C_2(\varphi) = 0 \tag{9.17}$$

From equations 9.6 and 9.15, using also equations 9.16 and 9.17,

$$\frac{\partial u}{\partial x} = -\frac{2}{Ehr}\frac{l^2}{\pi^2}\frac{4}{\pi}p\cos(\pi x/l)\cos\varphi$$

Therefore,

$$u = -\frac{2}{Ehr}\frac{l^3}{\pi^3}\frac{4}{\pi}p\sin(\pi x/l)\cos\varphi + C_3(\varphi) \tag{9.18}$$

where $C_3(\varphi)$ is an arbitrary constant as far as x is concerned. At the centre of the simply supported shell, when $x = 0$, u must be zero, therefore, from equation 9.18,

$$C_3(\varphi) = 0 \tag{9.19}$$

340   Shell roofs

From equations 9.7, 9.18 and 9.14, using also equations 9.16 and 9.19,

$$\frac{1}{r^2}\frac{2}{Eh}\frac{4}{\pi}p\frac{l^3}{\pi^3}\sin(\pi x/l)\sin\varphi + \frac{\partial v}{\partial x} = -\frac{4}{Eh}\frac{1}{\pi}\frac{4}{\pi}p\sin(\pi x/l)\sin\varphi$$

Therefore,

$$\frac{\partial v}{\partial x} = -\frac{2}{Eh}\frac{l}{\pi}\frac{4}{\pi}p\left(2 + \frac{1}{r^2}\frac{l^2}{\pi^2}\right)\sin(\pi x/l)\sin\varphi \tag{9.20}$$

whence,

$$v = \frac{2}{Eh}\frac{l^2}{\pi^2}\frac{4}{\pi}p\left(2 + \frac{1}{r^2}\frac{l^2}{\pi^2}\right)\cos(\pi x/l)\sin\varphi + C_4(\varphi) \tag{9.21}$$

where $C_4(\varphi)$ is an arbitrary constant as far as x is concerned. At the end of the simply supported shell, when $x = l/2$, the end diaphragm, or stiffening beam or frame, restrains the shell, so that v is zero. Therefore, from equation 9.21,

$$C_4(\varphi) = 0 \tag{9.22}$$

From equations 9.8, 9.21 and 9.12

$$\frac{2}{Ehr}\frac{l^2}{\pi^2}\frac{4}{\pi}p\left(2 + \frac{1}{r^2}\frac{l^2}{\pi^2}\right)\cos(\pi x/l)\cos\varphi - \frac{w}{r} = -\frac{r}{Eh}\frac{4}{\pi}p\cos(\pi x/l)\cos\varphi$$

Therefore,

$$w = \frac{2}{Eh}\frac{l^2}{\pi^2}\frac{4}{\pi}p\left(2 + \frac{1}{r^2}\frac{l^2}{\pi^2}\right)\cos(\pi x/l)\cos\varphi + \frac{r^2}{Eh}\frac{4}{\pi}p\cos(\pi x/l)\cos\varphi$$

whence,

$$w = \frac{1}{Eh}\frac{4}{\pi}p\left[r^2 + 2\frac{l^2}{\pi^2}\left(2 + \frac{1}{r^2}\frac{l^2}{\pi^2}\right)\right]\cos\varphi\cos(\pi x/l) \tag{9.23}$$

Thus, using the first term of equation 9.9 for the loading, equations 9.12 to 9.15, 9.18, 9.21 and 9.23 – using also equations 9.16, 9.17, 9.19 and 9.22 for the arbitrary constants (all zero) – give the values of N_φ, $\partial N_{\varphi x}/\partial x$, $N_{\varphi x}$, N_x, u, v and w, respectively. Allowing for the different notation and the opposite sign of $N_{\varphi x}$ between the sign conventions, these equations are exactly the same as those given by Tottenham.[27]

9.7.2 Shells of revolution

A surface of revolution or rotation (see also Section 9.2.2) is formed by rotating a plane curve, the 'generating curve' or 'meridian', about an axis (here considered vertical for convenience) in the plane of the curve, the 'meridian plane'. A differential element is shown in Figure 9.14; its sides are off meridians and its top and bottom are off horizontal circles, located by angles θ (the 'meridian angle') and φ, respectively, φ being the angle between the normal and the axis of rotation. The planes containing the element's principal curvatures, of radii r_1 and r_2, are the meridian plane and the plane perpendicular to it, respectively. The radius of the horizontal circle shown is r_0. Therefore, the lengths of the sides of the element are $r_1 \, d\varphi$ and $r_0 \, d\theta = r_2 \sin \varphi \, d\theta$ and its surface area is $r_0 r_1 \, d\varphi \, d\theta$ or $r_1 r_2 \sin \varphi \, d\varphi \, d\theta$.

The main loading on, say, a reinforced concrete dome comprises its self-weight and snow load, which together can be considered as a uniform load per unit of surface area (Section 9.5). Because of the assumed symmetry of loading, there will be no shearing forces, as these are asymmetric forces, on the sides of the element.

As shown in Figure 9.14, N_φ and N_θ are forces per unit length, and X and Y are external loads per unit area, all tangential to the element, whilst Z is the external loading per unit area normal to the element.

Figure 9.14 Differential element of a surface revolution.

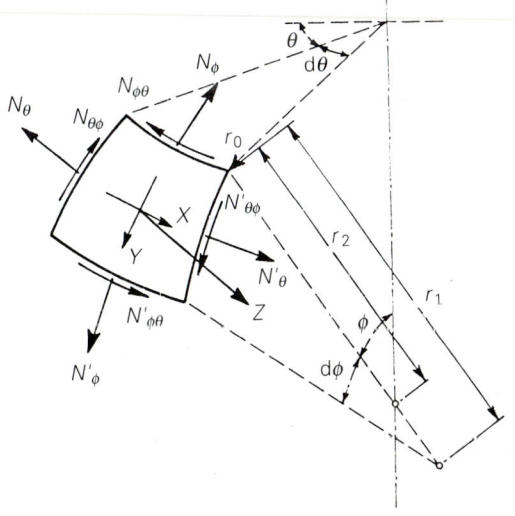

Shell roofs

Resolving the forces for the element in the direction of the tangent to the meridian, let us first consider the contribution due to N_φ:

$$-N_\varphi r_0 \, d\theta + \left(N_\varphi + \frac{\partial N_\varphi}{\partial \varphi} d\varphi\right)\left(r_0 + \frac{\partial r_0}{\partial \varphi} d\varphi\right) d\theta$$

$$= N_\varphi \frac{\partial r_0}{\partial \varphi} d\varphi \, d\theta + r_0 \frac{\partial N_\varphi}{\partial \varphi} d\varphi \, d\theta + \frac{\partial r_0}{\partial \varphi} \frac{\partial N_\varphi}{\partial \varphi} (d\varphi)^2 \, d\theta$$

Eventually, this will be divided by $d\varphi \, d\theta$, then the quantities $\partial r_0/\partial \varphi$, $\partial N_\varphi/\partial \varphi$ become in the limit zero/zero, that is normal quantities, e.g. rate of change of radius, rate of change of N_φ, but the last term of the above expression will contain $d\varphi$, which in the limit is zero. Therefore, the above becomes

$$N_\varphi \frac{\partial r_0}{\partial \varphi} d\varphi \, d\theta + r_0 \frac{\partial N_\varphi}{\partial \varphi} d\varphi \, d\theta = \frac{\partial}{\partial \varphi}(N_\varphi r_0) \, d\varphi \, d\theta$$

The contribution due to N_θ is a force $N_\theta r_1 \, d\varphi$ tangential to the horizontal circle of radius r_0. Resolving this and its incremental force

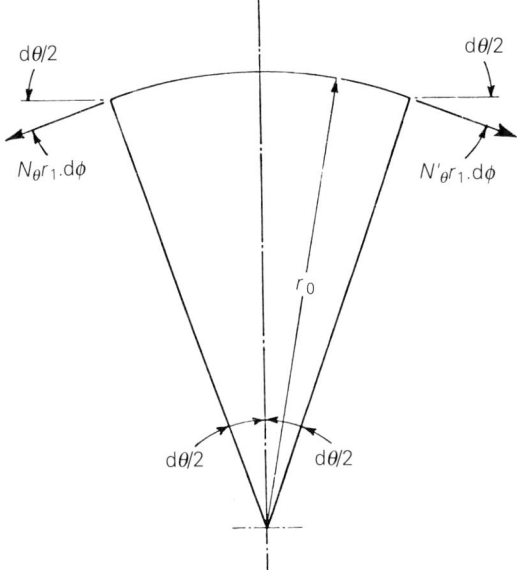

Figure 9.15. Tangential forces acting on element of surface of revolution.

Generalized membrane analysis

$N'_\theta (= N_\theta + (\partial N_\theta/\partial\theta) \, d\theta)$ at the other end of the element (Figure 9.15) in a horizontal direction gives

$$N_\theta r_1 \, d\varphi \sin(\tfrac{1}{2}d\theta) + N'_\theta r_1 \, d\varphi \sin(\tfrac{1}{2}d\theta)$$

and as in the limit when $d\theta \to 0$, $\sin(\tfrac{1}{2}d\theta) \to \tfrac{1}{2}d\theta$ this becomes

$$N_\theta r_1 \, d\varphi \frac{d\theta}{2} + \left(N_\theta + \frac{\partial N_\theta}{\partial \theta} d\theta \right) r_1 \, d\varphi \frac{d\theta}{2} = N_\theta r_1 \, d\varphi \, d\theta + \frac{r_1}{2} \frac{\partial N_\theta}{\partial \theta} d\varphi (d\theta)^2$$

Eventually, this expression will be divided by $d\theta \, d\varphi$, then the quantity $\partial N_\theta/\partial \theta$ becomes in the limit zero/zero, i.e. a normal quantity (the rate of change of N_θ with respect to θ), but the last term contains $d\theta$, which in the limit is zero. Therefore, the above expression becomes

$$N_\theta r_1 \, d\varphi \, d\theta$$

Then, referring to Figure 9.16, the component of this force in the direction of the tangent to the meridian is

$$(N_\theta r_1 \, d\varphi \, d\theta) \cos \varphi$$

The contribution due to $N_{\theta\varphi}$ is

$$-N_{\theta\varphi} r_1 \, d\varphi + \left(N_{\theta\varphi} + \frac{\partial N_{\theta\varphi}}{\partial \theta} d\theta \right) r_1 \, d\varphi = \frac{\partial N_{\theta\varphi}}{\partial \theta} r_1 \, d\theta \, d\varphi$$

Figure 9.16

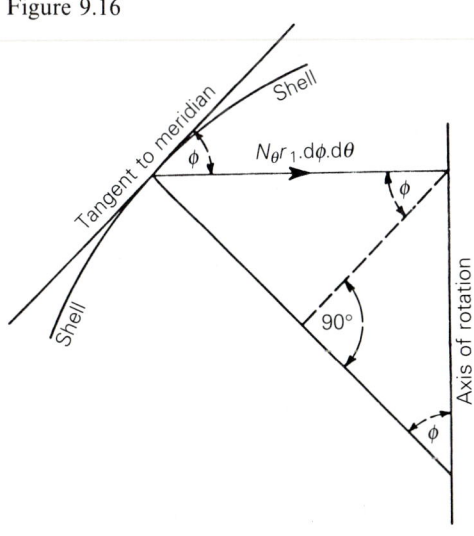

Adding the above forces and the component of the external force in the direction of the tangent to the meridian gives

$$\frac{\partial}{\partial \varphi}(N_\varphi r_0) \, d\varphi \, d\theta - N_\theta r_1 \cos\varphi \, d\varphi \, d\theta + \frac{\partial N_{\theta\varphi}}{\partial \theta} r_1 \, d\theta \, d\varphi + Y r_0 r_1 \, d\varphi \, d\theta = 0$$

Therefore,

$$\frac{\partial}{\partial \varphi}(N_\varphi r_0) - N_\theta r_1 \cos\varphi + \frac{\partial N_{\theta\varphi}}{\partial \theta} r_1 + Y r_0 r_1 = 0 \qquad (9.24)$$

Resolving the forces for the element tangentially to the horizontal circle of radius r_0, we see the contribution due to N_θ is

$$-N_\theta r_1 \, d\varphi + \left(N_\theta + \frac{\partial N_\theta}{\partial \theta} d\theta \right) r_1 \, d\varphi = \frac{\partial N_\theta}{\partial \theta} r_1 \, d\theta \, d\varphi$$

The contribution due to $N_{\varphi\theta}$ — a force of $N_{\varphi\theta} r_0 \, d\theta$ in the tangential direction to the horizontal circle of radius r_0 — is

$$-N_{\varphi\theta} r_0 \, d\theta + \left(N_{\varphi\theta} + \frac{\partial N_{\varphi\theta}}{\partial \varphi} d\varphi \right)\left(r_0 + \frac{\partial r_0}{\partial \varphi} d\varphi \right) d\theta$$

$$= N_{\varphi\theta} \frac{\partial r_0}{\partial \varphi} d\varphi \, d\theta + r_0 \frac{\partial N_{\varphi\theta}}{\partial \varphi} d\varphi \, d\theta + \frac{\partial N_{\varphi\theta}}{\partial \varphi} \frac{\partial r_0}{\partial \varphi} d\varphi \, (d\varphi)^2$$

Eventually, this expression will be divided by $d\theta \, d\varphi$, then the last term contains $d\varphi$, which in the limit is zero. Therefore, the expression becomes

$$\frac{\partial}{\partial \varphi}(r_0 N_{\varphi\theta}) \, d\varphi \, d\theta$$

There is a contribution due to $N_{\theta\varphi}$ because the horizontal components of $N_{\theta\varphi} r_1 \, d\varphi$ and $N'_{\theta\varphi} r_1 \, d\varphi$, namely, $N_{\theta\varphi} r_1 \, d\varphi \cos\varphi$ and $N'_{\theta\varphi} r_1 \, d\varphi \cos\varphi$ (Figure 9.16) contains a horizontal angle $d\theta$ (Figure 9.14). Therefore, the contribution is

$$\left(N_{\theta\varphi} + \frac{\partial N_{\theta\varphi}}{\partial \theta} d\theta \right) r_1 \, d\varphi \cos\varphi \sin d\theta$$

In the limit, when $d\theta \to 0$, $\sin d\theta \to d\theta$, and as the above expression will be eventually divided by $r_1 \, d\varphi$, the second term will then contain $d\theta$, which in the limit becomes zero. Therefore, the expression becomes

$$N_{\theta\varphi} r_1 \, d\varphi \, d\theta \cos\varphi$$

Generalized membrane analysis

Adding the above forces and the component of the external force in the direction tangential to the horizontal circle of radius r_0 gives

$$\frac{\partial N_\theta}{\partial \theta} r_1 \, d\theta \, d\varphi + \frac{\partial}{\partial \varphi} r_0 N_{\varphi\theta} \, d\varphi \, d\theta + N_{\theta\varphi} r_1 \, d\varphi \, d\theta \cos\varphi + X r_0 r_1 \, d\varphi \, d\theta = 0$$

Therefore,

$$\frac{\partial N_\theta}{\partial \theta} r_1 + \frac{\partial}{\partial \varphi}(r_0 N_{\varphi\theta}) + N_{\theta\varphi} r_1 \cos\varphi + X r_0 r_1 = 0 \tag{9.25}$$

Resolving the forces for the element normal to its plane, let us first consider the contribution due to N_φ. A figure similar to Figure 9.15 is relevant thus resolving normal to the element gives

$$(N_\varphi r_0 \, d\theta) \sin(\tfrac{1}{2} d\varphi) + \left(N_\varphi + \frac{\partial N_\varphi}{\partial \varphi} d\varphi\right)\left(r_0 + \frac{\partial r_0}{\partial \varphi} d\varphi\right) d\theta \sin(\tfrac{1}{2} d\varphi)$$

In the limit when $d\varphi \to 0$, $\sin(\tfrac{1}{2} d\varphi) \to \tfrac{1}{2} d\varphi$, so this expression becomes

$$N_\varphi r_0 \, d\theta \, d\varphi + \frac{N_\varphi}{2} \frac{\partial r_0}{\partial \varphi} d\theta (d\varphi)^2 + r_0 \frac{\partial N_\varphi}{\partial \varphi} d\theta (d\varphi)^2$$

Eventually, this expression will be divided by $d\theta \, d\varphi$, then the quantities $\partial r_0/\partial \varphi$ and $\partial N_\varphi/\partial \varphi$ become in the limit zero/zero, i.e. normal quantities (rates of change of r_0 and N_φ respectively with respect to φ) but the last two terms will contain $d\varphi$, which in the limit is zero. Therefore, the above expression becomes

$$N_\varphi r_0 \, d\theta \, d\varphi$$

The contribution due to N_θ – a force of $N_\theta r_1 \, d\varphi$ tangential to the horizontal circle of radius r_0 – gives, as previously, a resultant force of $N_\theta r_1 \, d\varphi \, d\theta$ in a horizontal direction. Referring to Figure 9.16 the component to the generating curve is

$$(N_\theta r_1 \, d\varphi \, d\theta) \sin\varphi$$

Adding the above forces and the component of the external force in the direction of the normal to the plane of the element gives

$$N_\varphi r_0 \, d\theta \, d\varphi + N_\theta r_1 \, d\varphi \, d\theta \sin\varphi + Z r_0 r_1 \, d\varphi \, d\theta = 0$$

Therefore,

$$N_\varphi r_0 + N_\theta \sin\varphi + Z r_0 r_1 = 0 \tag{9.26}$$

Taking moments about the normal to the centre of the element, or because of complementary shear stresses,

$$N_{\theta\varphi} = N_{\varphi\theta} \tag{9.27}$$

Thus, equations 9.24 to 9.27 inclusive enable N_φ, N_θ, $N_{\theta\varphi}$ and $N_{\varphi\theta}$ to be obtained. Allowing for very slightly different notation, these equations are the same as those derived by Flügge[23], who also applies them to various loading conditions. If the generating curve has a point of inflection, e.g. certain oriental domes. Flügge describes how φ presents difficulties and instead of it he uses the distance along the curve.

The author found that several works, excluding Flügge[23], used a diagram equivalent to Figure 9.14, but showing r_1 terminating where it hits r_2, which, of course, is incorrect, and/or made certain mistakes in precisely describing directions of forces or their components. The authors concerned undoubtedly know what they are doing but such deficiencies in communication could, without this warning, waste a considerable amount of the beginner's time.

If the loading is symmetric, the above equations can be used putting $N_{\theta\varphi} = N_{\varphi\theta} = 0$ because these are asymmetric stress resultants and thus cannot exist for a symmetric shell with symmetric loading. However, a much simpler method is to consider the equilibrium of a portion of a shell above a horizontal section defined by an angle φ (Figure 9.17). If the total loading on the portion is R, then resolving vertically gives

$$2\pi r_0 N_\varphi \sin\varphi + R = 0 \tag{9.28}$$

which yields N_φ. Then, N_θ can be obtained from equation 9.26.

Figure 9.17

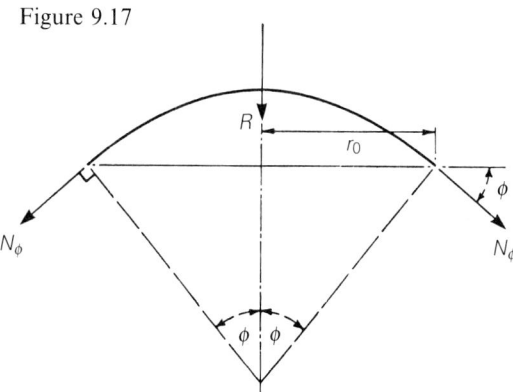

9.7.3 Shells of arbitrary shape

We must mention briefly a general membrane theory for shells of arbitrary shape. (In the preceding sections, the particular shapes treated allowed advantage to be taken of simpler theories.)

This was expounded by Flügge[23] and then applied to hypars, etc.

Two concepts are involved which have not been encountered so far in this chapter: firstly, the idea of taking and dealing with the components of all stress resultants in the horizontal plane; and secondly, the use of stress functions, which are a device to enable the solution of the partial differential equations.

9.8 Cylindrical shells

These can be designed and analysed as described in Sections 9.4 and 9.5. Now, with regard to the line load problem, this causes, as well as the in-planar (membrane) force, bending moments with accompanying shears and torsions in the shell. Various theories have been evolved for this edge load problem. The one which considers all such forces, moments and torsions is due to Donnell[29] in the U.S.A., Jenkins[30] in the U.K. and Vlassov[31] in the U.S.S.R. It is also described in Ref. 28. A theory which considers only in-planar forces and transverse bending moments with accompanying radial shears is due to Finsterwalder[32] and is also described in Ref. 26. Various assumptions were made by Schorer[33] to reduce the work involved in using this theory, because computers were not available to designers at the time. Tottenham[27] developed a method of using Schorer's theory, which tabulated the coefficients which needed to be obtained by using considerable arithmetical accuracy, and enabled the remainder of the calculations to be performed to slide-rule accuracy. Thus, a designer could use a slide rule and the tables for the whole of his design. Ref. 27 gives the tables and a design for a symmetric practical shell. It also derives Schorer's equation and discusses the practical range of applicability of Schorer's theory. About and prior to 1954 in the U.K., the theories used for designing practical shells were principally those given by Refs. 27, 30, 32 and 33. The author believes that to date more shells have been designed in the U.K. by these latter two methods than any other method. The author has shown, Ref. 34, how to use Tottenham's method for designing N-light shell roofs.

Schorer's theory has been criticized for making its assumptions. It is not suitable for very short shells, where the longitudinal bending moments with accompanying radial shears are important, i.e. the shell spans to some extent like a one-way spanning slab between the end diaphragms, arches or frames. The author regards it as the simplest of the reasonably respectable elastic theories and suitable for a considerable range of practical designs, borne out by experience in the U.K. Yet there are simpler methods than this, see Ref. 35.

The elastic analysis of practical cylindrical shells is explained in Ref. 19. This also gives the Schorer equations adapted for unit edge loads and gives all the algebraic equations (not published completely elsewhere) in Table 19 for the coefficients in these equations for both symmetric and asymmetric shells. These could easily be programmed for solution with a desk computer. Ref. 19 also deals with cylindrical shells with edge beams, valley beams and with numerous shells side by side.

9.8.1 Design tables and graphs

Without a computer program, the calculations are very arduous. The least arduous of the elastically respectable methods, by Tottenham[27], for an end (or external) shell and a valley, of a scheme of barrel shells, takes a minimum of 8 h and 4 h, respectively, of solid uninterrupted arithmetic before any checks are encountered. Usually, the checks show errors. Locating these is exceedingly time consuming and can easily take as long again. In the last scheme the author designed in this way, he had an assistant working independently with whom he cross-checked the arithmetic after every half hour or so. In this way, the above minimum times were slightly greater and with double labour, but the checks were correct and no searching for errors was involved. A minimum N-light[34] scheme takes about four times as long as the symmetric shell scheme just discussed, and the work is tedious rather than stimulating. All these times are for a skilled designer. A practising engineer having to learn about the method would need to allow much more time.

Hence the desirability of design tables.

The tables compiles by Schorer[33] are mainly not satisfactory in practice because of the large differences in the coefficients and the correct interpolation is non-linear. Tottenham's[27] tables overcome this difficulty. However, both of these give only partial assistance with the complete analysis. The tables in Ref. 18 are more complete in this respect, but there are significant differences between the coefficients and again the correct interpolation is non-linear. The author found that to try to use these for a particular shell design without interpolation was unsatisfactory and he realised that this would generally be the case.

The tables in Ref. 36 are published with adjacent German and English texts and common mathematical work. They are hard to follow and, because the notation is not collected in a single list, it is very time consuming trying to find out what various symbols mean. From the examples, considerable arithmetical work is obviously required. It is also obvious from the coefficients that large differences exist and one knows that correct interpolation is non-linear.

The tables in Ref. 37 give complete analyses. They deal with a limited range of symmetric shells. Unfortunately, they are for shells 76 mm (3 in) thick, whereas U.K. companies would generally use shells 64 mm (2.5 in) thick in most of the cases covered and designers in other countries would generally use shells thinner than 76 mm. However, an appendix gives some multiplying factors which may be used to convert the results to apply to shells 64 mm thick. In the author's opinion, some of the shells included have undesirable proportions in other practical respects.

Because of the shortcomings of past tables produced by authors who took on a tremendous task, including learning computing, the author and Dr Khwaja[3] have published tables which give complete solutions for a large range of shells of practical proportions for schemes of ordinary and north-light cylindrical shells. A complete example is given using the tables for a scheme of north-lights five side by side, each 30 ft (9.1 m) wide and all of span 60 ft (18.3 m). The example includes calculations of the necessary reinforcement and its practical disposition. Complete analyses are given for shells very near together so that, if a shell is required, say, 32 ft 4 in wide by 62 ft 8 in span, one consults the values of forces and moments in the following shells 32 ft wide by 62 ft span, 34 ft wide by 62 ft span, 32 ft wide by 64 ft span, and 34 ft wide by 64 ft span, and takes the highest values for designing the reinforcement. In this way, interpolation is not necessary and the solution is economically satisfactory. The shells have practical proportions known from experience to be satisfactory. One is limited to certain satisfactory geometries of the shells themselves but not to the areas they can cover, which can be of any rectangular shape or shapes.

The author[1] gives design graphs for complete designs of practical shell roofs. The range is much more limited than that of Wilby and Khwaja[3]. In practice, it has been found necessary to thicken shells towards their valleys and edge beams. Ref. 4 gives graphs for shells where this has been both taken and not taken into account. The former should be used and give a better practical design than previous tables. Ref. 4 also gives an example using the graphs for a single-bay barrel-shell, 8 m wide by 16 m span, which includes the design of edge beams and end stiffeners, and reinforcement is also calculated.

Ref. 2 shows, with an example, how continuous shells can be designed using simply supported solutions.

9.8.2 Analogue computer

Wilby and Bellamy[4,21,22] demonstrated that the analogue computer is a very useful tool for helping to design ordinary and north-light shells. It is much faster (instantaneous results) than the digital computer, particularly with regard to integrations. One simply has an 'integrator' unit whose input terminal can be connected to a cathode-ray oscilloscope or an x–y plotter to record a relationship between input voltage and time, and whose output terminal can be connected to a second oscilloscope or plotter to record the integral of this relationship between voltage and time. In addition, or as an alternative to the plotter, one can couple up a digital voltmeter which can easily be read at any time.

Time is taken as analogous to the angle φ and the voltages at different parts of the analogue circuit are taken as analogous to the values of the forces and moments in a shell.

The computer has slide rule accuracy, which is good enough for the results. Being instantaneously fast, it has a useful semi-design ability. If a shell is analysed and a certain force or moment is found to be undesirably great, ones inclination might be to alter a certain dimension, but this might not do much to solve the problem. The analogue computer, however, can be adjusted so that one can find the most powerful influence on the problem, which might be a peculiar relationship between certain dimensions. This gives a design as opposed to just an analysis facility.

The analogue computer was much cheaper than a digital computer. It still is much cheaper than a main-frame computer but doubtfully much cheaper than modern desk digital computers.

It enables[22] fewer assumptions to be made in the theory than the theories of Refs. 29–31, see Section 9.8.

9.8.3 Instability

Consider a thin, longitudinal strip of a shell, say 64 mm (2.5 in) thick and 18 m (60 ft) long; its length-to-thickness ratio is therefore 281:1. This would undoubtedly be unsuitable as a reinforced concrete column yet in some locations it would have to withstand compression. Its curvature saves it from buckling – the more curvature the better, from this point of view.

The shells of practical dimensions described in Refs. 2–4, 19 and 21 are known from experience to be satisfactory with regard to buckling. These are shells at least 64 mm thick and of reasonable curvatures and overall depth-to-span ratios. It is only when thinner shells of greater span-to-overall depth ratios and with less curvatures are used that buckling becomes a problem. Early guidance was given by Lundgren[20]. Professor Haas, at Delft, has been interested and has pursued experimental work on this problem. Ref. 38 outlines the problem and cites Refs. 20, 39 and 40. Furthermore, buckling is given significant treatment by Flügge[23].

9.9 Domes

Spherical domes are the most common for functional purposes because of the simpler shuttering.

For a spherical dome of radius a and a total load per unit area of curved surface, q, it is easy to obtain, Ref. 1, the following membrane forces from equation 9.28

$$N_\varphi = -aq/(1+\cos\varphi) \tag{9.29}$$

$$N_\theta = aq\left(\frac{1}{1+\cos\varphi} - \cos\varphi\right) \tag{9.30}$$

From these equations, N_φ will always be negative (i.e. compression) and increase as φ increases, whilst N_θ is negative for $\varphi < 50°50'$ but positive (tension) for $\varphi > 51°50'$. The latter condition is commonly the case in practice.

Because of the dome's double curvature, there is less worry in designing it by membrane theory only than there is in the case of the cylindrical shell; and many domes have been designed in the U.K. in this way[41]; that is, by providing edge supports designed strong enough to withstand the N_φ forces at the edge of the dome.

The membrane can be acceptable for a hemi-spherical dome supported vertically at its periphery. If the dome is less than a hemisphere and supported vertically, an edge ring force can be provided by post-tensioning so that the resultant of these vertical and horizontal forces is vectorially equal and opposite to N_φ.

In practice, the peripheries are often built in to a 'ring beam' which is supported vertically. If the design provides for the forces as already described, then the desire of the junction point between the ring beam and the dome to deflect differently for the two members will tend to cause cracking, unless the situation is as described in the last paragraph.

A more acceptable design is to consider unknown equal and opposite forces and moments at the junction point, and determine these (see Section 9.5) so that the above mentioned deflections agree. This is lucidly explained by Chatterjee[10] and a useful example is given. He does not derive the equations for the deflections, etc. These equations are derived in Ref. 1. Gol'denweiser[42] discusses the limitations of the membrane theory.

Instability (Section 9.8.3) is not as great a problem as with cylindrical shells because of the double curvature. It does not seem to cause trouble for normal domes, i.e. ones which are not particularly shallow. Buckling is considered analytically by Flügge[23], and summarized by Chatterjee[10].

9.10 Hypars

Figure 9.18 shows a surface of a hypar which can be expressed[3] as

$$z = kxy \sin \omega \tag{9.31}$$

When $x = a$ and $y = b$, $z = c$ and if $\omega = 90°$, then

$$z = \frac{c}{ab} xy \tag{9.32}$$

Figure 9.18 Surface of the hypar $z = kxy \sin \omega$ (From *Concrete Shell Roofs* by courtesy of the authors and Applied Sciences Publishers Ltd. London).

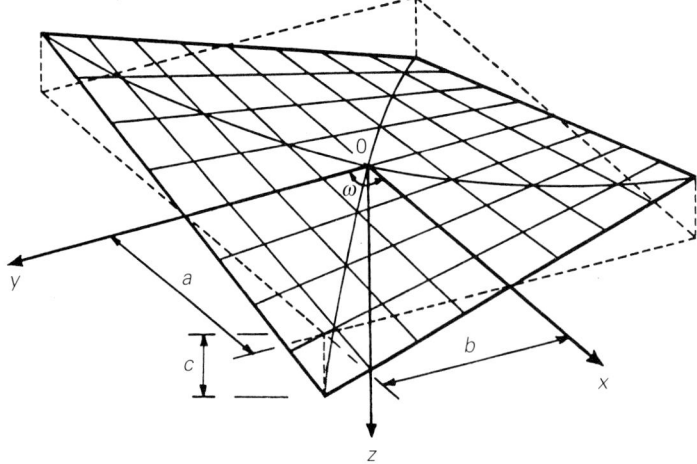

Its intersections with the vertical planes x = constant and y = constant are respectively the straight lines z = (const.) y and z = (const.) x, the 'generators'. Using equation 9.32 in the general membrane equations, Ref. 23, for shells of arbitrary shape (see Section 9.7.3) enables the membrane forces to be obtained. Timoshenko[1] considers a vertical load, such as snow load, of q per unit of horizontal area and this gives $N_x = N_y = 0$ and

$$N_{xy} = -\frac{q}{2k} \tag{9.33}$$

Then he considers a self-weight load of q_0 per unit surface area and this gives

$$N_{xy} = 0.5 q_0 (x^2 + y^2 + 1/k^2)^{0.5} \tag{9.34}$$

$$N_x = -\frac{q_0 y \cos \theta}{2 \sin \varphi} \log \left[\frac{x + (x^2 + y^2 + 1/k^2)^{0.5}}{(y^2 + 1/k^2)^{0.5}} \right] \tag{9.35}$$

$$N_y = -\frac{q_0 x \cos \varphi}{2 \cos \theta} \log \left[\frac{y + (x^2 + y^2 + 1/k^2)^{0.5}}{(x^2 + 1/k^2)^{0.5}} \right] \tag{9.36}$$

Billington[12] demonstrated that there is normally little error in assuming the self-weight, q_0, as an increase in the value of q so that the simpler solutions can be used. Essentially, this means that the shell mainly resists only the shear stress resultant of equation 9.33. Then, according to Pflüger[13]:

> Of course these shears must be received by appropriate stiffeners at the edges. The edge members are subject to certain strains while the shell is without strains in the x and y directions because of $N_x = N_y = 0$. This results in an incompatibility of the deformations that can be overcome only with the help of a bending theory.

Ref. 3 gives such a bending theory, and gives suitable computer program listings in the Appendices.

Many hypars have been designed, and some have cracked considerably, using membrane theory only. Worked examples indicating reinforcement details are given in Refs. 10 and 38.

The last paragraph of Section 9.9 on instability applies to hypars as well as domes.

9.11 Conoids

According to Wilby and Naqvi[13]

In the Cartesian system of coordinates, the middle surface of a shell is expressed by the equation

$$z = X(x) Y(y) \tag{9.37}$$

where the functions $X(x)$ and $Y(y)$ are independent of y and x respectively. If, however,

$$X(x) = Ax + B \tag{9.38}$$

where A and B are constants, then any section of the surface where y is constant is a straight line, and therefore a surface can be generated by straight lines. Such surfaces are called 'ruled surfaces'. A ruled surface is completely defined by the constants A and B, and the function $Y(y)$ is called the 'directrix' of the surface. A conoid belongs to this group of surfaces of which the most commonly used directrix is a parabola. Thus the equation of the parabolic conoid shown in Figure 9.19 is

$$z = -H\left(1 - \frac{x}{l}\right)\left(1 - \frac{y^2}{b^2}\right) \tag{9.39}$$

where $g = 1 - H_0/H$.

Figure 9.19 shows half of a conoid which might be used as a roof of width $2b$ and span a.

Ref. 13 goes on to consider edge beams and give a large number of computer-generated tables which may be used for designing conoids with edge beams. Then it gives two practical examples using the tables, namely, for unsupported and propped edge beams, and drawings give reinforcement details. In Ref. 2, the addendum and corrigenda to these tables are published and an example is given using the tables of a scheme of numerous conoidal shells covering a large area.

The background to Refs. 13, 44 and 45 is given in the PhD thesis[46] supervised by the author. Prior to this work, many conoids will have been designed using membrane theory[47] only. Because of their double curvature, there is less worry in designing conoids by membrane theory only than there is the case of cylindrical shells. However, one is likely to experience more cracking than using the method of Ref. 13.

The last paragraph of Section 9.9 on instability applies to conoids as well as domes.

9.12 Method of finite elements

Of the many methods of analysis (e.g. finite difference, relaxation), that of finite elements, given modern powerful computers, is at present generally considered to be a method applicable to a wide range of problems (in structures, hydraulics, soil mechanics, etc.) of the continuum mechanics type. It is a method only suitable for solution with computers.

Programs are commercially available world-wide. Some can be adapted for a variety of applications; for example, a program developed by a U.K. firm of civil engineering consultants is also widely used by mechanical and aeronautical engineers. They would claim to be able to adapt it adequately to any shape of shell. This is very time consuming and it is much faster to use tables such as given in Ref. 3.

Writing and adapting such programs becomes the work of specialists who usually have no experience of reinforced concrete shell design, construction or research, but are very experienced at elastic analysis using the finite element method and computers. For the beginner to set about learning how to write a finite element computer program for a shell design

Figure 9.19 Half of a parabolic conoid which might be used as a roof of width $2b$ and span a (From C. B. Wilby and M. M. Naqui, *Reinforced Concrete Conoidal Shell Roofs*, by permission of Eyre and Spottiswoode Publications Ltd. Leatherhead).

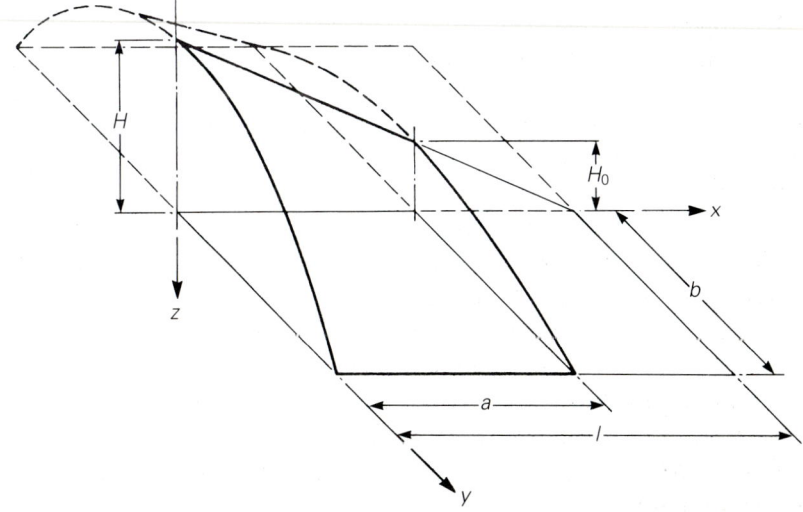

would not be practical if he were designing a job wanted in a normal time, because the program writing for this kind of work is very time consuming. The cost of producing the program would normally only be retrieved if the program were marketed for several designers to use.

For information on the finite element method (assisted by computers) applied to shells, the reader is referred to Refs. 24, 48 and 49.

9.13 Computer programs

In addition to the finite element packages just mentioned, other computer packages based on the methods previously outlined in this chapter may well be more available. For example, computer programs were used for cylindrical shells, hypars and conoids in Refs. 3 and 13, the program listing being published for the hypar in Ref. 3.

9.14 Design practicalities

Reinforced concrete shells, with or without prestressing, should be designed by those expert and experienced in structural concrete design, detailing and construction. It is also an advantage if the designer also has appropriate research or testing experience of how structural concrete behaves.

One difficulty is that many of the books and papers on shells are written by experts, with or without computing expertise, in the elastic analysis of shells for mechanical, aeronautical or structural problems, who have little or no experience of concrete as outlined above.

Experienced designers know that, to design many of the structures with which they are presented, or which they wish to create, they have to make assumptions which would occasionally horrify some analysts with little or no practical experience. Sometimes they have to make sure that forces are adequately resisted without the proper matching up of deflections. They then have to rely on some plastic action, which, from experience, they know will occur, and they will also have to detail reinforcement by experience to ensure adequate distribution of the cracks which will be caused. Often, a design is required within a very limited time and may involve situations which a pure analyst would take years to solve and then doubt whether his solution were really good enough to use for design.

A similar philosophy has to apply to the practical design of structural concrete shells. Often, architectural ideas present problems which have not previously been solved or for which the existing solutions are not presented in a form suitable for use in the time available to the designer.

For example, an industrialist may wish to utilize the whole area of a non-rectangular portion of his site, or want natural roof-lighting or special loads hung from the roof. Such demands will almost certainly pose special problems for the designer because it is unlikely that he can resort to ready-made solutions.

Laboratory and field tests made by the author and others show that shells are inherently strong and allow more liberties to be taken with strength than non-shell structures. Cracks can be troublesome, hence the use of special thickening and nominal reinforcement, see for example Refs. 2, 3, 4, 13 and 19.

References

1. Timoshenko, S. and Woinowsky-Krieger, S., *Theory of plates and shells*, 2nd Edn, McGraw–Hill, New York (1959).
2. Wilby, C. B., *Concrete for structural engineers*, Newnes/Butterworth, London (1977).
3. Wilby, C. B. and Khwaja, I., *Concrete shell roofs*, Applied Science Publishers, London (1977).
4. Wilby, C. B., *Design graphs for concrete shell roofs*, Applied Science Publishers, London (1980).
5. Gould, L. G., *Static analysis of shells*, Lexington Books, Massachusetts (1977).
6. Wilby, C. B., *Post-tensioned prestressed concrete*, Applied Science Publishers, London (1980).
7. Wilby, C. B., *Prestressed concrete beams*, Applied Science Publishers, London (1969).
8. Khwaja, I., 'Theoretical analysis and experimental behaviour of hyperbolic paraboloidal shells'. PhD Thesis, University of Bradford, England (1968).
9. Wilby, C. B., *Elastic stability of post-tensioned concrete members*, Edward Arnold, London (1964).
10. Chatterjee, B. K., *Theory and design of concrete shells*, Edward Arnold, London (1971).
11. Bharucha, J. N., Proposed classification of shell roofs, *Ind. Concr. J.*, Vol. 33, No. 12, Dec. (1959), pp. 419–21.
12. Billington, D. P., *Thin shell concrete structures*, McGraw–Hill, New York (1965).
13. Wilby, C. B. and Naqvi, M. M., *Reinforced concrete conoidal shell roofs*, Cement and Concrete Association, London (1973).
14. Ketchum, M. S., Economic factors in shell roof construction, *Proc. world conf. shell struct.*, San Francisco, 1962, National Academy of Sciences/National Research Council, Washington, DC, Publication No. 1187, 1964, pp. 97–102.
15. Whitney, C. S., Economics, *Proc. conf. thin concr. shells*, Massachusetts Institute of Technology, Cambridge, Mass., (1954), pp. 22–4.
16. Whitney, C. S., Cost of long span concrete shell roofs, *J. Am. Concr. Inst.*, Vol. 21, No. 10, June (1950) pp. 765–76.
17. Chatterjee, B. K., Economics of shell roof construction in India, *Ind. Concr. J.*, Vol. 33, No. 12, Dec. (1959) pp. 456–60.

18. American Society of Civil Engineers, *Design of cylindrical concrete shell roofs*, Manuals of engineering practice, No. 31, New York (1952).
19. Evans, R. H. and Wilby, C. B., *Concrete: plain, reinforced, prestressed and shell*, Edward Arnold, London (1963).
20. Lundgren, H., *Cylindrical shells*, Danish Technical Press, Copenhagen (1960).
21. Wilby, C. B. and Bellamy, N. W., *Elastic analysis of shells by electronic analogy*, Edward Arnold, London (1961).
22. Wilby, C. B., A proposed 'exact' theory for analysing shells, and its solution with an analogue computer, *Proc. Inst. Civ. Engrs*, Vol. 22, No. 3, July (1962) pp. 291–308.
23. Flügge, W., *Stresses in shells*, 2nd Edn, Springer-Verlag, New York, Heidelberg, Berlin (1973).
24. Gibson, J. E., *Thin shells*, Pergamon Press, Oxford (1980).
25. Paduart, A., *Shell roof analysis*, CR Books, London (1966).
26. Chronowicz, A., *The design of shells*, 3rd Edn, Crosby Lockwood, London (1968).
27. Tottenham, H., A simplified method of design for cylindrical shell roofs, *Struct. Eng.*, June (1954) pp. 161–80.
28. Gibson, J. E. and Cooper, D. W., *The design of cylindrical shell roofs*, Spon, London (1954).
29. Donnell, L. H., *Stability of thin-walled tubes under torsion*, NACA Report 479 (1934).
30. Jenkins, R. S., *Theory and design of cylindrical shell structures*, Ove Arup, London (1947).
31. Vlassov, V. Z., Some new problems on shells and thin structures, *Izv. Akad. Nauk SSSR*, No. 1 (1947) pp. 27–53. [English translation in NACA TM 1204].
32. Finsterwalder, U., Die Theorie der kreiszylindrischen Schalengewölbe System Zeiss-Dywidag, *J. Bridge Struct. Eng.* (1932) pp. 127–52.
33. Schorer, H., Line load on thin cylindrical shells, *Proc. Am. Soc. Civ. Eng.*, Vol. 61, No. 3, Mar. (1935) pp. 281–316.
34. Wilby, C. B., A method of designing north-light shell roofs, *Ind. Concr. J.*, Vol. 35, No. 1, Jan. (1961) pp. 6–10.
35. Paduart, A. and Dutron, R. (Eds), Simplified calculation methods of shell structures, *Proc. colloq. simplified calc. methods*, Brussels, 1961, North-Holland, Amsterdam; and Interscience, New York (1962).
36. Rüdiger, D. and Urban, J., *Circular cylindrical shells*, Teubner Leipzig (1955).
37. Gibson, J. E., *Computer analysis of cylindrical shells: design tables for cylindrical shell roofs calculated by automatic digital computer*, Spon, London (1961).
38. Ramaswamy, G. S., *Design and construction of concrete shell roofs*, McGraw-Hill, New York (1968).
39. Paduart, A., *Introduction au casul et à l'execution des voiles minces en béton armé*, Centre d'Information de l'Industrie Cimentière Belge, Eyrolles, Paris (1961).
40. Bradshaw, R. R., Some aspects of concrete shell buckling, *J. Am. Concr. Inst., Prov.*, Vol. 60, No. 3 (1963) pp. 313–28.
41. Reynolds, C. E., *Reinforced concrete designers' handbook*, Cement and Concrete Association, London (1971).
42. Gol'denweizer, A. L., *Theory of thin elastic shells*, GITTL, Moscow (1953).
43. Pflügger, A., *Elementary statics of shells*, 2nd Edn, English translation, Dodge Corporation, New York (1961).

References

44. Wilby, C. B. and Naqvi, M. M., Structural analysis of conoidal shells, *Struct. Eng.*, Vol. 50, No. 5, May (1972) pp. 97–201.
45. Wilby, C. B. and Naqvi, M. M., A flexural analysis of conoidal shells for reinforced concrete roofs, *Ind. Concr. J.*, Vol. 49, No. 7, July (1975) pp. 200–5.
46. Naqvi, M. M., 'Theoretical analysis and experimental behaviour of conoidal shells', PhD Thesis, University of Bradford, England (1969).
47. Soare, M., *Membrane theory of conoidal shells*, Library Translation No. 83, Cement and Concrete Association, London (1959).
48. Kraus, H., *Thin elastic shells*, Wiley, New York (1967).
49. Ashwell, D. G. and Gallagher, R. H. (Eds), *Finite elements for thin shells and curved members*, Wiley, London (1976).

10

Folded plate roofs

10.1 Folded plates

The economic and aesthetic advantages of curved shell roofs, when employed over large unobstructed floor areas, also accrue to a considerable extent to *folded plate* roofs, also known as *hipped plate* or *prismatic* structures. They have additional advantages over curved shell roofs in that shuttering, steel fixing, setting-out, concreting, the carrying of point loads and incorporation of large openings are all much easier, but the weight and total quantities of concrete and reinforcement are greater. Figure 10.1 shows a system of folded plates. Figure 10.2 shows a basically folded plate roof, designed by the author. The overall depth is greater at mid-span so strictly speaking the plates are slight hypars. It is useful to refer to Section 9.2.1.

Figure 10.1

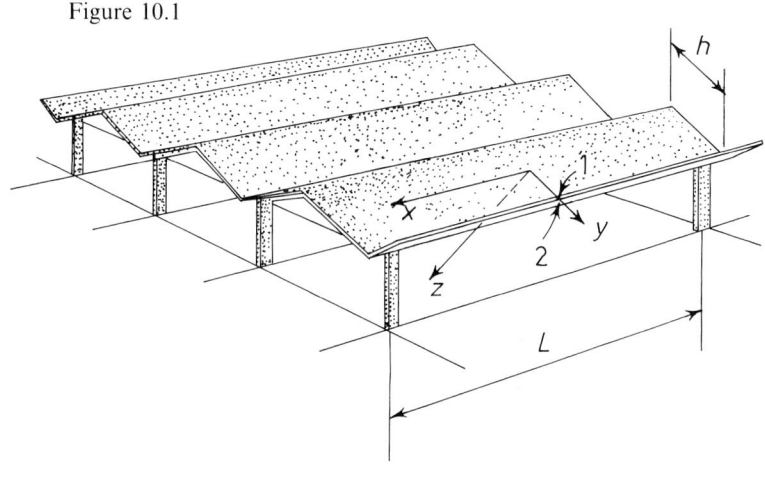

10.2 Design and analysis of folded plates

When they carry loads, internal forces and moments are induced in three dimensions as for cylindrical shells. Folded plates are designed by assessing their sizes by simple approximate design methods and/or experience, and then structurally analysing. The sizes are then altered if necessary and the process repeated. With experience it is often not necessary to alter the sizes and so in these cases the analysis gives confidence and allows economy of reinforcement compared to the original assessment.

10.3 Analysis of folded plates

A rigorous three-dimensional elastic analysis is given in Ref. 1. This has been programmed for digital computation[2]. Gibson[3] uses a program, based on the theory of Ref. 30 of Chapter 9 for analysing many cylindrical shells adjacent to one another, for analysing folded plates, making the curvatures of the shells very small. He defends this approximation of a shallow shell to a plate for a reasonable range of folded plates.

Figure 10.2

More approximate methods deal with less forces and moments. Resolving the loads in the directions y and z (see Figure 10.1) for each plate, then each plate can be considered as a beam of span L and depth h. If everything were symmetrical, that is if the folded plate system of Figure 10.1 continued *ad infinitum*, without terminating at either end, then the longitudinal flexural stresses in these plates would be the same where they touch one another and the design just mentioned would be adequate, and can be called a *beam analogy theory*. Each plate is very flexible in the z direction but relatively very stiff in the y direction. Hence each apex point and valley point is held in position, in that the two adjacent plates restrict its movement in their planes (i.e. y directions). Hence away from the end columns the slab can be regarded as spanning from apex to valley as a long continuous slab, assuming the slab is not two-way spanning, i.e. $L > 2h$. Thus with the beam analogy method we obtain longitudinal stresses in the direction x and transverse bending moments and shear forces. For a large number of similar folded plates the beam analogy could be reasonable for the central bays.

For a scheme like Figure 10.1 where there are end plates and not all the plates have the same dimension h, the beam analogy would mean that the longitudinal stresses calculated at say points 1 and 2 on Figure 10.1 would not be the same. Yet they must be, as the longitudinal strain at these points must be the same. Thus Winter and Pei[4] introduce shearing forces along each fold and then determine these shear forces so that the longitudinal stresses at adjacent points (such as 1 and 2) at each fold (valleys and apices) are the same. These unknown longitudinal shears can be obtained by solving simultaneous linear equations by computer, calculating machine, or by a relaxation method (analogous to moment distribution) given by Winter and Pei. The transverse moments are calculated as before.

Gaafar[5] showed that in many cases, particularly at a place like the edge valley of Figure 10.1 where two plates meet with very different values of h and they are discontinuous and continuous at their other ends respectively, the relative deflections of adjacent folds could make the Winter and Pei method inaccurate in assessment both of longitudinal stresses and of transverse moments. The relative deflections of the various folds (supports) have their effect upon the continuous slab (transverse) bending moments. This relative support settlement affects the reactions at the supports (folds) and thus the longitudinal stresses. This method is adequately accurate[6] for folded plates where $L > 3h$. In teaching students the presentation of this method by Simpson[7] is most useful, giving them a good insight into the principal behaviour of folded plates. Rapidly converging relaxation is used to eventually build up simultaneous equations, which essentially can be directly produced by Parme's method[8] by substituting numbers in equations (see Section 10.4).

An analysis making very similar assumptions is given by Thadani[9] in terms of tabulated coefficients based on a method by W. Tetzlaff. Computer programs are given for this analysis by Tamhankar and Jain[10].

10.4 Analysis due to Parme

Parme applies the slope deflection equations to the transverse moments. Then the support reactions (at folds) are resolved in the planes of the plates, and the deflections of these plates in their planes are then equated to the unknown deflections used in the slope deflection equations. This gives linear simultaneous equations which can be solved by calculating machine or computer. Without this equipment they can be solved by relaxation. Prior to the commonplace use of computers, the author used to teach MSc students at the University of Bradford, Simpson's Method and prior to that Gaafar's Method. When a computer with a matrix ROM became available the author then taught Parme's Method as this was ideal for quickly producing the simultaneous equations, merely by putting numbers in equations without study (and these equations could easily be programmed for evaluation on the micro computer). The simultaneous equations could be solved exceedingly quickly with the matrix ROM. Previously the author had preferred Simpson's (or Gaafar's) method because of the organised convergence (like that of Hardy Cross) to solutions, as opposed to Parme's Method which used a relaxation method which was not as systemised in converging upon the solutions to the simultaneous equations. So Parme's method which was not to be favoured in its day the by the author, suddenly became very attractive to him when he possessed a microcomputer with a matrix ROM.

Of course use can be made of commercially available finite element packages by adapting them to analyse folded plates. Still Parme's method can easily and rapidly be used with an inexpensive desk top microcomputer in the design office.

The following example, by courtesy of Dr A. L. L. Parme[11], is useful in indicating a procedure for a student or a practising engineer to pursue in his design office.

Table 10.1 in (a) shows the notation used for any type of folded plate. Then in (b) is shown the folded plate we have decided to analyse. This has been chosen to not look like the folded plate of (a), to indicate the general applicability of the method. When a plate is denoted by a number n, this number denotes the plate on its right-hand side. The length of the plates $= L = 70$ ft in this example, and W_n is determined from $W_n = h_n(w_D + w_L \cos \beta_n)$. In the example (density of reinforced concrete $= 150$ lb/ft^3) for $n = 0$, self-weight concrete $= 0.5 \times 150 = 75$ lb/ft$^2 = w_D$, $W_n = 3 \times 75 = 225$ lb/ft.

For $n = 1$, self-weight concrete $= \qquad 0.333 \times 150 = 50$

Roofing and insulation $\qquad = \underline{5}$

$$w_D = 55 \text{ lb/ft}^2$$

For roofs in general the superimposed load is due to snow. The loading for this is usually adequate to allow for the weight of maintenance workers and, if applicable, nominal grit from nearby industrial pollution, before it can be removed by wind or otherwise when the snow is not on the roof. In the U.K. the snow load of 0.75 kN/m^2 is rarely present because of the buildings being heated and the rare occurrence of snow. A loading of this small amount would be reasonable to use in geographical locations which never experience snow, so as to provide for loads due to maintenance staff, nominal grit/sand, etc. In North America approximate snow loads are as follows: Toronto (Canada) 2.4 kN/m^2, for various parts of the U.S.A. and Canada 3.2 kN/m^2 and for the U.S.A. and Canada towards the Rocky Mountains 4.0 generally, to 4.8 kN/m^2 in a small number of areas. Seemingly in the U.S.A. designers never use less than 20 lb/ft^2 (1.0 kN/m^2). In Parme's U.S.A. example, he uses U.S.A. (British Imperial) units and a.

Live load $= w_L = 25$ lb/ft^2
$W_1 = 10.01(55 + 25 \times 0.866) = 767.3$
W_1, W_2, W_3 and W_4 are taken as 800 lb/ft in this example due to Parme. There is no reason why accurate values should not be taken for W_1, W_2 etc.

The wind loading can be neglected as it is upwards and always less per unit area than the self-weight of each concrete plate. Tables 10.2, 10.2 and 10.3 are completed and it can be seen that for this example, as it is symmetrical, we do not need to consider all values of n. To illustrate how these tables are constructed, take for example the numbers in Column 35. They are obtained by adding the numbers in Columns 25, 26 and 36. The

Analysis due to Parme

Table 10.1

(a) [diagram showing α positive when clockwise, with points 0, 1, 2, n-1, n, n+1 and angles β, α]

(b) [diagram showing folded plate cross-section with centre line, points 0,1,2,3,4,5,6]

Given properties

(1)	(2)	(3)	(4)	(5)	(6)	(7)	(8)	(9)	(10)	(11)	(12)	(13)
Point or plate	α_n, deg	β_n, deg	t_{n-1}	t_n, ft	h_{n-2}	h_{n-1}	h_n, ft	h_{n+1}	W_{n-2}	W_{n-1}	W_n, lb	W_{n+1}
0		90										
1	60	30	0.5000	0.5000			3.000	10.01			225	800
2	20	10	0.3333	0.3333		3.000	10.01	10.07		225	800	800
3	20	−10	0.3333	0.3333	3.000	10.01	10.07	10.07	225	800	800	800
4	20	−30	0.3333	0.3333	10.01	10.07	10.01	10.01	800	800	800	800
5												
6												
7												
8												
9												
10												

Trigonometric values

(14)	(15)	(16)	(17)	(18)	(19)	(20)	(21)	(22)	(23)
$\cos\beta_{n-2}$	$\cos\beta_{n-1}$	$\cos\beta_n$	$\cos\beta_{n+1}$	$\sin\alpha_{n-1}$	$\sin\alpha_n$	$\sin\alpha_{n+1}$	$\cot\alpha_{n-1}$	$\cot\alpha_n$	$\cot\alpha_{n+1}$
			0.8660			0.8660			2.748
	0	0.8660	0.9848		0.8660	0.3420		0.5774	2.748
0	0.8660	0.9848	0.9848	0.8660	0.3420	0.3420	0.5774	2.748	2.748
0.8660	0.9848	0.9848	0.8660	0.3420	0.3420	0.3420	2.748	2.748	

Table 10.2

Computed values

Point or plate	㉔ $\frac{⑧}{⑦}\cdot\frac{⑲}{㉔}$	㉕ $\frac{⑲}{㉔}$	㉖ ㉒+㉓	㉗ ㉔²(㉑+㉒)	㉘ $1-\left(\frac{⑤}{④}\right)^2\times\frac{㉔}{④}$	㉙ $\frac{⑤\times㉔}{④}$	㉚ $\frac{⑤\times⑧}{⑥}\times\left(\frac{⑧}{l}\frac{\pi}{}\right)^2$	㉛ $\left(\frac{⑤}{l}\right)^3\times\left(\frac{⑧}{l}\frac{\pi}{}\right)^2$	㉜ $\frac{⑧\times㉔}{⑥\times⑱}$	㉝ ㉕+㉗+㉜	㉞ $2\times㉕+㉖+㉗$	㉟ ㉕+㉖+㊱	㊱ $\frac{⑨\times㉑}{⑧}$	㊲ $\frac{㉔}{㉘}$	㊳ $\frac{⑪\times⑮}{2}\cdot㉔$	㊴ $\frac{⑫\times⑯}{2}\cdot㉔$	㊵ $\frac{⑧\times㉚}{2}\cdot㉔$	㊶ $㉔\times⑭(⑩+⑪)\cdot\frac{⑱}{2}$	㊷ (⑪+⑫)$\left(\frac{⑲}{⑮+⑯\times㉔}\right)$	㊸ $\frac{⑳}{⑰}(⑫+⑬)$	㊹ $⑪\left(\frac{⑲}{⑮+⑯\times㉔}\right)+㊷$ $\frac{2\times}{㉔}⊛$	㊺	㊻ $\frac{⑰}{㉔}\left[⑫\times2+\left(\frac{⑨\times⑰}{⑧\times⑯}\right)+⑬\right]+\frac{⑳}{⑰}$
0	3.3337	3.853	3.325				220.6						0.3461				330.9						1250
1	1.006	2.942	5.496	3.365	0.9940	0.4495	8.908	0.01799	3.900	10.21	14.75	10.09	2.907	0.9881	0	346.4	44.58	0		3420	4607	4171	0
2	1.000	2.924	5.496	5.496	1.000	0.9940	8.750	0.01799	2.942	11.36	16.84	11.36	2.924	1.000	346.4	393.9	44.06	0.4051		8686	4607		
3						1.000	8.750			11.36		11.36	2.942		393.9	393.9	44.06			9215	4051		
4																							
5																							
6																							
7																							
8																							
9																							
10																							

Analysis due to Parme

Table 10.3

Coefficients for simultaneous equations

①	㊼	㊽	㊾	㊿	51	52	53	54	55	56	57
Point or plate	㉚ × ㊻	−㊵ × ㊻	㉙	2(1 + ㉙)	−㉚ × ㉟	㉚ × 36	㊵[−㊸ + ㊺(㉕ + ㉖)]	−㉛ × ㉜	㉛ × ㉝	2 × ㉘	−㉛ × ㉞
0	76.35	−413 600	0.4495	2.899	−89.88	25.90	−19 440	−0.07016	0.1837	1.988	−0.2654
1								−0.05293	0.2044	2.000	−0.3030
2											
3											
4											
5											
6											
7											
8											
9											
10											

Table 10.3 (*continued*)

①	Coefficients for simultaneous equations											
	㊽	㊾	㊿	㊶	㊷	㊸	㊹	㊺	㊻	㊼	㊽	㊾
Point or plate	㊳	㊴	㊵	㊶	㊷	㊸	㊹	㊺	㊻	㊼	㊽	㊾
	4(1+㉘)	㉛ × ㉟	−㉛ × ㊱	−⑧(㊲)(㊳ × ㊴ + ㊴)	㉚ × ㉜	㉙	−㉚ × ㉝	2(1+㉙)	㉚ × ㉞	−㉚ × ㉟	㉚ × ㊱	−㊵(㊶ − ㊷)(㊷ + ㊸)
0												
1												
2	7.976	0.2044	−0.05260	−7414	34.13	0.9940	−89.34	3.988	129.1	−99.40	25.58	179 700
3	8.000	0.2044	−0.05293	−7933	25.74	1.000	−99.40	4.000	147.4	−99.40	25.74	49 040
4												
5												
6												
7												
8												
9												
10												

method of constructing the necessary simultaneous equations for solution of the problem is shown in Table 10.4. α is the angle formed by the extention of one plate with the next one and is considered positive when the angle measured from the extention is clockwise. β is the angle formed by a plate and a horizontal line and is considered positive when the angle measured from the plate is clockwise. The equation numbers refer to mathematical equations in Parme's paper[3]. In all the tables so far given the shaded spaces eliminate quantities not required by the basic mathematical equations[8]. The numbers of the example are now inserted in Table 10.4 to give Table 10.5. The equations (rows) are renumbered 0 to 5, for simplicity. These equations can easily be solved with a high accuracy electronic calculator, or with a computer. With the former, the classical method of eliminating one unknown at once is better in practice than using determinants. Without such equipment an iteration method is recommended – a suitable one is demonstrated by Parme[8]. As mentioned previously the author finds a desk top microcomputer, with a program (better still a matrix ROM) for solving simultaneous equations very convenient.

The solutions are given in the last row of Table 10.5. Note that all units used are lb ft units, i.e. the stresses are units of lb/ft^2. The symbols used are as follows:

$M =$ Transverse bending moment at a fold and is considered positive when it creates tension on the underside of the plate (lb ft/ft length of plate)

$f =$ Longitudinal stress and is considered positive when it is compressive (lb/ft^2)

$W =$ Total vertical load acting on a plate, and is considered positive where acting downward (lb/ft length of plate)

Table 10.4.

Point	Eq.	f_0	f_1	f_2	M_2	f_3	M_3	= Load
0	29	2	1		(47)			= (48)
1	27	(49)	(50)	1	(51)		(52)	= (53)
2	8	(54)	(55)	(57) + (60)	(58)	(59)	2	= (61)
	33		(63)	(65)	(66) + (68)	1	(67)	= (69)
3	8		(54) + (60)	(55) + (59)	(56) + 2	(57)	(58)	= (61)
	33			(63) + 1	(64) + (67)	(65)	(66)	= (69)

Table 10.5

Point	Col.	0	1	2	3	4	5	6	7
	Row	f_0	f_1	f_2	M_2	f_3	M_3	= Load	Check column
0	0	2.000	1.000	0	76.35	0	0	= −413 600	−413 500
1	1	0.4495	2.899	1.000	−89.88	0	25.90	= −19 440	−19 500
2	2	−0.07016	0.1837	−0.3180	7.976	0.2044	2.000	= −7414	−7404
	3	0	0.9940	3.988	154.7	1.000	−99.40	= 179 700	179 800
3	4	0	−0.1059	0.4088	4.000	−0.3030	8.000	= −7933	−7921
	5	0	0	2.000	−198.8	4.000	147.4	= 49 040	48 990
Solutions		−169 800	−4554	33 890	−909.8	14 970	−1760		

Analysis due to Parme

These moments and stresses must be multiplied by $4/\pi$ to account for the difference in the sinusoidal load, assumed in the mathematical equations, and a uniformly distributed load. The use of a sinusoidal load is necessary for the mathematics and is a common device for many problems of elasticity.

Schemes of folded plates like the one in this example seem to be used in the U.S.A. In the U.K. the author has used schemes like the one shown in Figures 10.2 and 10.3.[12] The valleys act as gutters for rainwater as also do the upstand edge beams. The latter are often required to be vertical in line with the external walls of brickwork, blockwork or glazing, etc. The author favours sloping plates at the ends as shown in Figure 10.1 rather than vertical edge beams (if allowed by the architect) for structural reasons and also to protect the external walls/windows from rainfall. In southern Europe for example and hot countries these sloping plates at the ends are desirable to give shade to the windows from the sun.

The author set the following question on such a scheme as Figure 10.1 in the Christmas 1977 MSc Structural Design Examination:

> Design the reinforced concrete folded plate roof shown in Figure 10.4 to carry a vertical uniformly distributed loading of 0.75 kN/m^2 due to snow, built-up roofing felt weighing 0.215 kN/m^2, and its self weight assuming the weight density of reinforced concrete is 24 kN/m^3. Wind forces are to be ignored. Parme's method should be used and the small end cantilevers are to be ignored in this analysis. The weight of each valley thickening can be included by sharing this small weight between adjacent plates in proportion to their respective weights. Ignore the 10 mm chamfers and assume the plates to be concentrated at their centre lines. The lower extremities of the plates are 3.3 metres above ground floor level.

Figure 10.3

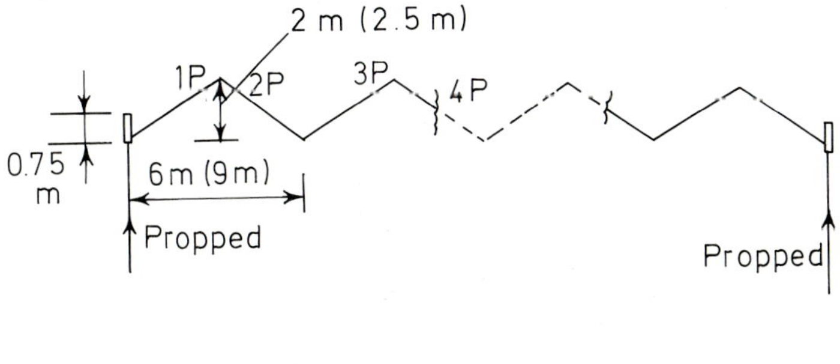

Design the complete structure including columns and bases assuming the soil one metre below ground floor level can be designed for an ultimate uniformly distributed pressure of 400 kN/m². Detail the reinforcement for one quarter of the structure (including columns and bases) so that the other quarters will be identical. Bar Bending Schedules should be produced but it is adequate to indicate the shape of each bar and link or stirrup

Figure 10.4

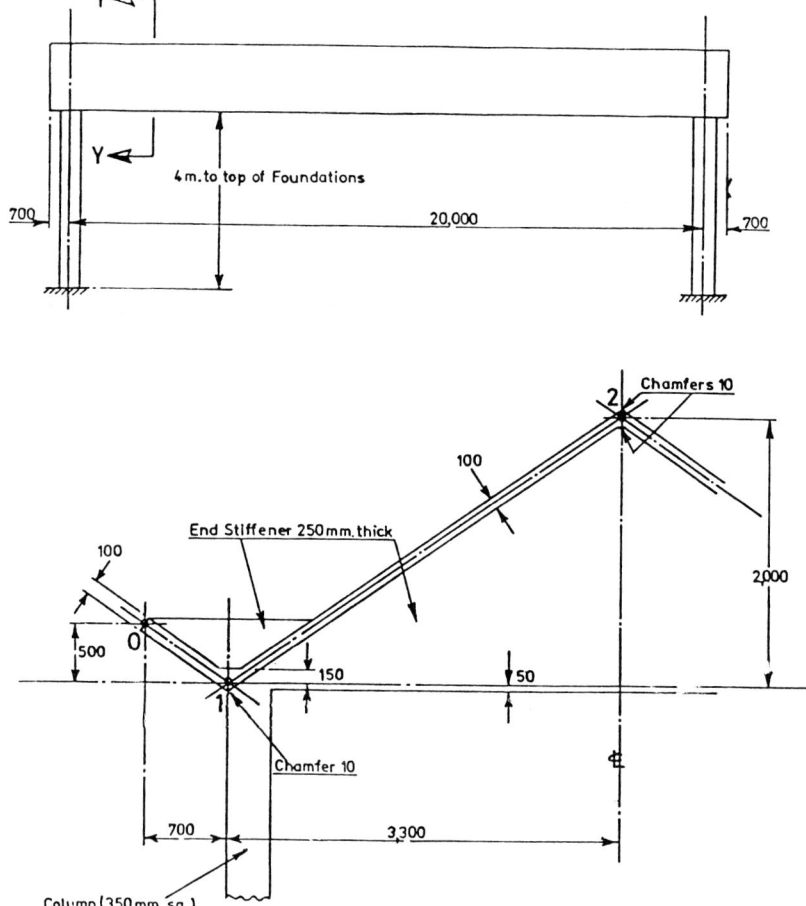

Section YY

and not give its dimensions. There are no cladding walls. All dimensions shown in Figure 10.4 are in mm units, and the structure is symmetrical.

The author then set this as coursework for the MSc and BEng/BTech/BSc final year students. The Parme solution produced by MSc students in 1981 is shown in Tables 10.6 to 10.9.

Table 10.6 in (a) shows the notation used for any type of folded plate. Then in (b) is shown the folded plate of the MSc question. This is of the type shown in Figure 10.3 (Ref. 12 gives Tables for designing such schemes with 8, 6 and 4 plates respectively, with vertical edge beams propped with intermediate columns adequately for them to be considered as providing line supports to the edge beams). In this present example:

Loading per m² of surface area:

self-weight of concrete	$= 0.1 \times 2.4 = 2.4$ kN/m²
roofing felt	$= 0.215$
total dead loading	$= 2.615$ kN/m²
live (snow) loading	$= 0.75$ kN/m²

From geometrical considerations, $h_0 = 0.86$, $h_1 = h_2 = 3.86$ m, $h_3 = 0.86$ m

$\alpha_1 = -66.76°$, $\alpha_2 = 62.44°$, $\beta_0 = -35.54°$, $\beta_1 = 31.22°$,

$\beta_2 = -31.22°$, $\beta_3 = 35.54°$

Using $W_n = h_n(w_D + w_L \cos \beta_n)$

$W_0 = 0.86(2.615 + 0.75 \cos 35.54°) = 2.77$ kN/m

$W_3 = W_0 = 2.77$ kN/m

$W_1 = 3.86(2.615 + 0.75 \cos 31.22°) = 12.57$ kN/m

$W_2 = W_1 = 12.57$ kN/m

Table 10.9 was based on Table 10.4.

For the remainder of this question concerning the design and detailing of the reinforcement for the plates, end stiffening beams and columns, the method of solution is similar to that shown in Refs. 3, 5 and 13 of Chapter 9, and is as taught to the students by the author and has been published in Ref. 13.

Folded plate roofs

Table 10.6

(a) [Diagram showing folded plate geometry with angles β_0, α_1, β_1, α_2, β_2, ..., α_n, β_n, α_{n+1}, β_{n+1} and thicknesses t_n, h_n. α positive when clockwise]

(b) [Diagram showing cross-section with points 0, 1, 2, 3, 4 and centre line at point 2]

		Given properties												Trigonometric values								
(1)	(2)	(3)	(4)	(5)	(6)	(7)	(8)	(9)	(10)	(11)	(12)	(13)	(14)	(15)	(16)	(17)	(18)	(19)	(20)	(21)	(22)	(23)
Point or plate	α_n deg	β_n deg	t_{n-1}	t_n ft	h_{n-2}	h_{n-1}	h_n metres	h_{n+1}	W_{n-2}	W_{n-1}	W_n kN	W_{n+1}	$\cos\beta_{n-2}$	$\cos\beta_{n-1}$	$\cos\beta_n$	$\cos\beta_{n+1}$	$\sin\beta_{n-1}$	$\sin\alpha_n$	$\sin\alpha_{n+1}$	$\cot\alpha_{n-1}$	$\cot\alpha_n$	$\cot\alpha_{n+1}$
0																						
1	−66.76	−35.54	0.1	0.1			0.86	3.86			2.77	12.57			0.8137	0.552			−0.9189		−0.4294	0.5219
2	62.44	31.22	0.1	0.1	0.86	0.86	3.86	3.86	2.77	2.77	12.57	12.57	0.8137	0.8137	0.8552	0.8552	−0.9189	−0.9189	0.8865	−0.4294	0.5219	−0.4294
3	−66.76	35.54			0.86	3.86	3.86	0.86	2.77	12.57	12.57	2.77		0.8552	0.8552	0.8137		0.8865	−0.9189	−0.4294		
4																						
5																						
6																						
7																						
8																						
9																						
10																						

Table 10.7 Analysis due to Parme

Computed values

Point of plate	㉔ $\frac{⑧}{⑦}$	㉕ $\frac{⑲}{㉔}$	㉖ ㉒ + ㉓	㉗ ㉔²(㉑ + ㉒)	㉘ $\frac{1}{\left(\frac{⑤}{④}\right)^3}$	㉙ $\frac{⑤ \times ㉔}{④}$	㉚ $\frac{⑤ \times ⑧}{⑥} \times \left(\frac{⑧}{7}\pi\right)^2$	㉛ $\left(\frac{⑧}{⑤}\right)^2 \times \left(\frac{7}{\pi}\right)^3$	㉜ $\frac{⑧ \times ㉔}{⑥ \times ⑱}$	㉝ ㉕ + ㉗ + ㉜	㉞ 2 × ㉕ + ㉖ + ㉗	㉟ ㉕ + ㉖ + 36	㊱ $\frac{⑨ \times ⑳}{㉘}$	㊲ $\frac{㉔}{⑪ \times ⑮}$	㊳ $\frac{2}{⑪ \times ⑮}$	㊴ $\frac{2}{⑫ \times ⑯}$	㊵ $\frac{⑧ \times ㉚}{2}$	㊶ $\frac{㉔ \times ⑭(⑩+⑪)}{⑱}$	㊷ (⑪+⑫) $\left(\frac{⑲}{⑮+⑯ \times ㉔}\right)$	㊸ $\frac{⑳}{⑰}(⑫+⑬)$	㊹ ⑪$\left(\frac{⑲}{⑮+⑯ \times ㉔}\right)$ + ㊷	㊺ $\frac{2 \times ㊳}{㉔}$	㊻ $\frac{⑰}{⑳}\left[⑬ + \left(\frac{⑧ \times ⑯}{⑨ \times ⑰}\right) \times 2\right]$
0	4.49	−4.89	0.0925				3822						−0.242				1644						−17.4
1	1.0	1.13	0.0925	0.0925	1.0	0.2227	42.3	0.0007	−4.88	−3.66	2.45	−3.67	1.128	1.00	1.13	5.38	81.64	−13.58	−77.69	24.25	−91.72	0.5	
2				1.0	1.0	42.3		−4.88	−3.67	−3.67	−4.88	5.38	5.38	81.64	48.5	−13.58							
3																							
4																							
5																							
6																							
7																							
8																							
9																							
10																							

376 Folded plate roofs

Table 10.8

Coefficients for simultaneous equations

(1) Point or plate	(47)	(48)	(49)	(50)	(51)	(52)	(53)	(54)	(55)	(56)	(57)
	㉚ × ㊱	−㊵ × ㊻	㉙	2(1 + ㉙)	−㉚ × ㉟	㉚ × ㊱	㊵[−㊸ + ㊹ + ㊺(㉕ + ㉖)]	−㉛ × ㉜	㉛ × ㉝	2 × ㉘	−㉛ × ㉞
0	−925	28 606									
1			0.2227	2.445	154.8	47.71	−9663.6	0.0034	−0.0026	2.0	−0.0017
2											
3											
4											
5											
6											
7											
8											
9											
10											

Table 10.8 (continued)

Analysis due to Parme

①	㊿	㊾	⑥⓪	⑥①	⑥②	⑥③	⑥④	⑥⑤	⑥⑥	⑥⑦	⑥⑧	⑥⑨
	4(1 + ㉘)	㉛ × ㉟	−㉛ × ㊱	−(⑧)(㊲ × ㊳ + ㊴)	㉚ × ㉜	㉙	−㉚ × ㉝	2(1 + ㉙)	㉚ × ㉞	−㉚ × ㉟	㉚ × ㊱	−(㊵)(㊶) − ㊷) + ㊸)

Point or plate

	Coefficients for simultaneous equations

Point or plate													
0												
1												
2	8.0	−0.0026	0.0034	−41.53	−206.6	1.0	154.9	4.0	103.6	154.8	−206.6	6177
3												
4												
5												
6												
7												
8												
9												
10												

Folded plate roofs

Table 10.9

Point	Col. Row	0 f_0	1 f_1	2 f_2	3 M_2	4 = Load
0	0	2.000	1.000	0	−925	= 28 606
1	1	0.2227	2.445	1.000	154.8	= −9663.6
2	2	0.0034	−0.0026	0.0017	8.0	= −41.53
2	3	0	1.0	4.0	−103.0	= 6177
Solutions		11 271	−5244	2540	−12.23	

The methods of this Chapter base design on elastic theory; the loads designed for are working loads and permissible working stresses as in CP 114 (Ref. 19 of Chapter 9) should be used. This is for similar reasons as given for shells in Section 9.4 and on page 205 of Ref. 12.

References

1. Goldberg, J. E. and Leve, H. L., *Theory of Prismatic Plate Structures*, International Assocn. Bridge and Structural Engineering (1957).
2. Goldberg, J. E., Glauz, W. D. and Setlur, A. V., *Computer Analysis of Folded Plate Structures*, International Asscn. Bridge and Structural Engineering, Rio de Janiero, 16 Aug. (1964).
3. Gibson, J. E. and Gardner, N. J., 'Investigation of Multi-folded Plate Structures', *Proc. I.C.E.*, May (1965).
4. Winter, G. and Pei, M., 'Hipped Plate Construction', *Journal of the American Concrete Institute*, Jan. (1947).
5. Gaafar, I., 'Hipped Plate Analysis Considering Joint Displacements', *Proceedings of the American Society of Civil Engineers*, Apr. (1953).
6. Rockey, K. C. and Evans, H. R., 'A Critical Review of the Methods of Analysis for Folded Plate Structures', *Proceedings of the Institute of Civil Engineers*, June (1971).
7. Simpson, H., 'Design of Folded Plate Roofs', *Journal of the Structural Division of the American Society of Civil Engineers*, Jan. (1958).
8. Parme, A. L. L., 'Direct Solution of Folded Plate Concrete Roofs', *Bulletin of the International Association of Shell Structures* No. 6 (1960).
9. Thadani, B. N., 'The Analysis of Hipped Plate Structures by Influence Coefficients', *Indian Concrete Journal*, Apr. (1957).
10. Tamhankar, M. G. and Jain, R. D., 'Computer Analysis of Folded Plates', *Indian Concrete Journal*, Oct. (1965).
11. Parme, A. L. L., 'Computational Arrangement for Analysis of Folded Plate by Direct Solution', *Advanced Engineering Bulletin*, 3a, Portland Cement Assoc., U.S.A. (1963).
12. Wilby, C. B., *Concrete for Structural Engineers*. Newnes–Butterworth, London–Boston (1977).
13. Westbrook, R., *Structural Engineering Design in Practice*. Longman Scientific and Technical (1984 and 1988).

APPENDIX 1

Tables and graphs for design

Throughout the text there are many tables, graphs and references which are useful for speeding up design. Many of these are similar to those all confined together at the end of the old Reynolds' Handbook. As these tables and graphs are scattered throughout the text, so that they occur where their basis is being described, the following list will enable easy reference to them for those engaged in design:

Mix design

Serviceability and safety	Chapter 1
Load combinations	Table 1.1
Values of γ_m for ultimate limit state	Table 1.2
Graphs plotting percentage passing against sieve aperture size for concrete aggregate	Figure 2.1
Table recommending suitable workabilities for various uses	Table 2.1
Graphs plotting average ultimate compressive stress against water-to-cement ratio for concretes of various ages	Figure 2.2
Table recommending minimum strength as percentage of average strength for various conditions of control of concreting	Table 2.2
Tables recommending aggregate-to-cement ratios for various gradings and types of aggregates, water-to-cement ratios, and workabilities	Table 2.3
Table showing how to determine a certain required grading from available sand and coarse aggregates	Table 2.4

380   Appendix 1

Relationship between standard deviation and characteristic strength (D.O.E.) — Figure 2.5

Fine aggregate, percentage passing various sieves by weight — Table 2.6

Approximate compressive strengths of concrete mixes made with a free-water-to-cement ratio of 0.5 — Table 2.7

Relationship between compressive strength and free-water-to-cement ratio — Figure 2.6

Approximate free-water contents required to give various levels of workability — Table 2.8

Estimated wet density of fully compacted concrete — Figure 2.7

Recommended proportions of fine aggregate according to percentage passing a 600 μm sieve — Figure 2.8

Completed concrete mix design form — Table 2.5

Weights of materials

Weights of materials in kN/m^3 and kN/m^2 — Table 7.4

Reinforcement

Table giving cross-sectional areas of numbers of bars and bars in slabs — Table 3.2

Tables giving BS 8110 values of f_y for various types of reinforcement bars — Table 2.10

Anchorage or bond lengths

Table giving tension anchorage lengths (l_b/d_b) for various values of f_{cu} and f_y — Table 2.9

Table giving compression anchorage lengths (l_b/d_b) for various values of f_{cu} and f_y — Table 2.11

Table giving anchorage length equivalents of hooks and nibs for various diameters of mild and high-yield steel bars — Table 2.12

Table giving compression and tension anchorage lengths for $f_{cu} = 20$ N/mm² and various values of f_y — Table 2.13

Table giving overall anchorage lengths using hooks and nibs for $f_y = 250$ N/mm² and $f_{cu} = 20$ N/mm² — Table 2.14

Tables and graphs for design

Table giving tension anchorage lengths for bars used in water-retaining structures, BS 2007	Table 2.16
Dimensioned drawings of hooks and nibs	Figure 2.16

Curtailment of bars in beams
Table giving points for stopping off or bending up tension reinforcement bars towards supports for simply supported, continuous and fixed beams	Table 2.15

Elastic theory
Table giving tension anchorage lengths for bars used in water-retaining structures, CP 2007	Table 2.16
Tables for calculating equivalent area, x and I	Table 3.1 and Table 3.4
Tables giving corresponding values of $K[= M/(bd^2)]$ and z_1 for $f_s = 85$ N/mm² and $\alpha_e = 15$	Table 3.3

Shear reinforcement
Table giving values of V/d for two-arm stirrups for various values of f_{yv}, d_b and s_v	Table 3.5
Table giving values of V for single bars bent up at 45° in single shear for various values of f_{yv} and d_b	Table 3.6

Plastic design of sections for bending moments
Table giving, for balanced design, values of $K_1[= M_u/(bd^2)]$ and $\rho\%[=100A_s/(bd)]$ for various values of f_y, f_s and f_{cu}	Table 3.7

Strength of steel in compression
Table giving f_{sc} (design ultimate compressive stress) for various values of f_y for compression steel	Table 3.8
Figure giving BS 8110, Fig. 3, stress–strain curve for 25 mm diameter alloy bar	Figure 8.6

Design of beams and slabs
Table to assist in the preliminary design of depths of beams and slabs of various spans	Table 7.1

Continuous beams and slabs

Tables giving bending moments and shear forces in continuous beams and slabs carrying dead and imposed loadings ⎰ Table 6.1 and ⎱ Table 6.2

Table giving bending moments and shear forces in continuous beams and slabs subjected to unit bending moment at one and both ends Table 7.3

Single-span beams with fixed and free end supports

Table giving bending moments, support reactions and deflections for beams with various loadings Table 7.2

APPENDIX 2

Units and Greek symbols

For the purpose of being absolutely clear internationally about the units used in this book, the following conversions (which should prove useful anyway to engineers internationally) are given.

British Imperial	U.S.A.	Metric	SI
1 ton	1 long ton	1016.0 kg	9.964 kN
2000 lb	1 short ton	907.1 kg	8.896 kN
0.9843 ton	0.9843 long tons	{1 tonne / 1000 kg}	9.807 kN
1 lb	1 lb	0.4536 kg	4.448 N
1000 lb	1 kip	453.6 kg	4.448 kN
1 inch	1 inch	2.54 cm	25.4 mm
1 foot	1 foot	30.48 cm	0.3048 m
1000 lb in	1 kip in	1.152 kg cm	0.1130 kN m
1000 lb/in	1 kip/in	178.6 kg/cm	175.1 kN/m
1 lb/in^2	1 psi	0.070309 kg/cm^2	6.895 kN/m^2
1000 lb/in^2	{1 kip/in^2 / 1000 psi}	70.309 kg/cm^2	6.895 N/mm^2
1 lb/ft^2	1 lb/ft^2	4.882 kg/m^2	0.04788 kN/m^2
1 ton/ft^2	1 long ton/ft^2	10940 kg/m^2	107.3 kN/m^2
1 lb/ft	1 lb/ft	1.488 kg/m	0.01459 kN/m
1 ton/ft	1 long ton/ft	3333 kg/m	32.69 kN/m
1 lb/ft^3	1 lb/ft^3	16.02 kg/m^3	0.15707 kN/m^3
145.0 lb/in^2	1 Pa	10.20 kg/cm^2	1 N/m^2
1 ton/in^2	1 long ton/in^2	75.97 kg/m^2	0.7451 kN/m^2

Notes

1. The terms 'force' and 'mass' have not been used above, and acceleration due to gravity = 9.807 m/s^2
2. p.s.i. = psi = lb/in^2 = pounds per square inch
3. kip = 1000 lb = 1000 pounds
4. kip/in^2 = 1000 psi = 1000 pounds per square inch
5. Pa = pascal

The Greek Alphabet

1.	A	α	alpha	13.	N	ν	nu
2.	B	β	beta	14.	Ξ	ξ	xi
3.	Γ	γ	gamma	15.	O	o	omicron
4.	Δ	δ	delta	16.	Π	π	pi
5.	E	ε	epsilon	17.	P	ρ	rho
6.	Z	ζ	zeta	18.	Σ	σ	sigma
7.	H	η	eta	19.	T	τ	tau
8.	Θ	θ	theta	20.	Y	υ	upsilon
9.	I	ι	iota	21.	Φ	φ	phi
10.	K	κ	kappa	22.	X	χ	chi
11.	Λ	λ	lambda	23.	Ψ	ψ	psi
12.	M	μ	mu	24.	Ω	ω	omega

Other symbols used in mathematics:

∇ del

∂ curly d (partial differential)

APPENDIX 3

Nomenclature

The following symbols are generally used in this present book. The symbols conform to an internationally agreed system of constructing symbols which can be used in creating further symbols.

A_c	Area of concrete
A_{cf}	Area of effective concrete flange
A_{ps}	Area of prestressing tendons
A'_s	Area of compression reinforcement
A'_{s1}	Area of compression reinforcement in the more highly compressed face
A_s	Area of tension reinforcement
A_{s2}	Area of reinforcement in other face
A_{sc}	Area of longitudinal reinforcement (for columns)
A_{s1}	Cross-sectional area of longitudinal reinforcement provided for torsion
$A_{s.prov.}$	Area of tension reinforcement provided
$A_{s.req.}$	Area of tension reinforcement required
A_{sv}	Cross-sectional area of the two legs of a link
a	Deflection
a'	Distance from compression face to the point at which the crack width is being calculated
a_b	Distance between bars
a_{cent}	Distance of the centroid of the concrete flange from the centroid of the composite section
a_{ct}	Distance from the point (crack) considered to the surface of the nearest longitudinal bar

Appendix 3

a_s	Distance of the centroid of the steel from the centroid of the net concrete section
a_v	Distance between the line of action of the load and the face of the supporting member
b	Width of section
b_c	Breadth of compression face
b_e	Width of contact surface (between *in situ* and precast components)
b_1	Breadth of section at level of tension reinforcement
b_w	Breadth of web or rib of a member
C	Torsional constant
c_{min}	Minimum cover to tension steel
D_c	Density of concrete at time of test
d	Effective depth of tension reinforcement
d'	Depth to compression reinforcement
d_c	Depth of concrete in compression
d_o	Depth to additional reinforcement to resist horizontal loading
d_t	Effective depth in shear
d_2	Depth to reinforcement
E_c	Static secant modulus of elasticity of concrete
E_{cf}	Modulus of elasticity of flange concrete
E_{cq}	Dynamic tangent modulus of elasticity of concrete
E_s	Modulus of elasticity of steel
e	Eccentricity
e	Base of Napierian logarithms
e_a	Additional eccentricity due to deflections in walls
e_x	Resultant eccentricity of load at right angles to plane of wall
e_{x1}	Resultant eccentricity calculated at top of wall
e_{x2}	Resultant eccentricity calculated at bottom of wall
F	Ultimate load
F_b	Anchorage value of reinforcement
F_{bst}	Tensile bursting force
F_{bt}	Tensile force due to ultimate loads in a bar or group of bars
F_h	Horizontal component of a load
F_k	Characteristic load
F_t	Tie force
F_v	Maximum vertical ultimate load
F_w	Horizontal force on stiffened section of wall
f_{bs}	Bond stress

Nomenclature

f_{ci}	Concrete strength at (initial) transfer
f_{co}	Stress in concrete at the level of the tendon due to initial prestress and dead load
f_{cp}	Compressive stress at the centroidal axis due to prestress
f_{cu}	Characteristic concrete cube strength
f_k	Characteristic strength
f_{pb}	Tensile stress in tendons at (beam) failure
f_{pe}	Effective prestress (in tendon)
f_{pt}	Stress due to prestress
f_{pu}	Characteristic strength of prestressing tendons
f_s	Service stress
f_{s2}	Stress in reinforcement
f_t	Maximum principal tensile stress
f_y	Characteristic strength of reinforcement
f_{y1}	Characteristic strength of longitudinal reinforcement
f_{yv}	Characteristic strength of link reinforcement
G	Shear modulus
G_k	Characteristic dead load
g	Distributed dead load
g_k	Characteristic dead load per unit area
h	Overall depth of section in plane of bending
h_{agg}	Maximum size of aggregate
h_c	Diameter of column head
h_e	Effective thickness
h_f	Thickness of flange
h_{max}	Larger dimension of section
h_{min}	Smaller dimension of section
I	Second moment of area
i	Radius of gyration
j	Number of days
j_c	Number of days of concrete hardening
j_i	Age at first loading
K	A constant (with appropriate subscripts)
k	A constant (with appropriate subscripts)
	Distance from face of support at the end of a cantilever
	or
	Effective span of a simply supported beam or slab
l_e	Effective height of a column or wall
l_{ex}	Effective height for bending about the major axis

388 Appendix 3

l_{ey}	Effective height for bending about the minor axis
l_m	Average of l_1 and l_2
l_c	Clear height of column between end restraints
l_{sb}	Length of straight reinforcement beyond the intersection with the stirrup
l_x	Length of the shorter side (of rectangular slab)
l_y	Length of the longer side (of rectangular slab)
l_1	Length of a slab panel in the direction of span measured from the centres of columns
l_2	Width of slab panel measured from the centres of columns
M	Bending moment due to ultimate loads
M_a	Increased moment in column
M_{add}	Maximum additional moment
M_{cs}	Hogging restraint moment at an internal support of a continuous composite beam and slab section due to differential shrinkage
M_{ds}	Design bending moments in flat slabs
M_i	Maximum initial moment in a column due to ultimate loads (but not less than $0.05\,Nh$)
M_{ix}	Initial moment about the major axis of a slender column due to ultimate loads
M_{iy}	Initial moment about the minor axis of a slender column due to ultimate loads
M_{sx}, M_{sy}	The bending moments at mid span on strips of unit width and spans l_x and l_y, respectively
M_t	Total moment in a column due to ultimate loads
M_{tx}	Total moment about the major axis of a slender column due to ultimate loads
M_{ty}	Total moment about the minor axis of a slender column due to ultimate loads
M_u	Ultimate resistance moment
M_{ux}	Maximum moment capacity in a short column assuming ultimate axial load and bending about the major axis only
M_{uy}	Maximum moment capacity in a short column assuming ultimate axial load and bending about the minor axis only
M_z, M_y	Moments about the major and minor axes of a short column due to ultimate loads
M_0	Moment necessary to produce zero stress
M_1	Smaller initial end moment due to ultimate loads (assumed negative if the column is bent in double curvature)

M_2	Larger initial end moment due to ultimate loads (assumed positive)
N	Ultimate axial load at section considered
N_{bal}	Axial load on a column corresponding to the balanced condition
N_{ux}	Axial load capacity of a column ignoring all bending
n	Total ultimate load per unit area ($1.4g_k + 1.6q_k$)
n_s	Number of storeys
n_w	Axial load per unit length of wall
P_k	Characteristic load in tendon
P_0	Prestressing force in the tendon at the jacking end (or at tangent point near jacking end)
P_x	Prestressing force at distance x from jack
Q_k	Characteristic imposed load
q	Distributed live load
q_k	Characteristic live load per unit area
r	Internal radius of bend
r_{ps}	Radius of curvature (of a prestressing tendon)
$\dfrac{1}{r_b}$	Curvature of a beam at mid span or, for cantilevers, at the support section
$\dfrac{1}{r_{cc}}$	Creep curvature
$\dfrac{1}{r_{cs}}$	Shrinkage curvature
$\dfrac{1}{r_x}$	Curvature of a beam at point x
S_c	First moment of area of the concrete to one side of the contact surface, about the neutral axis of the transformed composite section
s_b	Spacing of bars
s_v	Spacing of links along the member
T	Torsional moment due to ultimate loads
$T°$	Temperature in degrees
t	Time
u	Perimeter
u_{crit}	Length of a critical perimeter
V	Shear force due to ultimate loads
V_c	Ultimate shear resistance of concrete
V_{co}	Ultimate shear resistance of a section uncracked in flexure

V_{cr}	Ultimate shear resistance of a section cracked in flexure
V_d	Total vertical shear due to design service load
v	Shear stress
v_c	Ultimate shear stress in concrete
v_h	Horizontal shear stress per unit area of contact surface
v_t	Torsional shear stress
v_{tu}	Ultimate torsional shear stress
W_k	Characteristic wind load
x	Neutral axis depth
x_1	Smaller dimension of a link
y_0	Half the side of end block
y_{po}	Half the side of loaded area
y_1	Larger dimension of a link
z	Lever arm
α_c	A ratio of the sum of column stiffnesses to the sum of beam stiffnesses
α_{c1}	Value of α_c at lower end of column
α_{c2}	Value of α_c at upper end of column
α_{cmin}	Minimum value of α_{c1} and α_{c2}
α_e	Modular ratio
α_f	Angle of internal friction for concrete interfaces
α_n	Coefficient as a function of column axial loading
α_{xx}, α_{xy}	Bending moment coefficients for slabs with no provision to resist torsion at the corners or to prevent the corners from lifting
β_b	Ratio of beam moments with respect to service stress in beams
β_{cc}	Ratio of total creep to elastic deformation
β_{red}	Ratio of reduction in resistance moment
β_{sx}, β_{sy}	Bending moment coefficients for slabs with provision to resist torsion and to prevent corners from lifting
β_1	Ratio of the longer to shorter base sides
γ_f	Partial safety factor for load
γ_m	Partial safety factor for strength
δ_m	Degree of hardening at moment of loading
ε_{cs}	Shrinkage strain
ε_{c1}	Strain in concrete at the level of the tendon at time of loading
ε_{c2}	Strain in concrete at the centroid of the section at time of loading
ε_{diff}	Differential shrinkage strain
ε_m	Average strain
ε_1	Strain at the level considered

Nomenclature

η	Relaxation coefficient
θ_s	Angle between the compression face and the tension reinforcement
λ_w	Coefficient for walls dependent upon dimensions and concrete used
μ	Coefficient of friction
ξ_s	Depth of slab factor
ρ	$\rho = \dfrac{A_s}{bd}$
ρ_o	Coefficient which depends upon the percentage of tension and compression steel in the section
φ	Creep coefficient with appropriate subscripts
ΣA_{sv}	Area of shear reinforcement
Σu_s	Sum of the effective perimeters of the tension reinforcement
Δ_{cc}	Concrete creep deformation
Δ_{cs}	Concrete shrinkage deformation
Φ	Bar size

APPENDIX 4

Extracts from British Standard BS 8110:1985

Throughout this present book references are often made to BS 8110 and sometimes extracts are reproduced. Other extracts of BS 8110, which should be useful to readers of this book, are now reproduced in this Appendix.

BS 8110: Part 1: 1985

2.4.3.1.1 *General.* For load combinations 1 and 2 in table 2.1 (table 1.1 in this book) the 'adverse' partial factor is applied to any loads that tend to produce a more critical design condition while the 'beneficial' factor is applied to any loads that tend to produce a less critical design condition at the section considered.

2.5.2 Analysis of structure

For design service loads, the analysis by linear elastic methods will normally give a satisfactory set of moments and forces.

When linear elastic analysis is used, the relative stiffnesses of members may be based on any of the following.

(a) *The concrete section:* the entire concrete cross section, ignoring the reinforcement.

(b) *The gross section:* the entire concrete cross section, including the reinforcement on the basis of modular ratio.

(c) *The transformed section:* the compression area of the concrete cross section combined with the reinforcement on the basis of modular ratio.

In (b) and (c) a modular ratio of 15 may be assumed in the absence of better information.

A consistent approach should be used for all elements of the structure.

Extracts from British Standard BS 8110: 1985

2.5.3 Analysis of sections for the ultimate limit state

The strength of a cross section at the ULS under both short and long term loading may be assessed assuming the short term stress/strain curves derived from the design strengths of the materials as given in **2.4.4.1** (table 1.2 in this book) and figures 2.1 to 2.3 as appropriate.

Table 3.1 Strength of reinforcement	
Designation	Specified characteristic strength, f_y
	N/mm²
Hot rolled mild steel	250
High yield steel (hot rolled or cold worked)	460

3.3.1.2 *Bar size.* The nominal cover to all steel should be such that the resulting cover to a main bar should not be less than the size of the main bar or, where bars are in pairs or bundles, the size of a single bar of cross-sectional area equal to the sum of their cross-sectional areas. At the same time the nominal cover to any links should be preserved.

3.3.1.3 *Nominal maximum size of aggregate.* Nominal covers should not be less than the nominal maximum size of the aggregate.

3.3.3 Cover against corrosion

The cover required to protect the reinforcement against corrosion depends on the exposure conditions and the quality of the concrete as placed and cured immediately surrounding reinforcement. Table 3.4 gives limiting values for the nominal cover of concrete made with normal-weight aggregates as a function of these factors.

3.3.4 Exposure conditions

3.3.4.1 *General.* The exposure conditions in service listed in table 3.4 are described in table 3.2.

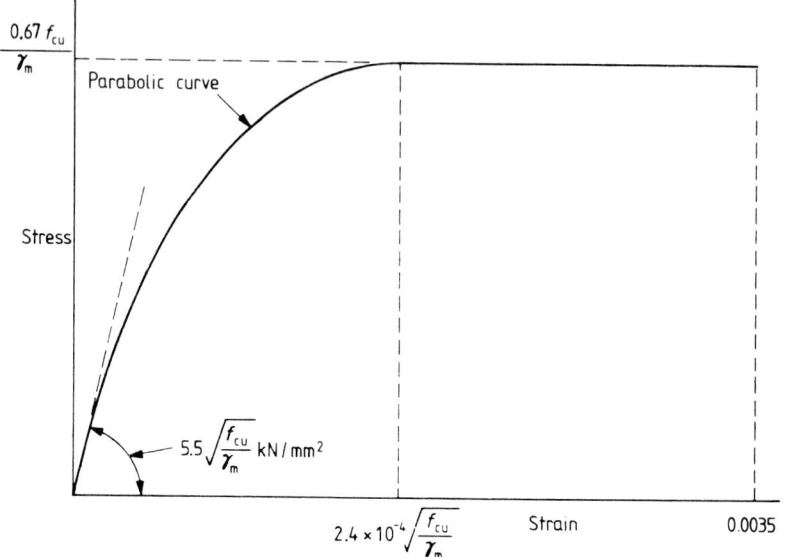

NOTE 1. 0.67 takes account of the relation between the cube strength and the bending strength in a flexural mem It is simply a coefficient and *not* a partial safety factor.
NOTE 2. f_{cu} is in N/mm^2.

Figure 2.1 Short term design stress-strain curve for normal-weight concrete

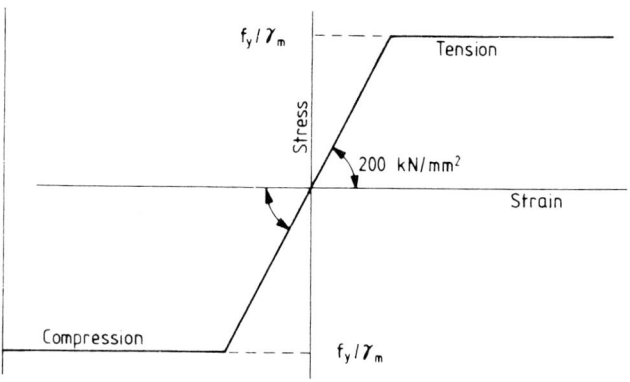

NOTE. f_y is in N/mm^2.

Figure 2.2 Short term design stress-strain curve for reinforcement

Extracts from British Standard BS 8110 : 1985

Table 3.2 Exposure conditions

Environment	Exposure conditions
Mild	Concrete surfaces protected against weather or aggressive conditions
Moderate	Concrete surfaces sheltered from severe rain or freezing whilst wet Concrete subject to condensation Concrete surfaces continuously under water
Severe	Concrete surfaces exposed to severe rain, alternate wetting and drying or occasional freezing or severe condensation
Very severe	Concrete surfaces exposed to sea water spray, de-icing salts (directly or indirectly), corrosive fumes or severe freezing conditions whilst wet
Extreme	Concrete surfaces exposed to abrasive action, e.g. sea water carrying solids or vehicles

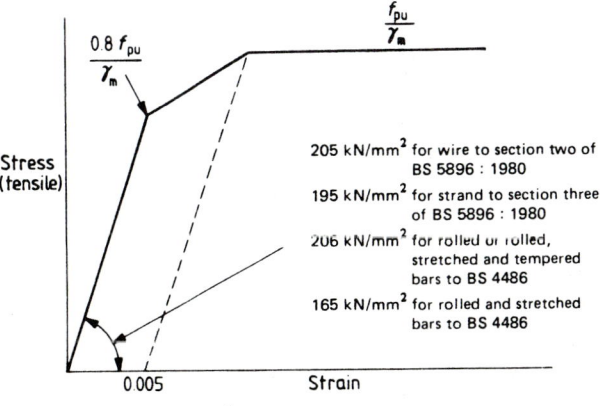

205 kN/mm² for wire to section two of BS 5896 : 1980
195 kN/mm² for strand to section three of BS 5896 : 1980
206 kN/mm² for rolled or rolled, stretched and tempered bars to BS 4486
165 kN/mm² for rolled and stretched bars to BS 4486

NOTE. f_{pu} is in N/mm².

Figure 2.3 Short term design stress-strain curve for prestressing tendons

Table 3.4 Nominal cover to all reinforcement (including links) to meet durability requirements

Conditions of exposure	Nominal cover				
	mm	mm	mm	mm	mm
Mild	25	20	20*	20*	20*
Moderate	—	35	30	25	20
Severe	—	—	40	30	25
Very severe	—	—	50†	40†	30
Extreme	—	—	—	60†	50
Maximum free water/cement ratio	0.65	0.60	0.55	0.50	0.45
Minimum cement content (kg/m³)	275	300	325	350	400
Lowest grade of concrete	C30	C35	C40	C45	C50

* These covers may be reduced to 15 mm provided that the nominal maximum size of aggregate does not exceed 15 mm.
† Where concrete is subject to freezing whilst wet, air-entrainment should be used.
NOTE 1. This table relates to normal-weight aggregate of 20 mm nominal maximum size.
NOTE 2. For grade of concrete C40 means $f_{cu} = 40$ N/mm², etc.

Extracts from British Standard BS 8110: 1985

Table 3.5 Nominal cover to all reinforcement (including links) to meet specified periods of fire resistance

Fire resistance	Nominal cover						
	Beams*		Floors		Ribs		Columns*
	Simply supported	Continuous	Simply supported	Continuous	Simply supported	Continuous	
h	mm	mm	mm	mm	mm	mm	mm
0.5	20†	20†	20†	20†	20†	20†	20†
1	20†	20†	20	20	20	20†	20†
1.5	20	20†	25	20	35	20	20
2	40	30	35	25	45	35	25
3	60	40	45	35	55	45	25
4	70	50	55	45	65	55	25

* For the purposes of assessing a nominal cover for beams and columns, the cover to main bars have been reduced by a notional allowance for stirrups of 10 mm to cover the range 8 mm to 12 mm.

† These covers may be reduced to 15 mm provided that the nominal maximum size of aggregate does not exceed 15 mm.

NOTE 1. The nominal covers given relate specifically to the minimum member dimensions given in figure 3.2.

NOTE 2. Cases that lie below the bold line require attention to the additional measures necessary to reduce the risks of spalling.

Figure 3.2 Minimum dimensions of reinforced concrete members for fire resistance

Fire resistance	Minimum beam width	Rib width of T-beam	Minimum thickness of floors	Column width		
				Fully exposed	50% exposed	One face exposed
h	mm	mm	mm	mm	mm	mm
0.5	200	125	75	150	125	100
1	200	125	95	200	160	120
1.5	200	125	110	250	200	140
2	200	125	125	300	200	160
3	240	150	150	400	300	200
4	280	175	170	450	350	240

3.4.1.2 *Effective span of simply-supported beams.* The effective span of a simply-supported beam may be taken as the smaller of the distance between the centres of bearings, or the clear distance between supports plus the effective depth.

3.4.1.3 *Effective span of a continuous member.* The effective span of a continuous member should be taken as the distance between centres of supports. The centre of action of support at an encastré should be taken to be at half the effective depth from the face of the support.

3.4.1.4 *Effective length of a cantilever.* The effective length of a cantilever should be taken as its length to the face of the support plus half its effective depth except where it forms the end of a continuous beam where the length to the centre of the support should be used.

3.4.1.5 *Effective width of flanged beam.* In the absence of any more accurate determination this should be taken as:

(a) *for T-beams:* web width *plus* $l_z/5$ or actual flange width if less;

(b) *for L-beams:* web width *plus* $l_z/10$ or actual flange width if less;

where

l_z is the distance between points of zero moment (which, for a continuous beam, may be taken as 0.7 times the effective span).

3.4.1.6 *Slenderness limits for beams, for lateral stability.* The clear distance between restraints should not exceed:

(a) *for simply-supported or continuous beams:* $60b_c$ or $250b_c^2/d$ if less;

(b) *for cantilevers with lateral restraint only at support:* $25b_c$ or $100b_c^2/d$ if less;

where

b_c is the breadth of the compression face of the beam, measured mid-way between restraints (or the breadth of the compression face of a cantilever);

d is the effective depth (which need not be greater than whatever effective depth would be necessary to withstand the design ultimate load with no compression reinforcement).

Extracts from British Standard BS 8110: 1985

3.4.3 Uniformly-loaded continuous beams with approximately equal spans: moments and shears

Table 3.6 may be used to calculate the design ultimate bending moments and shear forces, subject to the following provisos:

(a) characteristic imposed load Q_k may not exceed characteristic dead load G_k;

(b) loads should be substantially uniformly distributed over three or more spans;

(c) variations in span length should not exceed 15% of longest.

Table 3.6 Design ultimate bending moments and shear forces

	At outer support	Near middle of end span	At first interior support	At middle of interior spans	At interior supports
Moment	0	$0.09Fl$	$-0.11Fl$	$0.07Fl$	$-0.08Fl$
Shear	$0.45F$	—	$0.6F$	—	$0.55F$

NOTE. l is the effective span;
F is the total design ultimate load $(1.4G_k + 1.6Q_k)$.
No redistribution of the moments calculated from this table should be made.

3.4.4.1 *Analysis of sections.* In the analysis of a cross section to determine its ultimate moment of resistance the following assumption should be made.

Where a section is designed to resist only flexure, the lever arm should not be assumed to be greater than 0.95 times the effective depth.

Table 3.8 Form and area of shear reinforcement in beams

Value of v (N/mm^2)	Form of shear reinforcement to be provided	Area of shear reinforcement to be provided
Less than $0.5v_c$ throughout the beam	See note 1	
$0.5v_c < v < (v_c + 0.4)$	Minimum links for whole length of beam	$A_{sv} \geq 0.4 b_v s_v / 0.87 f_{yv}$ (see note 2)
$(v_c + 0.4) < v$ $0.8\sqrt{f_{cu}}$ or 5 N/mm^2	Links or links combined with bent-up bars. Not more than 50% of the shear resistance provided by the steel may be in the form of bent-up bars	

NOTE 1. While minimum links should be provided in all beams of structural importance, it will be satisfactory to omit them in members of minor structural importance such as lintels or where the maximum design shear stress is less than half v_c.

NOTE 2. Minimum links provide a design shear resistance of 0.4 N/mm^2.

3.4.5.5 *Spacing of links.* The spacing of links in the direction of the span should not exceed $0.75d$. At right-angles to the span, the horizontal spacing should be such that no longitudinal tension bar is more than 150 mm from a vertical leg; this spacing should in any case not exceed d.

3.4.6 Deflection of beams

3.4.6.2 *Symbols.* For the purposes of **3.4.6** the following symbols apply.

$A_{s,prov}$ area of tension reinforcement provided at mid-span (at support for a cantilever)

$A'_{s,prov}$ area of compression reinforcement

$A_{s,req}$ area of tension reinforcement required at mid-span to resist the moment due to design ultimate loads (at support for a cantilever)

b effective width of a rectangular beam, the effective flange width of a flanged beam or the total average width of the flanges

b_w average web width of a beam

d effective depth

f_s estimated design service stress in the tension reinforcement
M design ultimate moment at the centre of the span or, for a cantilever, at the support
β_b the ratio:

$$\frac{\text{(moment at the section after redistribution)}}{\text{(moment at the section before redistribution)}}$$

from the respective maximum moments diagram

3.4.6.3 *Span/effective depth for a rectangular or flanged beam.* The basic span/effective depth ratios for beams are given in table 3.10. These are based on limiting the total deflection to span/250 and this should normally ensure that the part of the deflection occurring after construction of finishes and partitions will be limited to span/500 or 20 mm, whichever is the lesser, for spans up to 10 m. For values of b_w/b greater than 0.3, linear interpolation between the values given in table 3.10 for rectangular sections and for flanged beams with b_w/b of 0.3 may be used.

Table 3.10 Basic span/effective depth ratios for rectangular or flanged beams

Support conditions	Rectangular sections	Flanged beams with $\dfrac{b_w}{b} \leqslant 0.3$
Cantilever	7	5.6
Simply supported	20	16.0
Continuous	26	20.8

3.4.6.4 *Long spans.* For spans exceeding 10 m, table 3.10 should be used only if it is not necessary to limit the increase in deflection after the construction of partitions and finishes. Where limitation is necessary, the values in table 3.10 should be multiplied by 10/span except for cantilevers where the design should be justified by calculation.

3.4.6.5 *Modification of span/depth ratios for tension reinforcement.* Deflection is influenced by the amount of tension reinforcement and its stress. The span/effective depth ratio should therefore be modified according to the area of reinforcement provided and its service stress at the centre of the span (or at the support in the case of a cantilever). Values of

Table 3.11 Modification factor for tension reinforcement

Service stress		M/bd^2								
		0.50	0.75	1.00	1.50	2.00	3.00	4.00	5.00	6.00
($f_y = 250$)	100	2.00	2.00	2.00	1.86	1.63	1.36	1.19	1.08	1.01
	150	2.00	2.00	1.98	1.69	1.49	1.25	1.11	1.01	0.94
	156	2.00	2.00	1.96	1.66	1.47	1.24	1.10	1.00	0.94
	200	2.00	1.95	1.76	1.51	1.35	1.14	1.02	0.94	0.88
	250	1.90	1.70	1.55	1.34	1.20	1.04	0.94	0.87	0.82
($f_y = 460$)	288	1.68	1.50	1.38	1.21	1.09	0.95	0.87	0.82	0.78
	300	1.60	1.44	1.33	1.16	1.06	0.93	0.85	0.80	0.76

NOTE. For a continuous beam, if the percentage of redistribution is not known but the design ultimate moment at mid-span is obviously the same as or greater than the elastic ultimate moment, the stress, f_s, in this table may be taken as $5/8 f_y$.

Table 3.12 Modification factor for compression reinforcement

$\dfrac{100 A'_{s,prov}}{bd}$	Factor
0.00	1.00
0.15	1.05
0.25	1.08
0.35	1.10
0.50	1.14
0.75	1.20
1.0	1.25
1.5	1.33
2.0	1.40
2.5	1.45
$\geqslant 3.0$	1.50

NOTE. The area of compression reinforcement $A'_{s,prov}$ used in this table may include all bars in the compression zone, even those not effectively tied with links.

Extracts from British Standard BS 8110: 1985

span/effective depth ratio obtained from table 3.10 should be multiplied by the appropriate factor obtained from table 3.11.

3.4.6.6 *Modification of span/depth ratios for compression reinforcement.* Compression reinforcement also influences deflection and the value of the span/effective depth ratio obtained from table 3.10 modified by the factor obtained from table 3.11 may be multiplied by a further factor obtained from table 3.12.

Table 3.13 Ultimate bending moment and shear forces in one-way spanning slabs

	At outer support	Near middle of end span	At first interior support	Middle of interior spans	Interior supports
Moment	0	0.086Fl	−0.86Fl	0.063Fl	−0.063Fl
Shear	0.4F	—	0.6F	—	0.5F

NOTE. F is the total design ultimate load $(1.4G_k + 1.6Q_k)$.
l is the effective span.

3.4.6.7 *Deflection due to creep and shrinkage.* Permissible span/effective depth ratios obtained from tables 3.10 to 3.12 take account of normal creep and shrinkage deflection.

3.8.1.6 *Effective height of a column*

3.8.1.6.1 *General.* The effective height, l_e, of a column in a given plane may be obtained from the following equation:

$$l_e = \beta l_0 \qquad \text{equation 30}$$

Values of β are given in tables 3.21 and 3.22 for braced and unbraced columns respectively as a function of the end conditions of the column.

In tables 3.21 and 3.22 the end conditions are defined in terms of a scale from 1 to 4. Increase in this scale corresponds to a decrease in end fixity.

3.8.1.6.2 *End conditions.* The four end conditions are as follows.

(a) *Condition 1.* The end of the column is connected monolithically to beams on either side which are at least as deep as the overall dimension of the column in the plane considered. Where the column is connected to a foundation structure, this should be of a form specifically designed to carry moment.

(b) *Condition 2.* The end of the column is connected monolithically to beams or slabs on either side which are shallower than the overall dimension of the column in the plane considered.

(c) *Condition 3.* The end of the column is connected to members which, while not specifically designed to provide restraint to rotation of the column will, nevertheless, provide some nominal restraint.

(d) *Condition 4.* The end of the column is unrestrained against both lateral movement and rotation (e.g. the free end of a cantilever column in an unbraced structure).

Table 3.21 Values of β for braced columns

End condition at top	End condition at bottom		
	1	2	3
1	0.75	0.80	0.90
2	0.80	0.85	0.95
3	0.90	0.95	1.00

Table 3.22 Values of β for unbraced columns

End condition at top	End condition at bottom		
	1	2	3
1	1.2	1.3	1.6
2	1.3	1.5	1.8
3	1.6	1.8	—
4	2.2	—	—

Extracts from British Standard BS 8110: 1985

3.8.1.7 *Slenderness limits for columns.* Generally, the clear distance, l_o, between end restraints should not exceed sixty times the minimum thickness of a column.

3.8.1.8 *Slenderness of unbraced columns.* If, in any given plane, one end of an unbraced column is unrestrained (e.g. a cantilever column), its clear height, l_o, should not exceed:

$$l_o = \frac{100b^2}{h} \leqslant 60b \qquad \text{equation 31}$$

NOTE. In equation 31 h and b are respectively the larger and smaller dimensions of the column.

The considerations of deflection may introduce further limitations.

3.8.2 Moments and forces in columns

3.8.2.4 *Minimum eccentricity.* At no section in a column should the design moment be taken as less than that produced by considering the design ultimate axial load as acting at a minimum eccentricity, e_{min}, equal to 0.05 times the overall dimension of the column in the plane of bending considered but not more than 20 mm. Where biaxial bending is considered, it is only necessary to ensure that the eccentricity exceeds the minimum about one axis at a time.

INDEX

adhesion, *see* bond
advanced method, Hillerborg's *see* slabs
age of concrete, 27, 32
aggregate:cement ratio, 36
aggregates, 21–4
  combining, 39
  fine, coarse, 21, 45, 49
  gradings, 37, 38, 40, 45
  lightweight, shape, 22
  rounded, irregular, angular, 23
air entrained concrete, 26
alkali–silica reaction, 21
analysis of sections for ultimate limit state, 393, 399
analysis of structure, 392
anchorage lengths *see* bond
anchorage of column bars into bases, 233
anchorage of shear bars, 77–8
anchorage of stirrups, 79–80
arches, 249
asbestos cement, 14
Aspdin, J., 13

balanced design, 119, 126, 128
basement wall, 15
bases, 230–3, 237–8
batching concrete by volume and weight, 23
beams
  breadth, 85
  continuous, 85, 129, 216–18
  depth, 85, 401
  elastic theory formulae, 95–6
  T-, 85, 86
bearing failure, 8

bearing stresses inside bends, 78–9
bending moments, redistribution, 216, 217
Bogue compounds, 14, 83
bond, 65–82, 87
bond failure, 8
bond stresses due to shear, 112, 113
bridges, 249
briquettes for tension test, 28
bulk volume, 23

calcium chloride, *see* chloride ion
carbonisation, 82–3
cathodic protection, 82–3
cement, 13–21
  chemical composition, 13, 14
  extra rapid hardening, 16
  for cold weather, 18
  high alumina (H.A.C.), 14, 16–18
  low-alkali, 21
  minimum content, 43, 44, 221, 223, 227, 231, 235–7
  Portland, 16
  Portland blast furnace, 20, 21
  Pozzolana, 20
  rapid hardening, 16, 36
  sulphate-resisting, 19
  super sulphated, 19
  testing of, 28
  water-repellant, 21
  with low coefficient of shrinkage, 19
  with low heat of setting, 20
characteristic load, 6, 7, 11
characteristic strength, 6, 11, 35
chemical conversion, 17
chloride ion, 18, 32

Index

Ciment Fondu, 14, 15
coefficient of shrinkage, 51
coefficient of variation, 35
coloured cements, 20
columns, 205–15
  axially loaded, 205–7
  circular, 213
  eccentrically loaded, 207–12
  effective height, 403
  end conditions, 403, 404
  moments in, 405
  prestressed, 283
  short, 205
  slender, 205, 214, 405
compacting factor, 25
composite construction, 9
compression failure, 8
compression steel, 129
compression steel near neutral axis, 130
concrete, 24–61
  abrasion, 28
  age of, 27, 32
  air-entrained, 26
  air-voids, 24
  compaction, 26
  curing of, 32, 33, 52
  density, 28
  design of mixes, 33–50
  grades, 34, 221, 396
  mean strength, 35
  mixes, 24
  no fines, 31
  porosity, 28
  quantities of materials, 49, 50
  segregation, 29
  strength (comp.) 6, 26–8, 46, 47
  strength (tensile), 57–61
  strength tests, 28, 32, 34
  use of, 25
  vacuum, 28
  vibrated, 29
  voids in, 24
  wet density, 48
connections, 9
continuous beams, *see* beams
continuous slabs, *see* slabs
corrosion of reinforcement, 82–3
cover of concrete, 393
cracks, cracking, 8, 10, 132
creation of structures, 63–4, 242–51
creep, 54–7, 257

curing of concrete, 32, 33
curtailment of reinforcement, 74–8
cylinder splitting test, 28

deflection, 4, 8, 10, 131, 400, 401–3
design calculations, 234–8
design load, 10
design of
  bases, 230–3, 237–8
  beams, 85, 223–7, 235
  columns, 227–30, 233, 236–7
  compression steel, 127–31
  concrete mixes, 33–49, 379
  floor of building, 221–3, 234, 235
  frames, 219–21
  prestressed concrete members, 259–94
  shear reinforcement, 105–12, 330–1
  slabs, 134–204, 221–3, 235
  structures, 219–51
design philosophy (BS 8110), 5–12
design strength, 11, 12
designers' tables and graphs, 379–82
  anchorage lengths, 69, 70, 72, 81
  areas of bars, 97
  areas of bars for slabs, 97
  balanced design values K_1 and ρ, 123
  bending moments in continuous beams and slabs, 217, 240, 399, 403
  bending moments, support reactions and deflections for beams with fixed and free supports, 239
  bent-up bars, 111, 400
  cambers for cylindrical shells, 330
  coefficients for elastic design, 98
  columns, 404
  concrete mix design, 22, 25, 27, 33, 34, 37, 38, 42–9
  cover, 396, 397
  curtailment of bars, 75
  exposure conditions, 395
  fire resistance, 397
  folded plate roof, 365–70, 374–8
  grading of aggregates, 22
  hooks and nibs, 70, 72
  load combinations, 9
  moment of inertia, 92
  ratios of span to overall depth, beams and slabs, 221, 401, 402
  reinforcement strength, 276, 393
  shear forces in continuous beams and slabs, 218, 240, 399

408 Index

shear stresses, 105, 123
span to depth ratios (beams and slabs), 221
stirrups, 110, 400
stress in compression steel, 128, 394
stress/strain curves for
  concrete, 394
  prestressing wires and strand, 276, 395
  reinforcement, 394
torsion stresses, 115
two-way spanning slabs, 139–41
values of γ_m for ultimate limit state, 9
values of z_1 and $K = M/(bd^2)$ for elastic theory, 98
weights of materials, 241
diagonal tensile stresses, 103, 106
dilatency, 66
D.O.E. mix design method, 41–50
durability, 11, 44

economics, 63–4
effective spans, 398
effective widths, 398
elastic modulus, 55–7, 88
elastic theory/analysis/design, 2, 3, 4, 5, 86–104, 131, 205
elastic theory formulae for slabs and beams, 95
end anchorages, 70–3
end blocks, 8, 261
epoxy resin, 253
equivalent area, 89
exposure conditions, 393–6
extra rapid hardening cement, 16

factor of safety, 2, 4
factor γ_f, 10, 11
factor γ_m, 9, 11
fatigue, 11
fire resistance, 11, 397
flash set, 14
flat slabs, 64, 141, 201–3, 234, 244, 248
  dropped panels, 64
flexural failure, 8, 119
flexural bond, 112, 113
fly ash, 20
folded plate roofs, 64, 242–3, 249, 250, 303, 304, 360–78
  analysis due to Parme, 363–78
  analysis of, 361–78
  design of, 361–78

frames, 213–4, 216–7
frost resistance, 26, 252

gap-graded concrete, 29–31
grades of concrete, 24
grading, combining aggregates, 39–41
granolithic concrete, 22, 57–60
gravity, acceleration due to, 384
greek alphabet, 384
grip, 66
grouting ducts, 252, 253

hardening, rate of, 15
high alumina cement (H.A.C.), 14, 16, 17, 32
Hillerborg's strip/advanced methods, *see* slabs
history, 13, 57–60
hollow tile floors and roofs, 64, 250
hooks, 70, 71

In situ R.C. construction, 64, 248
  design, 219–51
  slabs, 247–8
instability, 398

Johansen, *see* yield line

laps in reinforcement, 73–4
lightness of weight, 64
limit state design, 8, 11
Lin T. Y., load-balancing, 296–8, 300
links, *see* stirrups
load
  combinations, 9
  dead, 5, 9
  factor, 3, 4, 5
  imposed, 5, 9
  wind, 9
loading tests, 167

mean strength, 34
mix design, 33–49, 379
modular ratio, 89, 91, 93, 95, 96, 99
modulus of elasticity, *see* elastic modulus
modulus of rupture, 88
moment arm, 4
moment of inertia, 88, 91–4
moment of resistance, 91

neutral axis, 91–4
nibs, 70–2

Index

no fines concrete, 31, 32
nomenclature, 301–3, 385–91

over-reinforced sections, 119

Parme, *see* folded plate roofs
partial factor of safety, 2, 4, 7, 11
partial prestressing, 260
permissible stresses, 2, 3, 4
petrographic analysis, 21
P.F.A., 20
philosophy of design (BS 8110), 5, 6, 11
plastic analysis, 118–31
plastic analysis assumptions, 118–19
plastic collapse mechanisms, 5
plastic design, 119–31, 205
plasticisers, 26
plastic strain, *see* creep
Poisson's ratio, 87
polyester, 253
Portland blast furnace cement, 20, 21
Portland cement, 13, 15
   ordinary, 16, 27
   rapid hardening, 16, 27
post-compressing, 299, 300
post-tensioning, 252–3, 258, 299
   for shear, 253
precast concrete floors, roofs and frames, 64, 247–50
prescribed mixes, 24
prestress, losses due to
   creep, 255, 257, 264, 272
   elastic deformation, 256, 257
   friction, 258, 259
   relaxation (creep) of steel, 255, 256, 270
   slip of anchorage, 257
   shrinkage, 257, 272
   stream curing, 259
prestressed concrete, 64, 249, 252–300
   additional untensioned steel, 281–2
   assumptions for elastic design, 264
   advantages and disadvantages, 253–4
   classes of structures, 260
   columns, 283
   composite construction, 287–94
   compression steel, 282
   continuity, 294–5
   deflections, 265, 266, 268, 269
   end blocks, 8, 261
   end splitting forces, 295
   elastic design, 265

   inclined tendons, 287
   limit state design, 259
   limit states of deflections and stresses, 2
   load balancing (Lin), 296–8
   losses, 255–9, 263, 267, 269, 270–4
   materials, 255, 276
   non-prestressed steel, 282
   post-compressing (Reiffenstuhl), 299, 300
   post-tensioning, 252–3, 258, 299
   segmental construction, 253
   shear resistance, 284–7
   steam curing, 259
   stress corrosion, 255
   stresses, 265
   tanks, pipes, domes, shells and piles, 295, 296
   tendons, 276
   ties, 283
   torsional resistance, 296
   ultimate limit state due to flexure, 274
   unbonded tendons, 282
prestressing, 252, 254
prestressing beds, 252, 284
prestressing dams, 253
prestressing forces, 253
prestressing wires, transmission length, 261
pressure compaction, 29
pretensioning, 252
probability, 7
proof stress, 2
punching shear stress, 244

Reiffenstuhl, post-compressing, 299–300
reinforcement, 61–3, 144–6, 276, 393, 401
   areas, 97
relaxation, 255
Road Note No 4, 22, 25, 33, 39, 41

safety, 1, 2
sealing water leaks, 17
serviceability limit state, 11
set, initial, 15
   final, 15
setting time, initial, 15
   final, 15
shear failure, 8
shear reinforcement, 105, 112
shear stresses (elastic), 100–4
shell roofs, 64, 86, 242–3, 249, 250, 251, 320–59

analogue computer, 350–1
analysis, 321–4
applications, 311–12
computer programs, 356
conoidal, 309, 310, 354, 355
construction, 324–31
cylindrical, 305, 332, 347–51
design, 319–21, 356
design tables and graphs, 348–50
designation, 310, 311
domes, 308, 346, 351, 352
doubly curved, 308
economics, 313–18
finite elements, 355–6
hyperbolic paraboloids, 309, 352
membrane analysis, 332–46
north-light, 307, 347, 349
notation, 301–3
of arbitrary shape, 347
of revolution, 341–6
types, 303–10
shrinkage, 15, 50–3, 64, 257
shrinkage stresses, 50–3, 87
slabs
  affine, transformations, 162–6
  continuous, 216–18, 250
  design of, 201–3
  elastic theory formulae, 95
  flat, 64, 141, 201–3, 234, 244, 248
  Hillerborg's advanced method, 141, 169–200
  Hillerborg elements with edge shear forces, 196–200
  Hillerborg's strip method, 141, 166–9
  holes in, 190–6
  isotropically reinforced, 142
  Johansen's yield line, *see* yield line
  on solid, 60–1
  orthogonally anisotropically reinforced, 144
  orthotropically reinforced, 144
  skew, 163, 165
  spanning one way, 134, 246
  spanning two ways, 64, 134–41, 201–3, 234, 245
  traditional U.K. design office methods, 201
  waffle, 64
slip, 66, 257, 273
slump test, 25
space frames, 283

speed of loading, 55
standard deviation, 35, 44
statistics, 5, 35
steam curing, 17, 18, 259
stirrups, 79, 80, 400
stopping-off bars, order of, 75
strain loss, 256, 257
strand, 63, 276
strength tests, 28
stress-block, 120–2
stress: strain relationship, 54–7
strip method, Hillerborg's, *see* slabs
  discontinuity lines, 167, 168
sulphate-resisting cement, 19
super-plasticisers, 26
super sulphated cement, 19

T-beams, 64, 129
tables for designers, *see* designers tables
tank for water, 96
tanking, 16
temperature stresses, 87
time flow, *see* creep
tobermorite gel, 15
torsion, 113–18
torsion failure, 8
torsion test, 28
truss-analogy, 106

ultimate limit state, 8, 11
under-reinforced, 119, 120, 127, 279
units, conversion British Imperial, U.S.A., metric and S.I., 383, 384

vacuum concrete, 28
variation, coefficient of, 35
VB consistometer test, 25
vibrated concrete, 25, 29
vibration, 8, 65

waffle slabs, 244
walls, 212
water-repellant cements, 21
water-retaining structures, 3, 96
water-to-cement ratio, 26, 27, 47
wedge action, 66
weights of materials, 241
Whitney, 123, 275, 278
workability, 23, 25, 26, 48
working loads, 2, 6
working stress, 2

yield line, Johansen's, *also see* slabs, 141–66, 168–9
  affine slab transformations, 162–6
  combination of equilibrium and virtual-work methods, 154–62
  corner levers, 147
  design of bases, 233
  equilibrium method of analysis, 142, 148–52
  kinking of reinforcement, 146
  lower-bound solutions, 147, 148
  upper-bound solutions, 147, 148
  virtual-work method of analysis, 143, 153–62